IEE ELECTROMAGNETIC WAVES SERIES 49

Series Editors: Professor P. J. B. Clarricoats
Professor E. V. Jull

Ridge waveguides and passive microwave components

Other volumes in this series:

Volume 10 **Aperture antennas and diffraction theory** E. V. Jull
Volume 11 **Adaptive array principles** J. E. Hudson
Volume 12 **Microstrip antenna theory and design** J. R. James, P. S. Hall and C. Wood
Volume 13 **Energy in electromagnetism** H. G. Booker
Volume 14 **Leaky feeders and subsurface radio communications** P. Delogne
Volume 15 **The handbook of antenna design, Volume 1** A. W. Rudge, K. Milne, A. D. Olver, P. Knight (Editors)
Volume 16 **The handbook of antenna design, Volume 2** A. W. Rudge, K. Milne, A. D. Olver, P. Knight (Editors)
Volume 17 **Surveillance radar performance prediction** P. Rohan
Volume 18 **Corrugated horns for microwave antennas** P. J. B. Clarricoats and A. D. Olver
Volume 19 **Microwave antenna theory and design** S. Silver (Editor)
Volume 20 **Advances in radar techniques** J. Clarke (Editor)
Volume 21 **Waveguide handbook** N. Marcuvitz
Volume 22 **Target adaptive matched illumination radar** D. T. Gjessing
Volume 23 **Ferrites at microwave frequencies** A. J. Baden Fuller
Volume 24 **Propagation of short radio waves** D. E. Kerr (Editor)
Volume 25 **Principles of microwave circuits** C. G. Montgomery, R. H. Dicke, E. M. Purcell (Editors)
Volume 26 **Spherical near-field antenna measurements** J. E. Hansen (Editor)
Volume 27 **Electromagnetic radiation from cylindrical structures** J. R. Wait
Volume 28 **Handbook of microstrip antennas** J. R. James and P. S. Hall (Editors)
Volume 29 **Satellite-to-ground radiowave propagation** J. E. Allnutt
Volume 30 **Radiowave propagation** M. P. M. Hall and L. W. Barclay (Editors)
Volume 31 **Ionospheric radio** K. Davies
Volume 32 **Electromagnetic waveguides: theory and application** S. F. Mahmoud
Volume 33 **Radio direction finding and superresolution** P. J. D. Gething
Volume 34 **Electrodynamic theory of superconductors** S.-A. Zhou
Volume 35 **VHF and UHF antennas** R. A. Burberry
Volume 36 **Propagation, scattering and dissipation of electromagnetic waves** A. S. Ilyinski, G. Ya. Slepyan and A. Ya. Slepyan
Volume 37 **Geometrical theory of diffraction** V. A. Borovikov and B. Ye. Kinber
Volume 38 **Analysis of metallic antennas and scatterers** B. D. Popović and B. M. Kolundžija
Volume 39 **Microwave horns and feeds** A. D. Olver, P. J. B. Clarricoats, A. Kishk and L. Shafai
Volume 40 **Uniform stationary phase method** V. A. Borovikov
Volume 41 **Approximate boundary conditions in electromagnetics** T. B. A. Senior and J. L. Volakis
Volume 42 **Spectral theory and excitation of open structures** V. P. Shestopalov and Y. V. Shestopalov
Volume 43 **Open electromagnetic waveguides** T. Rozzi and M. Mongiardo
Volume 44 **Theory of nonuniform waveguides** B. Z. Katsenelenbaum, M. K. A. Thumm, L. Mercader Del Río, M. Pereyaslavets and M. Sorolla Ay
Volume 45 **Parabolic equation methods for electromagnetic wave propagation** M. Levy

A complete list of publications in this series may be obtained from the IEE

Ridge waveguides and passive microwave components

J. Helszajn

The Institution of Electrical Engineers

Published by: The Institution of Electrical Engineers, London, United Kingdom

The Institution of Electrical Engineers,
Michael Faraday House,
Six Hills Way, Stevenage,
Herts. SG1 2AY, United Kingdom

British Library Cataloguing in Publication Data

A CIP catalogue record for this book
is available from the British Library

ISBN 0 85296 794 2

Printed in England by MPG Books Limited, Bodmin, Cornwall

Toujours, Ille

Contents

Preface

An important transmission line met in microwave engineering is the ridge waveguide. It consists of a regular rectangular waveguide with one or more metal inserts or ridges. The main advantage of this type of waveguide over a conventional one is the wider separation between the dominant mode and the first order one. Important quantities that enter into the description of any waveguide are the definitions of its propagation constant, attenuation and mode spectrum. The power-voltage, power-current and voltage-current definitions of its impedance are other salient properties. The text includes closed form descriptions and finite element calculations on each of these quantities. Another quantity of some importance in the design of non-reciprocal devices is the existence of circular polarisation of the alternating magnetic field. Its study is given special attention. Propagation in the ridge waveguide with a dielectric filler between the ridges is separately investigated using the finite element method. Such fillers support planes of counter-rotating circular polarisation on either side of the interfaces between the dielectric and free space regions and everywhere outside. Another ridge structure is the circular or square waveguide with more than one ridge. Propagation in the quadruple ridge waveguide with and without a gyro-magnetic filler is separately attended to. The latter arrangement supports so-called Faraday rotation, which plays an important role in the design of a number of nonreciprocal ferrite devices. It is also given special attention. The important problem of a step discontinuity between a regular and single ridge waveguide is separately addressed. One canonical representation of this sort of step is a shunt susceptance in cascade with an ideal transformer. Most standard passive components met in microwave engineering can, in practice, be realised in ridge topology. One typical device dealt with is the cross-guide directional coupler. An introduction to the design of lowpass and bandpass filters based on the mode matching method is also included. Nonreciprocal ferrite devices such as isolators, phase shifters, differential phase shift and junction circulators are also readily constructed in this waveguide. The text includes typical realisations of some of these components. A closely related waveguide to the ridge one is the finline geometry about which much has already been written. Only a brief introduction to it is included here for completeness' sake. One typical component is the 3-port finline circulator. Since much of the material presented in this text relies in its formulation on the finite element method an introductory chapter on the origin of this method is included for completeness.

November 2000

Chapter 1

The ridge waveguide

1.1 Introduction

A waveguide used in many broadband microwave equipments is the ridge geometry. An important feature of this sort of waveguide compared to the conventional rectangular waveguide is the wider separation between the cut-off numbers of its dominant and first higher order mode. Another is the fact that its impedance is bracketed between that of the regular rectangular waveguide (377 Ω) and those of coaxial and stripline structures (50 Ω). The original ridge waveguide consisted of a regular rectangular waveguide with one or two ridge inserts. Most passive components that may be realised in conventional rectangular waveguides are also available in ridge geometry. This chapter includes some typical arrangements by way of introduction. Some more recent configurations have various arrangements of two or more ridges. Typical geometries are illustrated in Figure 1.1. Square or round waveguides with one or more ridges have also been described. Figure 1.2 depicts some such structures. Figure 1.3 illustrates the field patterns for the first quasi-TE mode for two different geometries in this sort of waveguide.

1.2 Cut-off space of ridge waveguide

An important fundamental quantity entering into the description of any waveguide is its cut-off number. The influence of the ridge inserts on the cut-off space of the dominant mode in the double ridge arrangement is illustrated in Figure 1.4 by way of example. One property of this sort of waveguide is that its cut-off number can be varied by adjusting the details of the ridge without altering its outside dimensions. The guide wavelength

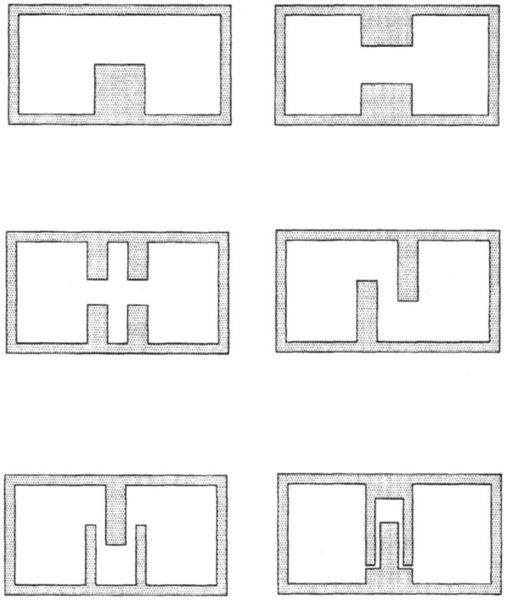

Figure 1.1 Schematic diagrams of rectangular ridge waveguides

(λ_g) is related to the cut-off (λ_c) and free space (λ_0) wavelengths in the usual way by

$$\left(\frac{2\pi}{\lambda_g}\right)^2 = \left(\frac{2\pi}{\lambda_0}\right)^2 - \left(\frac{2\pi}{\lambda_c}\right)^2$$

Figure 1.2 Schematic diagrams of square and round ridge waveguides

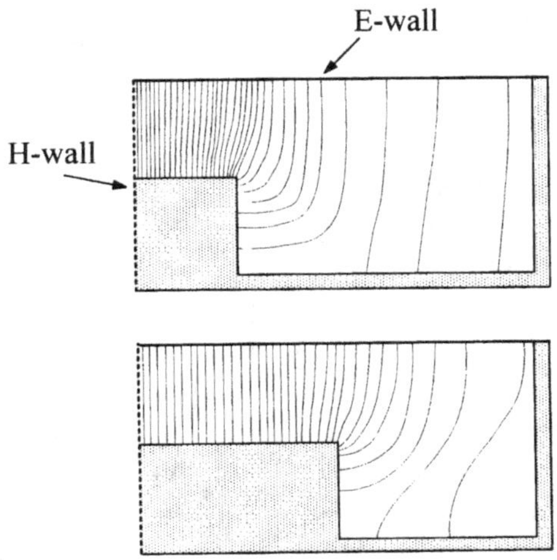

Figure 1.3 Standing wave solutions of dominant TE mode of double ridge waveguide for two different values of s/a (b/a = 0.5, d/b = 0.5)

Reprinted with permission, Helszajn & McKay (1998, *IEE Proc.*)

A table of standard commercial double ridge waveguides for various frequency bands is reproduced in Table 1.1.

The cut-off space of the ridge waveguide may be deduced by making use of the mode matching method (MMM), the finite element method (FEM) or other numerical techniques.

1.3 Impedance of ridge waveguide

A feature of the ridge waveguide is that the value of its characteristic impedance can be readily adjusted between that of a regular rectangular waveguide and of a coaxial cable. Another property of the ridge waveguide, in keeping with the conventional one, is that it has no unique definition of impedance. The three usual choices are based on voltage-current (Z_{vi}), power-voltage (Z_{pv}) and power-current (Z_{pi}) definitions. Still another property of this waveguide is that its impedance at finite frequency is related to that at infinite frequency by

$$Z(\omega) = Z(\infty)\left(\frac{\lambda_g}{\lambda_0}\right)$$

This identity holds equally for each of the definitions of impedance in common usage. Figure 1.5 depicts the relationship between the power-voltage definition of impedance and the details of the double ridge waveguide.

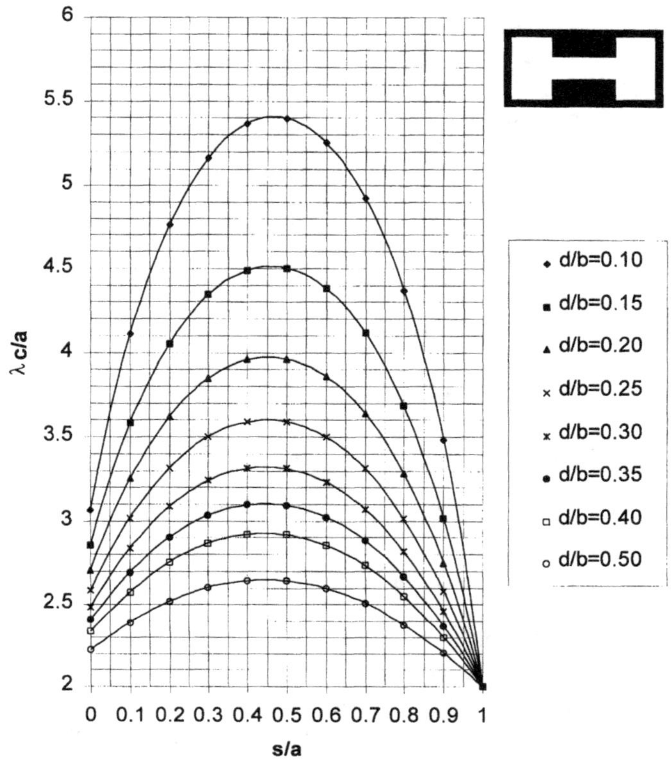

Figure 1.4 Cut-off space of double ridge waveguide ($b/a = 0.50$)
After Cohn (1947)

1.4 Attenuation of ridge waveguide

An important quantity in the specification of any transmission line is its attenuation per unit length (α).

If the dissipation per unit length is small, the time averaged power transmitted through the waveguide may be approximately written as

$$P_t \rightarrow P_t \exp(-2\alpha z)$$

The power loss per unit length is separately defined by

$$P_\ell = -\frac{\partial P_t}{\partial z} = 2\alpha P_t$$

The attenuation per unit length (α) is therefore described by

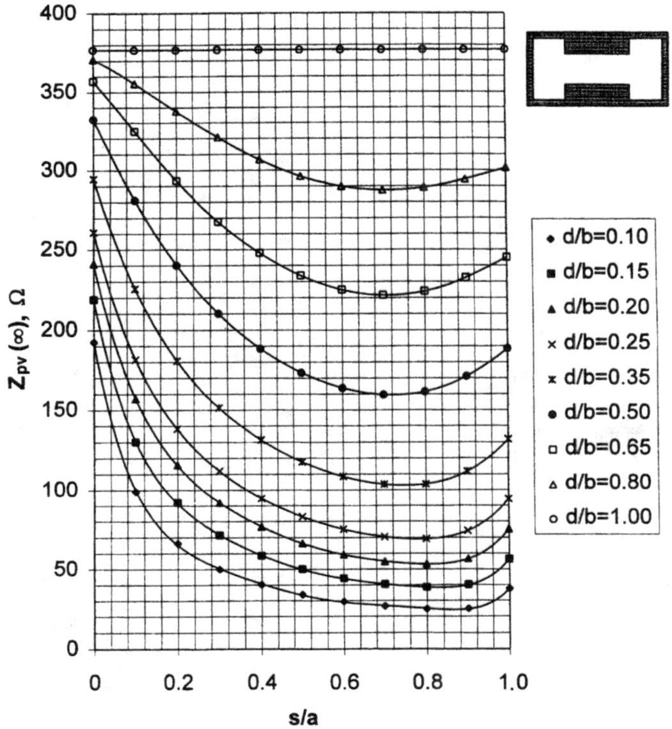

Figure 1.5 Power-voltage impedance of double ridge waveguide ($b/a = 0.50$)

After Hopfer (1955)

$$\alpha = \frac{P_\ell}{2P_t} \text{ dB/m}$$

Figure 1.6 summarises the relationships between frequency and attenuation per unit length of various waveguides and lines.

1.5 Ridge waveguide junctions

Most commercial passive microwave components which are available in regular rectangular waveguides can also be fabricated in ridge geometry. One practical component that embodies both rectangular and circular ridge waveguides is the turnstile junction. It is a 6-port network with four

Table 1.1 Standard commercial double ridge waveguides

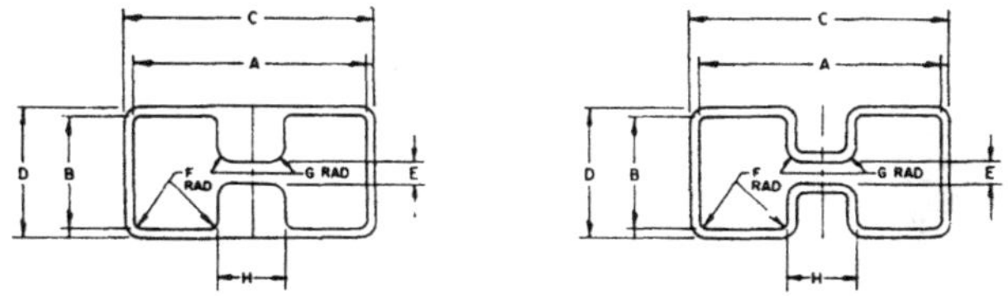

MIL-W-23351/4B — Dimensions in inches; mm on second line

WRD	Frequency range GHz	Theoretical attenuation for copper at 1.73fc dB/FT	A	B	C	D	E	F max	G ±10%	H
200	2.00–4.80	0.0089	2.590 ± 0.004	1.205 ± 0.004	2.750 ± 0.004	1.365 ± 0.004	0.512 ± 0.002	0.050	0.102	0.648 ± 0.002
			65.79 ± 0.10	30.61 ± 0.10	69.85 ± 0.10	34.67 ± 0.10	13.00 ± 0.05	1.27	2.59	16.46 ± 0.05
350	3.50–8.20	0.0204	1.480 ± 0.003	0.688 ± 0.003	1.608 ± 0.004	0.816 ± 0.004	0.292 ± 0.002	0.030	0.053	0.370 ± 0.002
			37.59 ± 0.08	17.48 ± 0.08	40.84 ± 0.10	20.73 ± 0.10	7.42 ± 0.05	0.76	1.47	9.40 ± 0.05
475	4.75–11.00	0.0324	1.09 ± 0.003	0.506 ± 0.003	1.190 ± 0.003	0.606 ± 0.003	0.215 ± 0.002	0.030	0.043	0.272 ± 0.002
			27.69 ± 0.08	12.85 ± 0.08	30.23 ± 0.08	15.39 ± 0.08	5.46 ± 0.05	0.76	1.09	6.91 ± 0.05
580	5.80–16.00	0.065	0.780 ± 0.003	0.370 ± 0.003	0.880 ± 0.003	0.470 ± 0.003	0.120 ± 0.002	0.020	0.043	0.200 ± 0.002
			19.82 ± 0.08	9.40 ± 0.08	22.35 ± 0.08	11.94 ± 0.08	3.05 ± 0.05	0.51	1.09	5.08 ± 0.05
650	6.50–18.00	0.080	0.720 ± 0.003	0.321 ± 0.003	0.820 ± 0.003	0.421 ± 0.003	0.101 ± 0.002	0.020	0.022	0.173 ± 0.002
			18.29 ± 0.08	8.15 ± 0.08	20.83 ± 0.08	10.69 ± 0.08	2.57 ± 0.05	0.51	0.56	4.39 ± 0.05
750	7.50–18.00	0.0641	0.691 ± 0.003	0.321 ± 0.003	0.791 ± 0.003	0.421 ± 0.003	0.136 ± 0.002	0.020	0.027	0.173 ± 0.002
			17.55 ± 0.08	8.15 ± 0.08	20.09 ± 0.08	10.69 ± 0.08	3.45 ± 0.05	0.51	0.69	4.39 ± 0.05
110	11.00–26.50	0.114	0.471 ± 0.003	0.219 ± 0.003	0.551 ± 0.003	0.299 ± 0.003	0.093 ± 0.002	0.015	0.019	0.118 ± 0.002
			11.96 ± 0.08	5.56 ± 0.08	14.00 ± 0.08	7.59 ± 0.08	2.362 ± 0.05	0.38	0.48	2.997 ± 0.05
180	18.00–40.00	0.238	0.288 ± 0.003	0.134 ± 0.003	0.368 ± 0.003	0.214 ± 0.003	0.057 ± 0.002	0.015	0.011	0.072 ± 0.002
			7.32 ± 0.08	3.40 ± 0.08	9.35 ± 0.08	5.44 ± 0.08	1.448 ± 0.05	0.38	0.28	1.829 ± 0.05

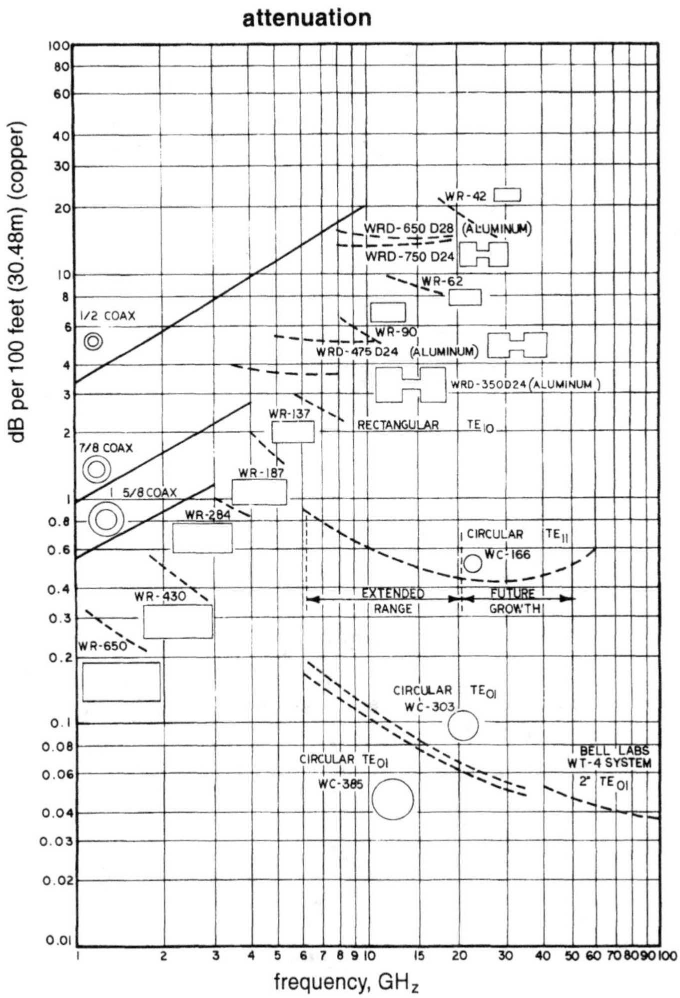

Figure 1.6 Attenuation in ridge and various other waveguides

Courtesy of Litton Inc.

ports in the rectangular waveguide and two in the circular one. Figure 1.7 depicts its geometry. The scattering matrix of this junction is described by

$$[S] = \begin{pmatrix} \alpha & \gamma & \delta & \gamma & \vdots & \varepsilon & 0 \\ \gamma & \alpha & \gamma & \delta & \vdots & 0 & \varepsilon \\ \delta & \gamma & \alpha & \gamma & \vdots & -\varepsilon & 0 \\ \gamma & \delta & \gamma & \alpha & \vdots & 0 & -\varepsilon \\ \cdots & \cdots & \cdots & \cdots & \cdots & \cdots & \cdots \\ \varepsilon & 0 & -\varepsilon & 0 & \vdots & \beta & 0 \\ 0 & \varepsilon & 0 & -\varepsilon & \vdots & 0 & \beta \end{pmatrix}$$

If the junction is matched, then

$$\alpha = 0$$
$$\beta = 0$$
$$\delta = 0$$
$$|\gamma| = \frac{1}{\sqrt{2}}$$
$$|\varepsilon| = \frac{1}{\sqrt{2}}$$

Another classic component is the cross-guide directional coupler. It is defined as a matched 4-port network with one adjacent port decoupled from any input port. Single and double ridge geometries are illustrated in

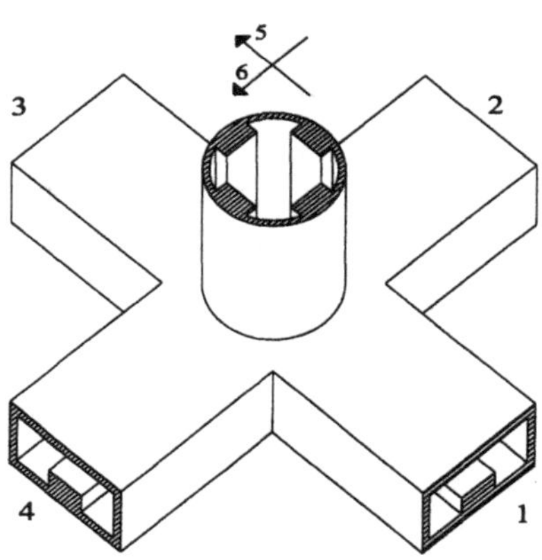

Figure 1.7 Ridge waveguide turnstile – a symmetrical 6-port junction

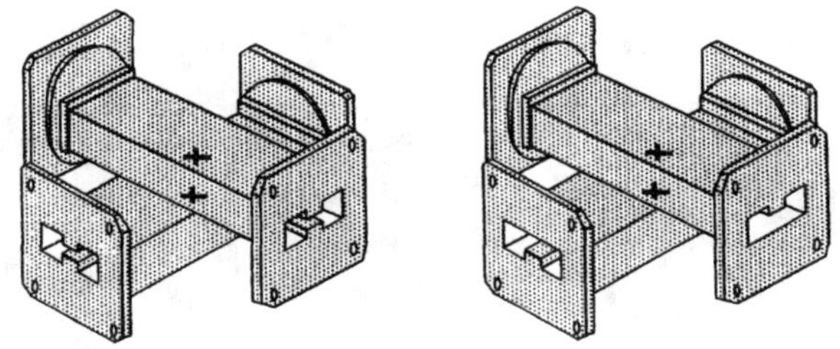

Figure 1.8 Schematic diagrams of single and double ridge cross-guide directional coupler

Figure 1.8. The waveguide section containing ports 1 and 3 is sometimes referred to as the primary waveguide, while that containing ports 2 and 4 is denoted the secondary waveguide. The scattering matrix of the ideal, symmetric and lossless network with port 2 decoupled from port 1 is given in the usual way by

$$[S] = \begin{bmatrix} 0 & 0 & S_{31} & S_{41} \\ 0 & 0 & S_{41} & S_{31} \\ S_{31} & S_{41} & 0 & 0 \\ S_{41} & S_{31} & 0 & 0 \end{bmatrix}$$

The relationship between the transmitted (S_{31}) and coupled (S_{41}) coefficients is given by the unitary condition

$$S_{31}S_{31}^* + S_{41}S_{41}^* = 1$$

$$S_{31}S_{41}^* + S_{31}^*S_{41} = 0$$

One solution is

$$S_{31} = \alpha$$

$$S_{41} = j\beta$$

Another much used ridge component is the magic-tee. A commercial version is reproduced in Figure 1.9. It may be visualised as a combination of H- and E-plane tee junctions. A wave incident at the H-plane port of the junction divides equally between the symmetric ports; a wave at the E-plane port produces out-of-phase equal amplitude waves at the two symmetric ports. In-phase or out-of-phase waves at the symmetric ports recombine at the H- and E-plane ports. The scattering matrix of the circuit is

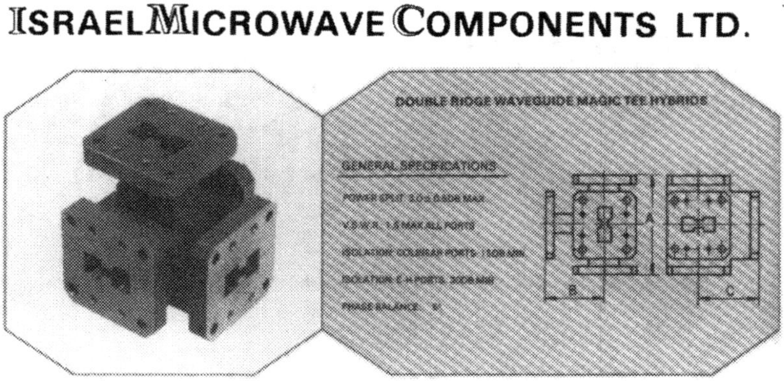

Figure 1.9 Commercial ridge magic-tee

Courtesy of Israel Microwave Components Ltd.

$$[S] = \begin{bmatrix} 0 & 0 & \alpha & \beta \\ 0 & 0 & \beta & \alpha \\ \alpha & \beta & 0 & 0 \\ \beta & -\alpha & 0 & 0 \end{bmatrix}$$

where

$$\alpha = \beta = \frac{1}{\sqrt{2}}$$

1.6 Waveguide transitions

It is not unusual, in any piece of equipment, to have some of its components in one sort of waveguide and some in another. Transitions between various types of transmission lines are therefore necessary in practice. One common transition is that between the standard and the ridge waveguide; another is between a coaxial cable and a ridge. The waveguide transition consists of one or more quarter-wave long impedance transformers with intermediate cross-sections to those being joined. A single section arrangement is indicated in Figure 1.10*a* and a coaxial to ridge waveguide transition in Figure 1.10*b*.

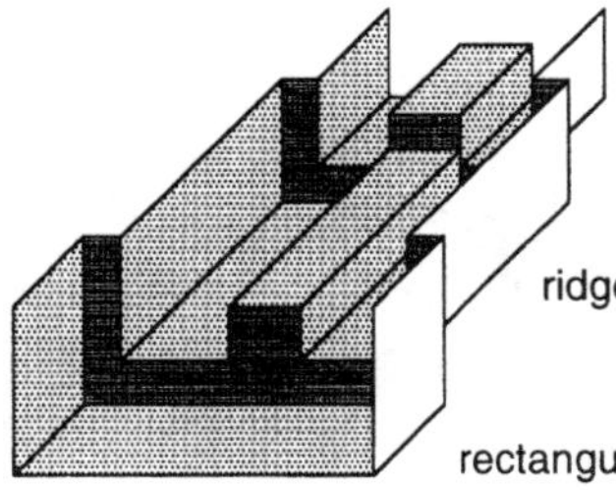

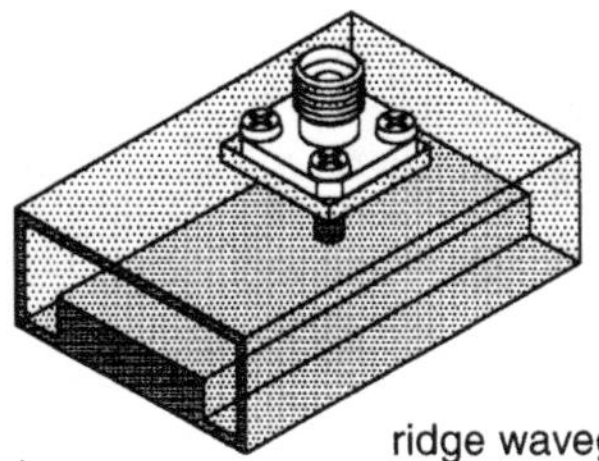

Figure 1.10 Schematic diagrams of ridge transitions

a Single ridge section; *b* coaxial to ridge

1.7 Filter circuits

No passive piece of microwave equipment is complete without one or more filter circuits. The ridge waveguide is again appropriate in this instance. One possible topology is a bandpass filter made up of half-wave long UEs separated by suitable immittance inverters. One realisation of a typical inverter is a short section of cut-off rectangular waveguide. Figure 1.11 depicts the overall structure in question.

1.8 Turnstile junction circulator

Another important class of ridge waveguide components is the non-reciprocal one. One possible assembly is the 4-port turnstile circulator. It is realised from the 6-port turnstile junction by a 45 degree Faraday rotation bit in the round waveguide. This arrangement is shown in Figure 1.12. The operation of the circulator in question may be understood by considering a typical input wave at port 1. Such a wave produces no reflection at port 1, decouples ports 3 and 6, establishes equal in-phase waves at ports 2 and 4

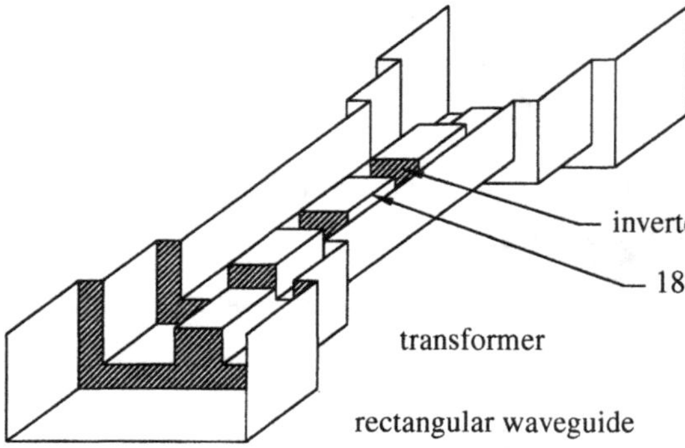

Figure 1.11 Schematic diagram of ridge bandpass filter using immittance inverters

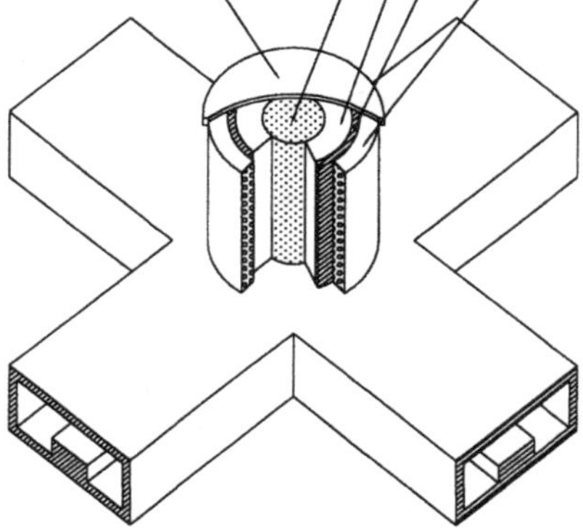

Figure 1.12 Turnstile circulator: polarisation 5 is rotated 45 degrees, reflected, then rotated 45 degrees further to polarisation 6

After Allen (1956)

and produces a component at port 5. The wave at port 5, upon traversing up and down the 45 degree rotator section, is now aligned with port 6 at the plane of the rectangular waveguide. Such a wave decouples ports 1, 3 and 5 and produces out-of-phase waves at ports 2 and 4 which have equal amplitudes to those established by the original incident wave. The net effect is to produce a single output at port 2. Similar considerations indicate that a wave at port 2 is emergent at port 3 and so on in a cyclic manner. The port notation of the circulator is that introduced in connection with the description of the turnstile junction in Figure 1.7.

Chapter 2

Propagation and impedance in rectangular waveguides

2.1 Introduction

The description of any waveguide includes its cut-off number, its propagation constant, one or more definitions of impedance, power flow and attenuation and a description of its field pattern. Since the rectangular waveguide embodies all the salient properties of the ridge waveguide it is apt to review some of its more important features before tackling the ridge structure. The fields in this sort of waveguide are usually deduced by obtaining a solution for E_z or H_z or both which satisfy the wave equation. The other field components are then obtained by using Maxwell's equations. A feature of particular importance in this sort of waveguide is the lack of uniqueness in the definition of its characteristic impedance. Another is the existence of planes of circular polarisation between the top and bottom walls of the waveguide on either side of its symmetry plane. Counter-rotating alternating magnetic fields are also displayed with different hands in a ridge waveguide. The chapter is restricted to a description of the dominant mode in the geometry but the existence of higher order modes is understood. The orthogonal property of any two modes of the waveguide is separately established.

2.2 The wave equation

The property of an inhomogeneous waveguide is that the transverse components of the electric and magnetic fields may be written in terms of the longitudinal ones. This property is a classic result in the literature. The required result in Cartesian co-ordinates is

$$H_x = \frac{-\gamma}{k_c^2}\left(\frac{\partial H_z}{\partial x}\right) + \frac{j\omega\varepsilon_0}{k_c^2}\left(\frac{\partial E_z}{\partial y}\right) \tag{1a}$$

$$H_y = \frac{-\gamma}{k_c^2}\left(\frac{\partial H_z}{\partial y}\right) - \frac{j\omega\varepsilon_0}{k_c^2}\left(\frac{\partial E_z}{\partial x}\right) \tag{1b}$$

$$E_x = \frac{-j\omega\mu_0}{k_c^2}\left(\frac{\partial H_z}{\partial y}\right) - \frac{\gamma}{k_c^2}\left(\frac{\partial E_z}{\partial x}\right) \tag{1c}$$

$$E_y = \frac{j\omega\mu_0}{k_c^2}\left(\frac{\partial H_z}{\partial x}\right) - \frac{\gamma}{k_c^2}\left(\frac{\partial E_z}{\partial y}\right) \tag{1d}$$

E_z and H_z satisfy both the wave equation and the boundary conditions of the problem region:

$$(\nabla_t^2 + k_c^2)H_z = 0 \tag{2a}$$

$$(\nabla_t^2 + k_c^2)E_z = 0 \tag{2b}$$

The solution to this sort of problem therefore amounts to finding E_z or H_z (or both) which satisfies both the wave equation and the boundary conditions and thereafter deducing the other components of the field by using equation (1). The modes in this and other waveguides are labelled TE, TM or EH according to whether H_z, E_z or E_z and H_z exist. The work is, however, restricted to the dominant TE_{10} mode.

2.3 Dominant mode in rectangular waveguides

The dominant mode in the rectangular waveguide is designated TE_{10}. It is described by

$$H_z = A_{10}\cos\left(\frac{\pi x}{a}\right) \tag{3a}$$

$$H_y = 0 \tag{3b}$$

$$H_x = jA_{10}\left(\frac{\lambda_c}{\lambda_g}\right)\sin\left(\frac{\pi x}{a}\right) \tag{3c}$$

$$E_z = 0 \tag{3d}$$

$$E_y = -jA_{10}\left(\frac{\lambda_c}{\lambda_0}\right)\sqrt{\frac{\mu_0}{\varepsilon_0}}\sin\left(\frac{\pi x}{a}\right) \tag{3e}$$

$$E_x = 0 \tag{3f}$$

The guide wavelength (λ_g) is separately defined in the usual way by

$$\left(\frac{2\pi}{\lambda_g}\right)^2 = \left(\frac{2\pi}{\lambda_0}\right)^2 - \left(\frac{2\pi}{\lambda_c}\right)^2 \tag{4}$$

The cut-off wavelength (λ_c) is related to the wide dimension of the waveguide (a) and the separation constant (k_c) by

$$k_c = \frac{2\pi}{\lambda_c} \tag{5a}$$

For the TE_{10} mode with $m = 1$, $n = 0$,

$$\lambda_c = 2a \tag{5b}$$

The phase constant of the waveguide (β) is related to λ_g by

$$\beta = \frac{2\pi}{\lambda_g} \tag{6}$$

The transverse wave impedance of the waveguide is denoted by Z_{TE},

$$Z_{TE} = \frac{E_y}{H_x} = \left(\frac{\lambda_g}{\lambda_0}\right)\sqrt{\frac{\mu_0}{\varepsilon_0}} \tag{7}$$

The free space wave impedance η_0 is given by

$$\eta_0 = \sqrt{\frac{\mu_0}{\varepsilon_0}} = 120\pi \tag{8}$$

The rectangular waveguide considered here is indicated in Figure 2.1 and a typical field pattern is shown in Figure 2.2.

2.4 Impedance in waveguides

The characteristic impedance of a uniform transmission line supporting TEM propagation may be defined in one of three possible ways:

$$Z_{PV} = \frac{VV^*}{2P_t} \tag{9a}$$

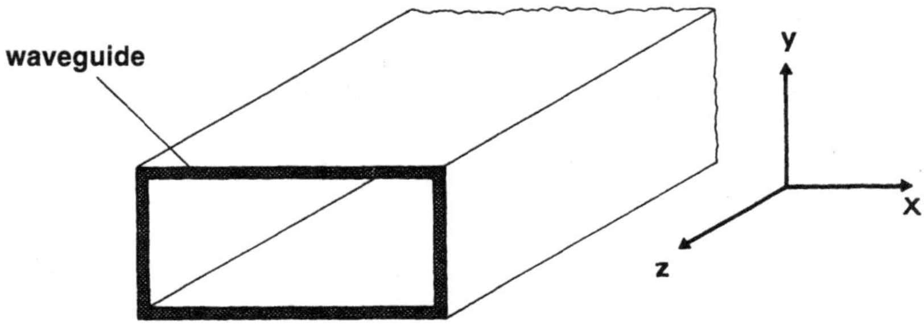

Figure 2.1 Schematic diagram of rectangular waveguide

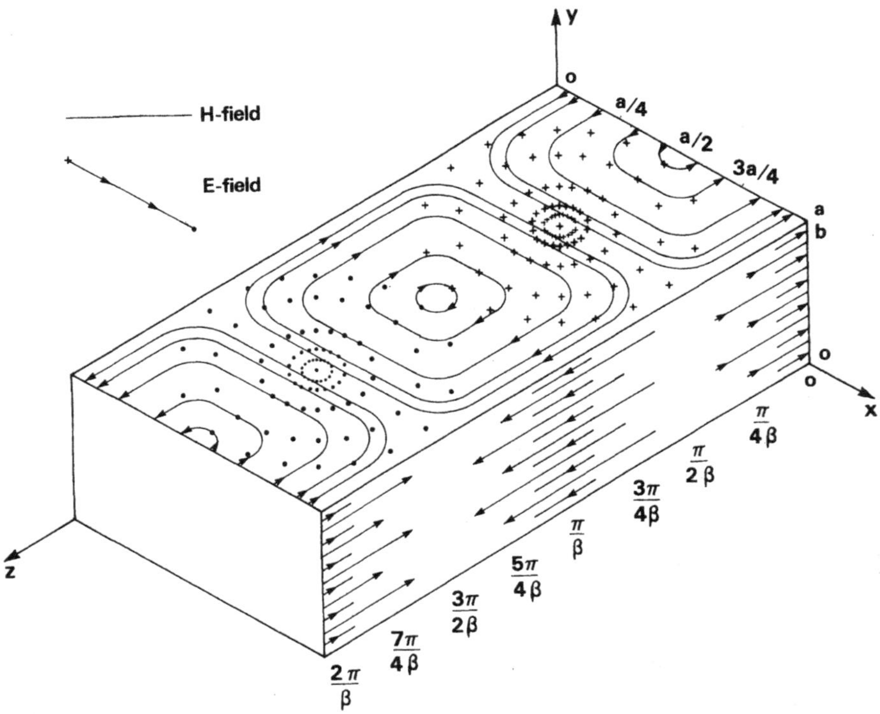

Figure 2.2 Field pattern of dominant TE_{10} mode in rectangular waveguide

$$Z_{\mathrm{PI}} = \frac{2P_{\mathrm{t}}}{II^*} \tag{9b}$$

$$Z_{\mathrm{VI}} = \frac{V}{I} \tag{9c}$$

and

$$Z_{\mathrm{PV}} = Z_{\mathrm{PI}} = Z_{\mathrm{VI}} \tag{9d}$$

However, in rectangular or circular waveguides, the definitions of voltage and current are not unique, so that

$$Z_{\mathrm{PV}} \neq Z_{\mathrm{PI}} \neq Z_{\mathrm{VI}} \tag{10}$$

It is readily observed that one relationship between the various definitions of impedance is

$$Z_{\mathrm{VI}} = \sqrt{Z_{\mathrm{PV}} Z_{\mathrm{PI}}} \tag{11}$$

2.5 Power transmission through rectangular waveguides

An important quantity in the description of any waveguide is the average power (P_{t}) transmitted through it. It may be evaluated by making use of the complex Poynting theorem,

$$P_{\mathrm{t}} = \frac{1}{2} \int\!\!\int_s \mathrm{Re}(\bar{E} x \bar{H}^*)\,\mathrm{d}\bar{S} \tag{12}$$

where $\bar{S}$ is the total surface perpendicular to the direction of propagation.

For the dominant TE_{10} mode in a rectangular waveguide

$$\frac{1}{2}(\bar{E} x \bar{H}^*) = \frac{1}{2} \begin{bmatrix} \bar{a}_x & \bar{a}_y & \bar{a}_z \\ 0 & E_y & 0 \\ H_x^* & 0 & H_z^* \end{bmatrix} \tag{13a}$$

and

$$\tfrac{1}{2}(\bar{E} x \bar{H}^*) = \tfrac{1}{2}[\bar{a}_x(E_y H_z^*) + \bar{a}_y(0) + \bar{a}_z(-E_y H_x^*)] \tag{13b}$$

Noting that $E_y H_z^*$ is a pure imaginary quantity gives

$$\frac{1}{2}\,\mathrm{Re}(\bar{E} x \bar{H}^*) = \frac{1}{2}\left(\frac{\lambda_{\mathrm{c}}}{\lambda_{\mathrm{g}}}\right)\left(\frac{\lambda_{\mathrm{c}}}{\lambda_0}\right)\sqrt{\frac{\mu_0}{\varepsilon_0}}\,\sin^2\!\left(\frac{\pi x}{a}\right) \tag{14}$$

The derivation now proceeds by integrating this quantity over the waveguide cross-section:

$$P_t = \frac{1}{2}\left(\frac{\lambda_c}{\lambda_g}\right)\left(\frac{\lambda_c}{\lambda_0}\right)\sqrt{\frac{\mu_0}{\varepsilon_0}}\int_0^a\int_0^b \sin^2\left(\frac{\pi x}{a}\right) \mathrm{d}x\, \mathrm{d}y \tag{15}$$

The required result is

$$P_t = \frac{ab}{4}\left(\frac{\lambda_c}{\lambda_g}\right)\left(\frac{\lambda_c}{\lambda_0}\right)\sqrt{\frac{\mu_0}{\varepsilon_0}} \tag{16}$$

The amplitude term A_{10}^2 in the description of P_t is understood.

2.6 Impedance in rectangular waveguides

The derivations of each of the three definitions of impedance met in the case of a rectangular waveguide propagating the TE_{10} mode will now be illustrated by way of an example. The voltage V across the waveguide is defined as the line integral of the electric field at the midpoint of the waveguide and the current I as the total longitudinal current flowing in the wide surface of one of the waveguide walls. The power flow P_t is given by equation (16).

The derivation of the power-voltage definition of impedance Z_{PV} starts by forming V at the symmetry plane of the waveguide:

$$V = \int_0^b E_y \, \mathrm{d}y \tag{17}$$

Evaluating this quantity at $x = a/2$ gives

$$V = -j\left(\frac{2ab}{\lambda_0}\right)\sqrt{\frac{\mu_0}{\varepsilon_0}} \tag{18}$$

Combining this result with the description of the power flow in the waveguide produces the required result:

$$Z_{PV} = \left(\frac{2b}{a}\right)Z_{TE} \tag{19}$$

The power-current definition for the impedance (Z_{PI}) is established by evaluating the total longitudinal current flowing in the wide dimension of the waveguide:

$$I = \int_0^a J_z \, \mathrm{d}x \tag{20}$$

This may be done by noting the relationship between the current density and the magnetic field:

$$|J_z| = |H_x| \tag{21}$$

The total longitudinal current is therefore

$$I = j\frac{\lambda_c^2}{\pi\lambda_g} \tag{22}$$

The corresponding power-current definition of impedance is

$$Z_{\mathrm{PI}} = \left(\frac{\pi^2 b}{8a}\right) Z_{\mathrm{TE}} \tag{23}$$

The voltage-current definition of impedance in a rectangular waveguide is now deduced by making use of the relationship between it and the power-voltage and power-current definitions in equation (11). The ensuing result is

$$Z_{\mathrm{VI}} = \left(\frac{\pi b}{2a}\right) Z_{\mathrm{TE}} \tag{24}$$

A scrutiny of the three descriptions of impedance in the waveguide suggests that Z_{PV} is of the order of Z_{TE} in a standard rectangular waveguide.

2.7 Circular polarisation in rectangular waveguides

An important property of a rectangular waveguide propagating the dominant TE_{10} mode is that it displays regions on either side of the symmetry plane at which the alternating magnetic field is circularly polarised with opposite senses of rotation. Such polarisation is defined by two equal amplitude waves in the time-space quadrature. Furthermore, if propagation is in the negative z direction the two hands of polarisation are interchanged. These features of a rectangular waveguide are of particular significance in that the operation of a number of nonreciprocal ferrite devices relies on such polarisations. The positions at which the magnetic field is circularly polarised can be derived without any difficulty by first putting down the three field components for the waveguide:

$$H_z = \cos\left(\frac{\pi x}{a}\right)\exp j(\omega t - \beta z) \tag{25a}$$

$$H_x = j\left(\frac{\lambda_c}{\lambda_g}\right)\sin\left(\frac{\pi x}{a}\right)\exp j(\omega t - \beta z) \tag{25b}$$

$$E_y = -j\left(\frac{\lambda_c}{\lambda_0}\right)\sqrt{\frac{\mu_0}{\varepsilon_0}}\sin\left(\frac{\pi x}{a}\right)\exp j(\omega t - \beta z) \tag{25c}$$

where propagation is assumed along the positive z direction.

Scrutiny of the preceding equations indicates that H_x and H_z are in time-space quadrature. If a region can now be located where the amplitudes are also equal, then it would exhibit circular polarisation there. This condition is in fact satisfied on either side of the centre line of the waveguide provided that

$$\tan\left(\frac{\pi x}{a}\right) = \left(\frac{\lambda_g}{\lambda_c}\right) \tag{26}$$

The two possible solutions to the preceding equation are satisfied in the vicinity of

$$x \approx \frac{a}{4} \tag{27a}$$

$$x \approx \frac{3a}{4} \tag{27b}$$

The nature of the circular polarisation in either direction of propagation in a rectangular waveguide may now be examined by taking the real parts of H_x and H_z along each direction. Taking the solution at $x = a/4$ by way of example gives

$$H_z = 0.707\cos(\omega t - \beta z), \qquad x = \frac{a}{4} \tag{28a}$$

$$H_x = -0.707\sin(\omega t - \beta z), \qquad x = \frac{a}{4} \tag{28b}$$

and

$$H_z = 0.707\cos(\omega t + \beta z), \qquad x = \frac{a}{4} \tag{28c}$$

$$H_x = 0.707\sin(\omega t + \beta z), \qquad x = \frac{a}{4} \tag{28d}$$

respectively.

Taking the solution at $x = 3a/4$ gives

$$H_z = -0.707\cos(\omega t - \beta z), \qquad x = \frac{3a}{4} \tag{29a}$$

$$H_x = -0.707\sin(\omega t - \beta z), \qquad x = \frac{3a}{4} \tag{29b}$$

and

$$H_z = -0.707\cos(\omega t + \beta z), \qquad x = \frac{3a}{4} \tag{29c}$$

$$H_x = 0.707\sin(\omega t + \beta z), \qquad x = \frac{3a}{4} \tag{29d}$$

respectively. Figure 2.3 shows the required results at $\omega t = 0$ and $\beta z = 0$, $\pi/2$, π and $3\pi/2$. Figure 2.4 indicates the corresponding result with $\beta z = 0$, $-\pi/2$, $-\pi$ and $-3\pi/2$.

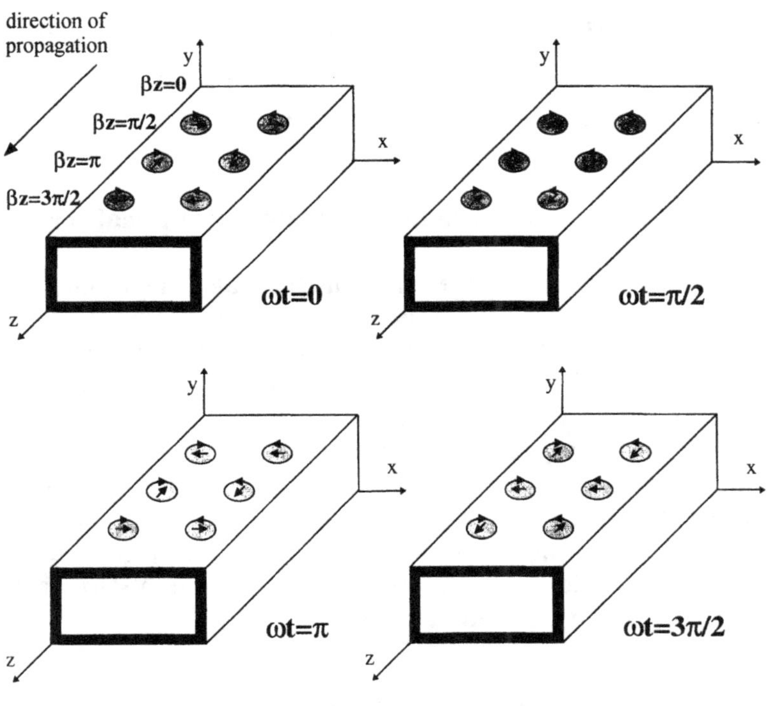

Figure 2.3 Planes of circular polarisation in rectangular waveguide for forward direction of propagation

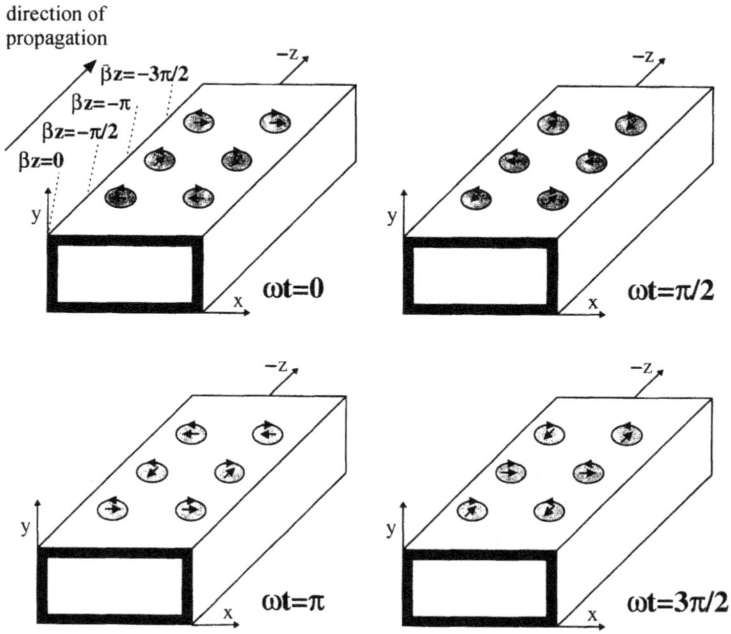

Figure 2.4 Planes of circular polarisation in rectangular waveguide for reverse direction of propagation

2.8 Calculation of impedance based on a mathematical technique

One mathematical technique that may be used to calculate the impedance of any homogeneous waveguide will now be outlined. It relies for its implementation on the fact that the amplitude distributions of the electric and magnetic fields are identical at both cut-off and infinite frequency and on the fact that the fields are related by the wave impedance at infinite frequency. This means that a knowledge of the amplitude distribution of either the electric or magnetic field at cut-off is sufficient to evaluate the impedance of this class of waveguide. The required procedure may be understood by construction of the voltage-current definition of impedance in a rectangular waveguide by way of an example. It amounts to writing the impedance $Z_{VI}(\omega)$ at finite frequency as

$$Z_{VI}(\omega) = \left(\frac{\lambda_g}{\lambda_0}\right) Z_{VI}(\infty) \tag{30}$$

where

$$Z_{\mathrm{VI}}(\infty) = \frac{V(\infty)}{I(\infty)} \tag{31}$$

Forming $V(\infty)$ and $I(\infty)$ in terms of the electric field $E(\infty)$ gives

$$V(\infty) = \int_0^b E_y(\infty)\,\mathrm{d}y \tag{32}$$

$$\eta_0 I(\infty) = \int_0^a E_y(\infty)\,\mathrm{d}x \tag{33}$$

The calculation is completed once the electric field distribution in the waveguide is at hand. For a standard rectangular waveguide propagating the dominant TE_{10} mode the distribution of the electric field is described by

$$E_y = A \sin\left(\frac{\pi x}{a}\right) \tag{34}$$

Introducing this relationship in the preceding equations readily gives

$$V(\infty) = Ab \tag{35}$$

$$\eta_0 I(\infty) = A\left(\frac{2a}{\pi}\right) \tag{36}$$

The impedances at infinite and finite frequencies are

$$Z_{\mathrm{VI}}(\infty) = \eta_0\left(\frac{\pi b}{2a}\right) \tag{37}$$

and

$$Z_{\mathrm{VI}}(\omega) = \eta_0\left(\frac{\lambda_g}{\lambda_0}\right)\left(\frac{\pi b}{2a}\right) \tag{38}$$

in agreement with the original result. While λ_0 and λ_g tend to zero at infinite frequency the ratio of the two quantities is finite there.

The other definitions of impedance encountered in the description of this sort of waveguide may also be expressed in a like manner:

$$Z_{\mathrm{PV}}(\omega) = \left(\frac{\lambda_g}{\lambda_0}\right) Z_{\mathrm{PV}}(\infty) \tag{39}$$

$$Z_{\mathrm{PI}}(\omega) = \left(\frac{\lambda_g}{\lambda_0}\right) Z_{\mathrm{PI}}(\infty) \tag{40}$$

The original choice of fields in the waveguide does not readily permit the power at finite frequency to be expressed in terms of that at infinite frequency. To do so it is necessary to assume that E_y rather than H_z is frequency independent:

$$E_y = A_{10} \sin\left(\frac{\pi x}{a}\right) \tag{41}$$

$$H_x = -A_{10}\left(\frac{\lambda_0}{\lambda_g}\right)\sqrt{\frac{\varepsilon_0}{\mu_0}} \sin\left(\frac{\pi x}{a}\right) \tag{42}$$

$$H_z = jA_{10}\left(\frac{\lambda_0}{\lambda_c}\right)\sqrt{\frac{\varepsilon_0}{\mu_0}} \cos\left(\frac{\pi x}{a}\right) \tag{43}$$

The power flow at finite frequency is then given in terms of that at infinite frequency by

$$P_t(\omega) = P_t(\infty)\left(\frac{\lambda_0}{\lambda_g}\right) \tag{44}$$

where

$$P_t(\infty) = A_{10}^2\left(\frac{ab}{4}\right)\sqrt{\frac{\varepsilon_0}{\mu_0}} \tag{45}$$

A_{10} is again an arbitrary constant that may be used to set the power flow along the waveguide to unity or some other suitable value.

2.9 Orthogonal properties of waveguide modes

An important feature of any two modes of a waveguide that enters into a number of calculations is its orthogonal properties. The required relationships are readily demonstrated in the cases of the longitudinal components E_z or H_z and with a little more difficulty in the case of the transverse components. The desired derivation in the case of either H_z or E_z starts by noting the wave equations for two typical solutions ϕ_i and ϕ_j:

$$\nabla_t^2 \phi_i + k_i^2 \phi_i = 0 \tag{46}$$

$$\nabla_t^2 \phi_j + k_i^2 \phi_j = 0 \tag{47}$$

It proceeds by multiplying the first equation by ϕ_j and the second one by ϕ_i, forming the difference between the two, and integrating each quantity over the cross-section of the waveguide. This gives

$$(k_i^2 - k_j^2) \iint_s (\phi_i \phi_j) \, \mathrm{d}s = \iint_s (\phi_i \nabla_t^2 \phi_j - \phi_j \nabla_t^2 \phi_i) \, \mathrm{d}s \tag{48}$$

The derivation now proceeds by recalling Green's second identity

$$\iint_s (A \nabla_t^2 B) \, \mathrm{d}s = \oint_c \left(A \frac{\mathrm{d}B}{\mathrm{d}n} \right) \mathrm{d}c$$

s denotes the cross-sectional surface of the waveguide and c the guide boundary. Introducing this identity in the preceding relationship gives

$$(k_i^2 - k_j^2) \iint_s (\phi_i \phi_j) \, \mathrm{d}s = \oint_c \left(\phi_i \frac{\mathrm{d}\phi_j}{\mathrm{d}n} - \phi_j \frac{\mathrm{d}\phi_i}{\mathrm{d}n} \right) \mathrm{d}c \tag{49}$$

Since ϕ_i and ϕ_j are zero for TE type modes and $\mathrm{d}\phi_i/\mathrm{d}n$ and $\mathrm{d}\phi_j/\mathrm{d}n$ for TM modes then, as asserted,

$$\iint_s (\phi_i \phi_j) \, \mathrm{d}s = 0, \quad i \neq j \tag{50}$$

It may be separately demonstrated that this condition is equally valid in the case of degenerate modes.

Chapter 3

Impedance and propagation in ridge waveguides using the transverse resonance method

J. Helszajn and M. Caplin

3.1 Introduction

Important quantities in the description of any waveguide are the definition of its propagation constant and the voltage-current, power-voltage and power-current definitions of its characteristic impedance. One purpose of this chapter is to summarise some closed form descriptions of propagation, power flow and impedance in the ridge waveguide based on the transverse resonance method (TRM). Since the different notations introduced in its characterisation are on occasion difficult to follow readily, another purpose of this chapter is to reproduce the existing literature in a single nomenclature. Still another is to summarise graphically its voltage-current, power-voltage and power-current definitions of impedance in a unified way. No view is, however, taken about which definition of impedance is appropriate in any single application nor any attempt made to derive any result from first principles. The topologies of the single and double ridge versions of the ridge waveguide are illustrated in Figure 3.1. It is convenient in this sort of waveguide to write the impedance at finite frequency in terms of that at infinite frequency in that it avoids the need to make separate calculations at each and every frequency. The approach adopted here is in keeping with this convention. The cut-off space of the waveguide is established by using the transverse resonance condition.

3.2 Cut-off space of ridge waveguide

One quantity that enters into the description of any waveguide is its cut-off number. Its knowledge is sufficient for the description of the propagation

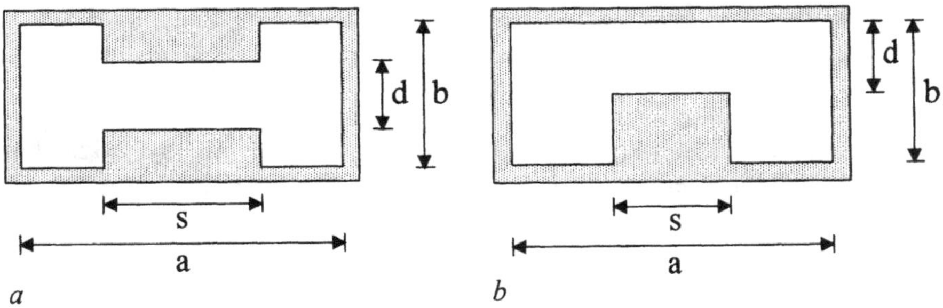

Figure 3.1 Schematic diagrams of single and double ridge waveguides

constant of the waveguide. Figure 3.1 illustrates the two possible ridge arrangements considered here. The physical variables entering into the descriptions of these waveguides are separately indicated on these diagrams. Figure 3.2 indicates the nature of the electric fields of the dominant modes in these sorts of waveguides.

One early calculation of the cut-off condition for either the single or the double ridge waveguide for the even TE_{no} family of modes is based on the transverse resonance method. It is determined by

$$-\cot\theta_1 + \left(\frac{Y_{02}}{Y_{01}}\right)\tan\theta_2 + \frac{B}{Y_{01}} = 0 \tag{1}$$

where

$$Y_{01} = \frac{k_c}{\omega\mu_0}\left(\frac{1}{b}\right) \tag{2a}$$

$$Y_{02} = \frac{k_c}{\omega\mu_0}\left(\frac{1}{d}\right) \tag{2b}$$

and

$$\theta_1 = \frac{\pi(a-s)}{\lambda_c} = \pi\left(1-\frac{s}{a}\right)\left(\frac{a}{\lambda_c}\right) \tag{3a}$$

$$\theta_2 = \frac{\pi s}{\lambda_c} = \pi\left(\frac{s}{a}\right)\left(\frac{a}{\lambda_c}\right) \tag{3b}$$

Figure 3.2 Electric fields in single and double ridge waveguides

The cut-off wavenumber k_c is related to the corresponding wavelength λ_c by

$$k_c = \frac{2\pi}{\lambda_c} \tag{4}$$

B/Y_{01} represents the step discontinuity on either side of the ridge. One approximation in the case of the single ridge is (Marcuvitz, 1964)

$$\frac{B}{Y_{01}} \approx 4\left(\frac{b}{a}\right)\left(\frac{a}{\lambda_c}\right)\ln \operatorname{cosec}\left(\frac{\pi d}{2b}\right) \tag{5}$$

The corresponding result for a double ridge is

$$\frac{B}{Y_{01}} \approx 2\left(\frac{b}{a}\right)\left(\frac{a}{\lambda_c}\right)\ln \operatorname{cosec}\left(\frac{\pi d}{2b}\right) \tag{6}$$

The relationship between the fringing capacitances of the single and double ridge waveguides may be understood by introducing an electric wall at the plane of symmetry of the latter arrangement. A scrutiny of this arrangement indicates that the two capacitors formed in this way are in series due to the fact that the top and bottom ridges are at different potentials in order to support the electric field. The normalised capacitance of the single ridge with the nomenclature used to label the details of the waveguide is therefore twice that of the double one as asserted. The equivalent circuit met in this problem region is indicated in Figure 3.3. The natural log term entering in these relationships is also sometimes written as

$$\ln \operatorname{cosec}\left(\frac{\pi\alpha}{2}\right) \approx \frac{1}{2}\left[\left(\frac{1+\alpha^2}{\alpha}\right)\ln\left(\frac{1+\alpha}{1-\alpha}\right) - 2\ln\left(\frac{4\alpha}{1-\alpha^2}\right)\right] \tag{7}$$

where

$$\alpha = \frac{d}{b}$$

Figure 3.4 illustrates the cut-off space of the single ridge structure and Figure 3.5 that of the double ridge arrangement. The cut-off condition at s/a equal to zero corresponds to that of an infinitely thin finline; at s/a equal to unity it reduces to that of a standard rectangular waveguide.

One closed form approximation for the cut-off space of the dominant mode in the double ridge waveguide is (Hoefer and Burton, 1982)

$$\frac{a}{\lambda_c} = \frac{a}{2(a-s)}\left[1 + \frac{4}{\pi}\left(1 + 0.2\sqrt{\frac{b}{a-s}}\right)\left(\frac{b}{a-s}\right)\ln \operatorname{cosec}\left(\frac{\pi d}{2b}\right)\right.$$
$$\left. + \left(2.45 + 0.2\frac{s}{a}\right)\left(\frac{sb}{d(a-s)}\right)\right]^{-\frac{1}{2}} \tag{8}$$

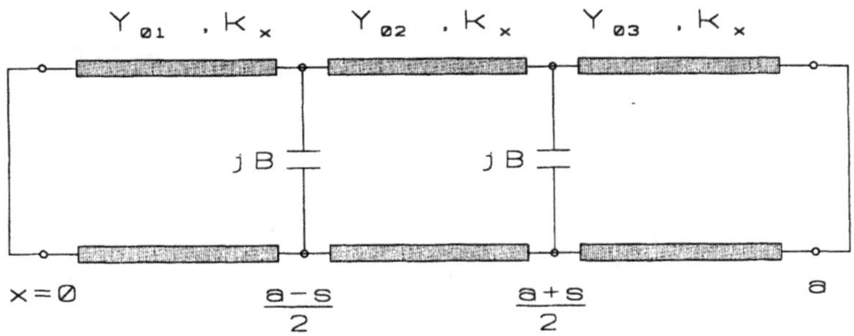

Figure 3.3 Equivalent circuit of ridge waveguide

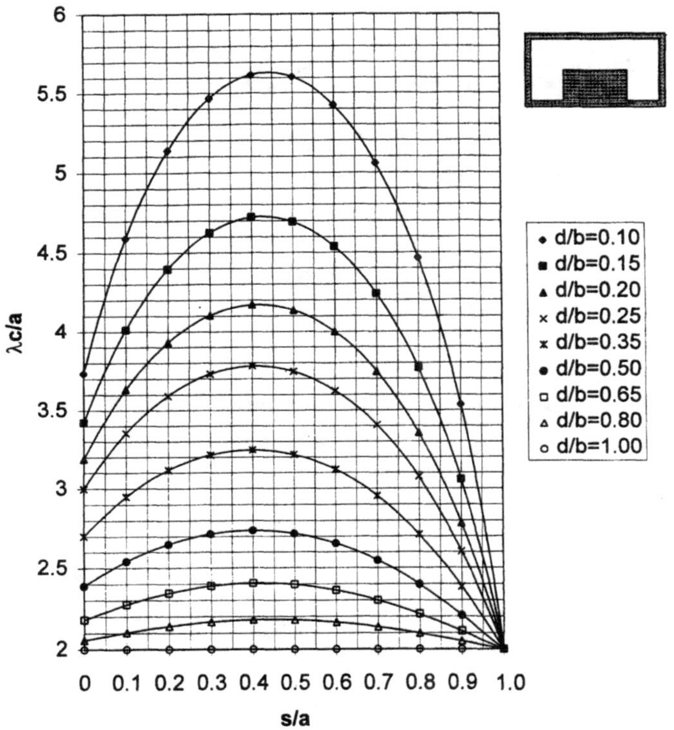

Figure 3.4 Cut-off space of single ridge waveguide (b/a = 0.45)

After Cohn (1947)

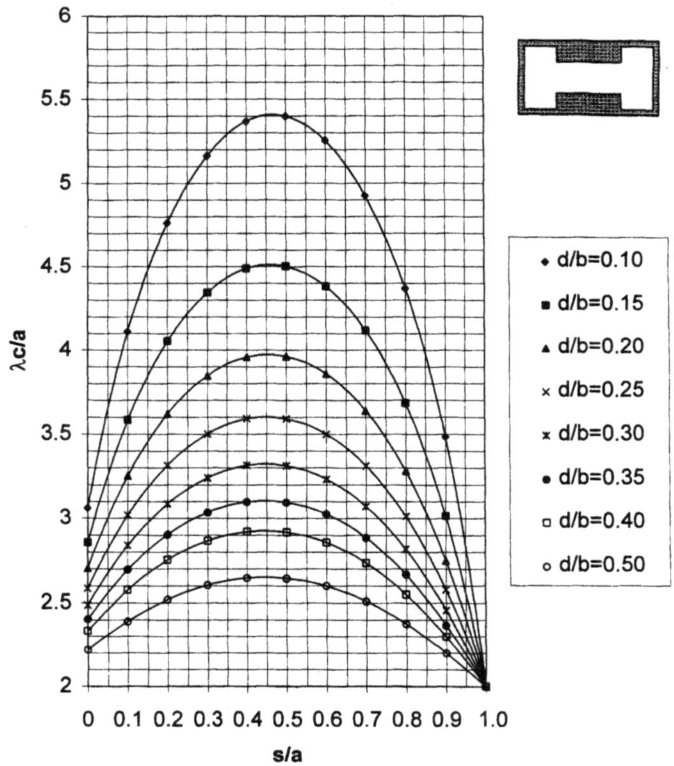

Figure 3.5 Cut-off space of double ridge waveguide ($b/a = 0.50$)

After Cohn (1947)

This representation agrees with numerical methods within 1% provided that

$$0.01 \leqslant \frac{d}{b} \leqslant 1$$

$$0 \leqslant \frac{b}{a} \leqslant 1$$

$$0 \leqslant \frac{s}{a} \leqslant 0.45$$

The corresponding approximate single ridge solution is obtained from that of the double one by replacing b by $2b$ in the coefficient multiplying the log term, which may be taken to represent the susceptance one.

One possible computation procedure is to initialise the root finding subroutine for a/λ_c in the exact transverse resonance condition of the problem region by the closed form solution for a/λ_c.

3.3 Power flow in ridge waveguide

The power flow (P_t) in the waveguide at finite frequency is also usually expressed in terms of that at infinite frequency. One description of the power flow at infinite frequency applicable to both single and double ridge waveguides is (Hopfer, 1955)

$$P_t(\infty) = \left(\frac{E_0^2 d^2}{2\pi\eta_0}\right)\left\{\left(\frac{d}{b}\right)\left(\frac{b}{a}\right)\left(\frac{2ma}{\lambda_c}\right)\ln\operatorname{cosec}\left(\frac{\pi d}{2b}\right)\cos^2\theta_2 + \frac{\theta_2}{2}\right.$$
$$\left. + \frac{\sin 2\theta_2}{4} + \left(\frac{d}{b}\right)\left(\frac{\cos\theta_2}{\sin\theta_1}\right)^2\left[\frac{\theta_1}{2} - \frac{\sin 2\theta_1}{4}\right]\right\}\left(\frac{a}{b}\right)\left(\frac{b}{d}\right)\left(\frac{\lambda_c}{a}\right) \tag{9}$$

The power flow at finite frequency is then given by

$$P_t(\omega) = P_t(\infty)\left(\frac{\lambda_0}{\lambda_g}\right) \tag{10}$$

To cater for the lumped element susceptances of single and double ridge waveguides m takes the value 1 for double ridge and 2 for single ridge waveguides, and E_0 is the peak electric field intensity (V/m) at the centre of the waveguide,

$$E_0 = \frac{V}{d} \tag{11}$$

When $s = a$ and $d = b$, $\theta_1 = 0$ and $\theta_2 = \pi/2$ and $P(\omega)$ reduces to the result for the ordinary waveguide in Chapter 2, as is readily verified.

3.4 Voltage-current definition of impedance in ridge waveguide

The approach used to calculate the impedance of the ridge waveguide consists of initially forming this quantity at infinite frequency prior to recovering it at finite frequency by introducing the dispersion factor λ_g/λ_0. The impedance in a ridge waveguide is again not unique. Taking the voltage-current definition of impedance by way of example gives

$$Z_{VI}(\omega) = Z_{VI}(\infty)\left(\frac{\lambda_g}{\lambda_0}\right) \tag{12}$$

The calculation of impedance at infinite frequency relies on the fact that the field distributions at the cut-off frequency and at infinite frequency are identical and that E_y and H_x are related there by the wave impedance of free space. Knowledge of the distribution of E_y either at cut-off or at infinite frequency is, therefore, sufficient for the solution of this sort of problem.

One approximate voltage-current definition for this problem region based on this observation but which omits the step discontinuities on either side of the ridge has been historically described by (Cohn, 1947)

$$Z_{\mathrm{VI}}(\infty) = \frac{\pi\eta_0}{\sin\theta_2 + \left(\dfrac{d}{b}\right)\tan\dfrac{\theta_1}{2}\cos\theta_2}\left(\frac{b}{a}\right)\left(\frac{d}{b}\right)\left(\frac{a}{\lambda_c}\right) \tag{13}$$

In obtaining this result the voltage has been taken as the line integral over the electric field between the ridges and the current has been defined as the total longitudinal surface current in the three regions of the structure. θ_1 and θ_2 are described in terms of λ_c in equations (3a) and (3b) and the physical variables are specified in Figure 3.1.

One solution which supersedes that in equation (13), in that it caters for the fringing effect at the steps, is (Hoefer and Burton, 1982)

$$Z_{\mathrm{VI}}(\infty) = \frac{\pi\eta_0}{\sin\theta_2 + \left(\dfrac{d}{b}\right)\left[\dfrac{B}{Y_{01}} + \tan\dfrac{\theta_1}{2}\right]\cos\theta_2}\left(\frac{b}{a}\right)\left(\frac{d}{b}\right)\left(\frac{a}{\lambda_c}\right) \tag{14}$$

The value of the normalised susceptance given in equation (5) is used in evaluating the impedance for a single ridge and that given in equation (6) is utilised in calculating that of a double ridge. The normalised cut-off frequency a/λ_c is given by either equation (1) or (8). The result for a single ridge is depicted in Figure 3.6 and that for a double ridge in Figure 3.7.

3.5 Power-voltage definition of impedance in ridge waveguide

The power-voltage definition of impedance at infinite frequency is given by using its definition,

$$Z_{\mathrm{PV}}(\infty) = \frac{V(\infty)V^*(\infty)}{2P_t(\infty)} \tag{15}$$

The power-voltage definition of impedance at infinite frequency is extracted from this relationship as

$$Z_{\mathrm{PV}}(\infty) = \frac{\pi\eta_0\left(\dfrac{b}{a}\right)\left(\dfrac{d}{b}\right)\left(\dfrac{a}{\lambda_c}\right)}{\left\{\left(\dfrac{d}{b}\right)\left(\dfrac{b}{a}\right)\left(\dfrac{2ma}{\lambda_c}\right)\ln\operatorname{cosec}\left(\dfrac{\pi d}{2b}\right)\cos^2\theta_2 + \dfrac{\theta_2}{2} + \dfrac{\sin 2\theta_2}{4} + \left(\dfrac{d}{b}\right)\left(\dfrac{\cos\theta_2}{\sin\theta_1}\right)^2\left[\dfrac{\theta_1}{2} - \dfrac{\sin 2\theta_1}{4}\right]\right\}} \tag{16}$$

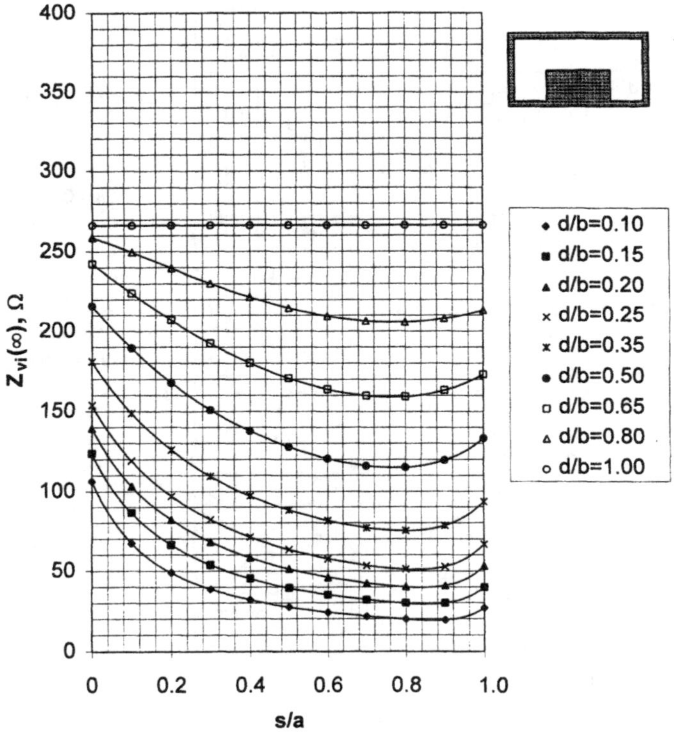

Figure 3.6 Voltage-current impedance of single ridge waveguide ($b/a = 0.45$)

After Hoefer & Burton (1982)

$m = 2$ for a single ridge and $m = 1$ for a double ridge. Figure 3.8 indicates the result in the case of a single ridge structure and Figure 3.9 that of the double structure.

$Z_{PV}(\omega)$ is connected to $Z_{PV}(\infty)$ by a similar relationship to that between $Z_{VI}(\omega)$ and $Z_{VI}(\infty)$ in equation (12).

3.6 Power-current definition of impedance in ridge waveguide

The power-current definition of impedance at infinite frequency in a ridge waveguide may also be deduced once those of the voltage-current and power-voltage are at hand. It is given by using the relationship between the three possible descriptions,

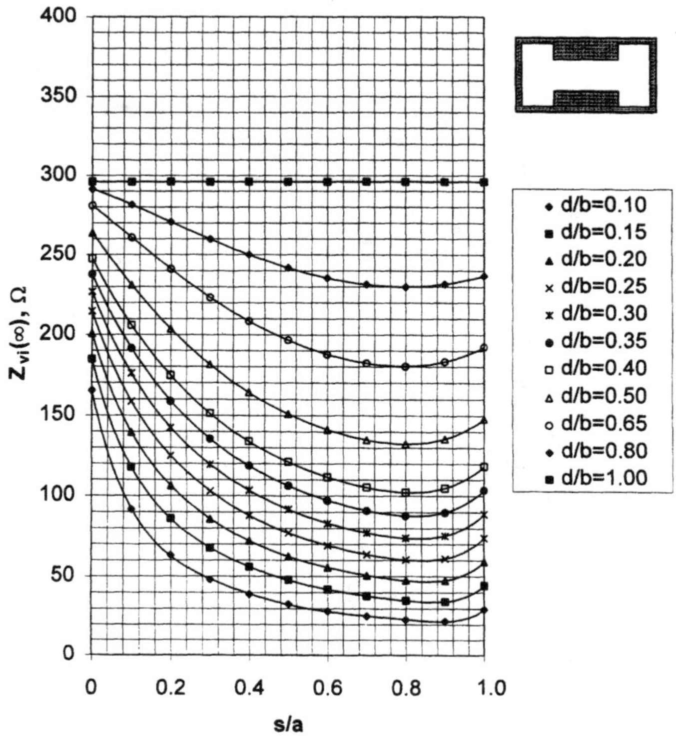

Figure 3.7 Voltage-current impedance of double ridge waveguide ($b/a = 0.50$)

After Hoefer & Burton (1982)

$$Z_{PI}(\infty) = \frac{Z_{VI}^2(\infty)}{Z_{PV}(\infty)} \tag{17}$$

The nature of this impedance is indicated in Figure 3.10 in the case of a single ridge configuration and in Figure 3.11 in that of a double ridge waveguide.

3.7 Admittances of double ridge waveguide

It is sometimes advantageous to work with admittance instead of impedance. The purpose of this section is to reproduce some data on each admittance definition met in this sort of waveguide. Since the double ridge wave is the more common commercial geometry the data are restricted to that situation. Figures 3.12–3.14 illustrate the required results.

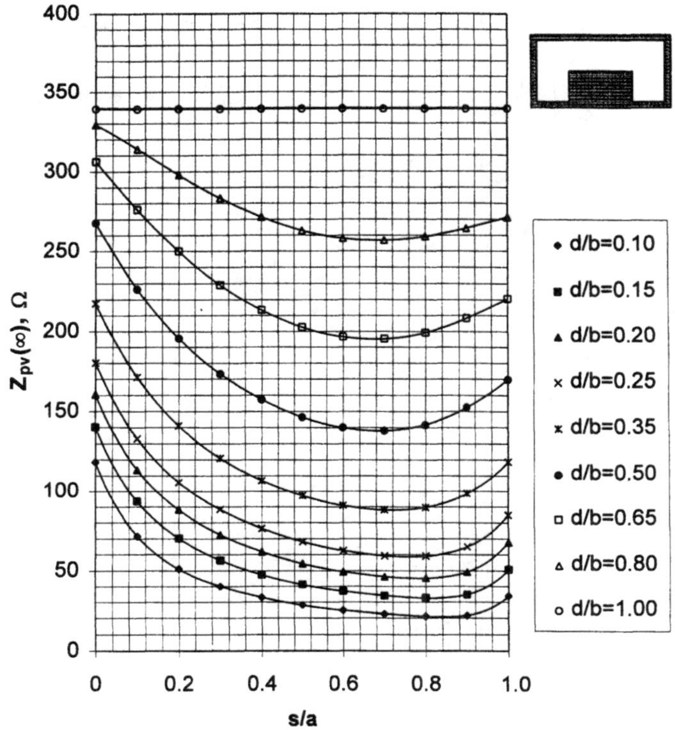

Figure 3.8 Power-voltage impedance of single ridge waveguide ($b/a = 0.45$)

After Hopfer (1955)

3.8 Closed form polynomials for single and double ridge waveguides

Some closed form polynomials for the cut-off wavelength and impedance of single ridge waveguides in terms of the gap-factor (d/b) with the aspect ratio (s/a) equal to 0.50 are summarised below:

$$\frac{a}{\lambda_c} = 0.109 + 0.735\left(\frac{d}{b}\right) - 0.596\left(\frac{d}{b}\right)^2 + 0.273\left(\frac{d}{b}\right)^3 \tag{18}$$

$$Z_{VI}(\infty) = -226.57\left(\frac{d}{b}\right)^4 + 414\left(\frac{d}{b}\right)^3 - 201.51\left(\frac{d}{b}\right)^2 + 279.86\left(\frac{d}{b}\right) + 0.6237 \tag{19}$$

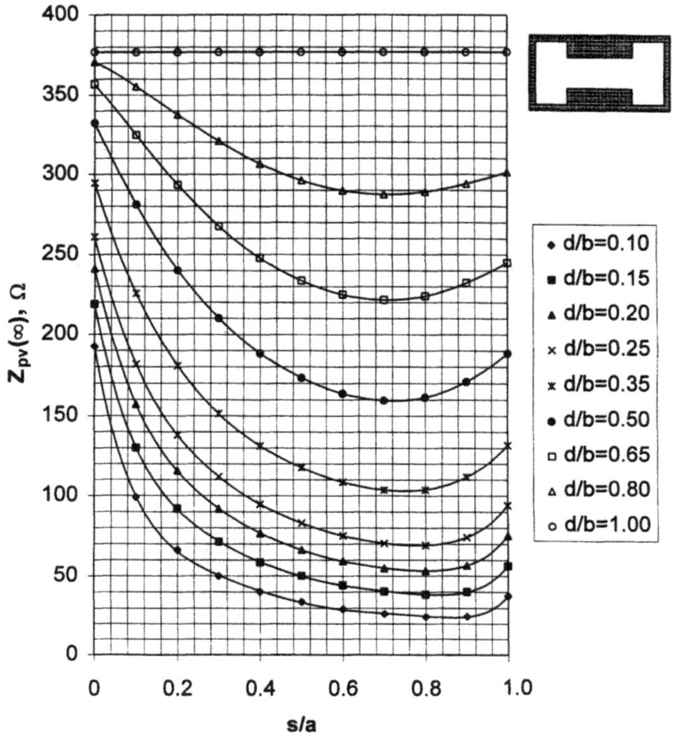

Figure 3.9 Power-voltage impedance of double ridge waveguide ($b/a = 0.50$)
After Hopfer (1955)

$$Z_{PV}(\infty) = -274.75\left(\frac{d}{b}\right)^4 + 495.25\left(\frac{d}{b}\right)^3 - 168.5\left(\frac{d}{b}\right)^2 + 286.73\left(\frac{d}{b}\right) + 0.5054 \qquad (20)$$

$$Z_{PI}(\infty) = -192.67\left(\frac{d}{b}\right)^4 + 368.41\left(\frac{d}{b}\right)^3 - 241.38\left(\frac{d}{b}\right)^2 + 274.11\left(\frac{d}{b}\right) + 0.717 \qquad (21)$$

The graphical solutions are indicated in Figures 3.15 and 3.16.

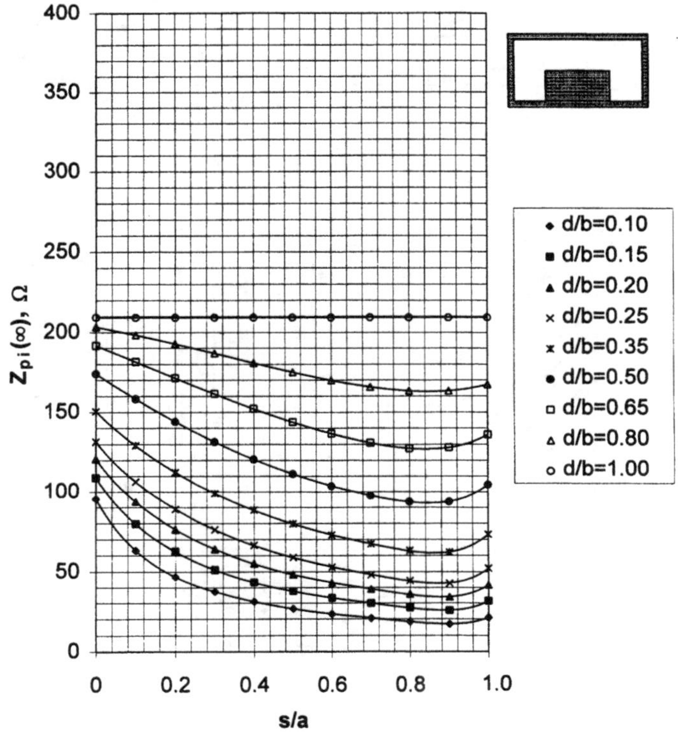

Figure 3.10 Power-current impedance of single ridge waveguide ($b/a = 0.45$)

The corresponding quantities for a double ridge are:

$$\frac{a}{\lambda_c} = 0.103 + 0.924\left(\frac{d}{b}\right) - 1.090\left(\frac{d}{b}\right)^2 + 0.667\left(\frac{d}{b}\right)^3 \tag{22}$$

$$Z_{VI}(\infty) = -23.958\left(\frac{d}{b}\right)^3 + 25.422\left(\frac{d}{b}\right)^2 + 293.6\left(\frac{d}{b}\right) + 1.9553 \tag{23}$$

$$Z_{PV}(\infty) = -45.635\left(\frac{d}{b}\right)^3 + 130.27\left(\frac{d}{b}\right)^2 + 291.37\left(\frac{d}{b}\right) + 2.1205 \tag{24}$$

$$Z_{PI}(\infty) = 3.8777\left(\frac{d}{b}\right)^3 - 65.419\left(\frac{d}{b}\right)^2 + 293.02\left(\frac{d}{b}\right) + 1.8943 \tag{25}$$

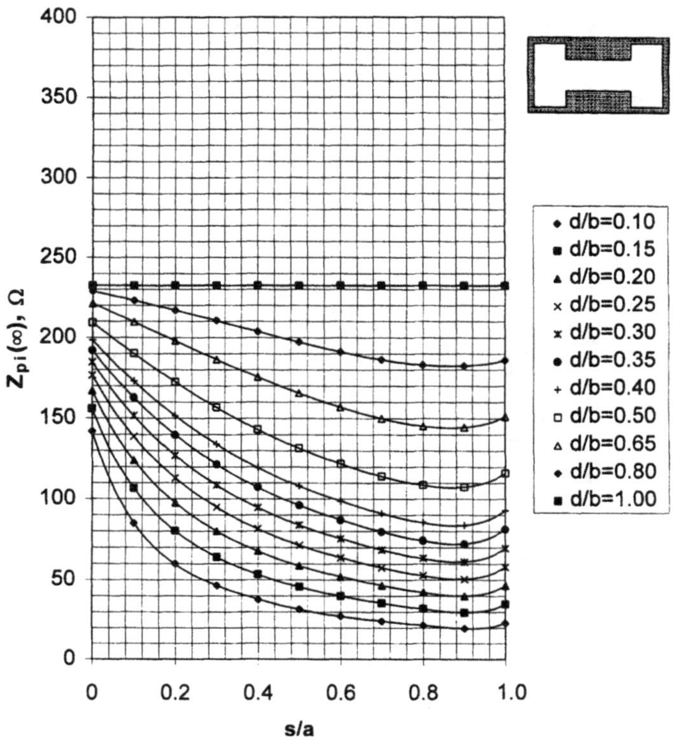

Figure 3.11 Power-current impedance of double ridge waveguide ($b/a = 0.50$)

Figures 3.17 and 3.18 summarise the graphical solutions.

Figure 3.19 separately compares the various definitions of impedance for $b/a = 0.5$ and $s/a = 0.45$ for different values of d/b.

3.9 Synthesis of quarter-wave ridge transformers

Provided s/a is fixed, the design procedure for a ridge impedance transformer involves the solution of a transcendental equation in d/b. The background to this calculation will now be demonstrated by way of an example. It starts by using the usual relationship between the immittances met in connection with an ideal quarter-wave long impedance transformer,

$$Y_r^2(\omega) = rY_0(\omega)G(\omega)$$

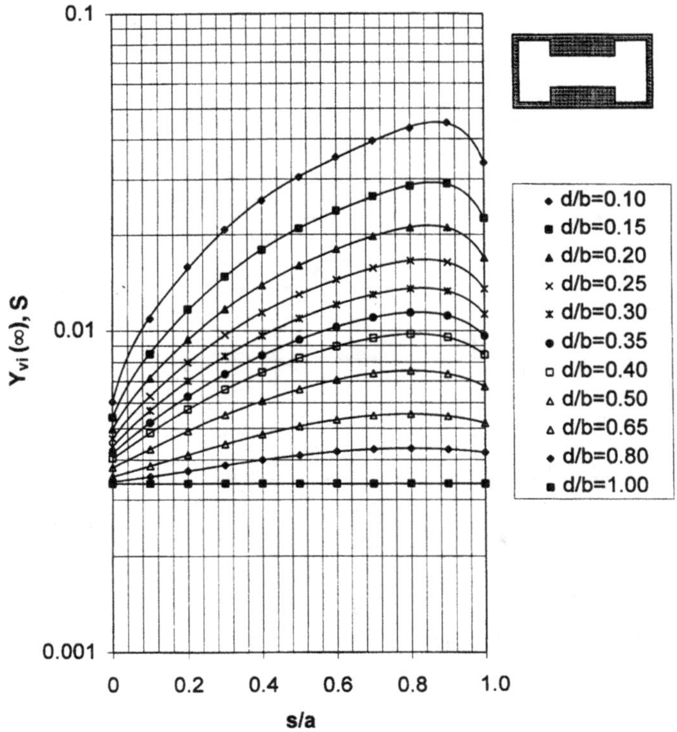

Figure 3.12 Voltage-current admittance of double ridge waveguide ($b/a = 0.50$)

After Hoefer & Burton (1982)

$G(\omega)$ is the load conductance and

$$Y_r(\omega) = Y_r(\infty)\left(\frac{\lambda_0}{\lambda_{gr}}\right)$$

$$Y_0(\omega) = Y_0(\infty)\left(\frac{\lambda_0}{\lambda_{g0}}\right)$$

r is the VSWR at the input terminals of the transformer. $Y_r(\infty)$ represents either the voltage-current, power-voltage or power-current definition of impedance of the ridge transformer. The characteristic equation for d/b is therefore specified by

$$Y_r(\infty)\left(\frac{\lambda_0}{\lambda_{gr}}\right) - \sqrt{r}\,Y_0(\infty)\left(\frac{\lambda_0}{\lambda_{g0}}\right)\sqrt{G(\omega)} = 0$$

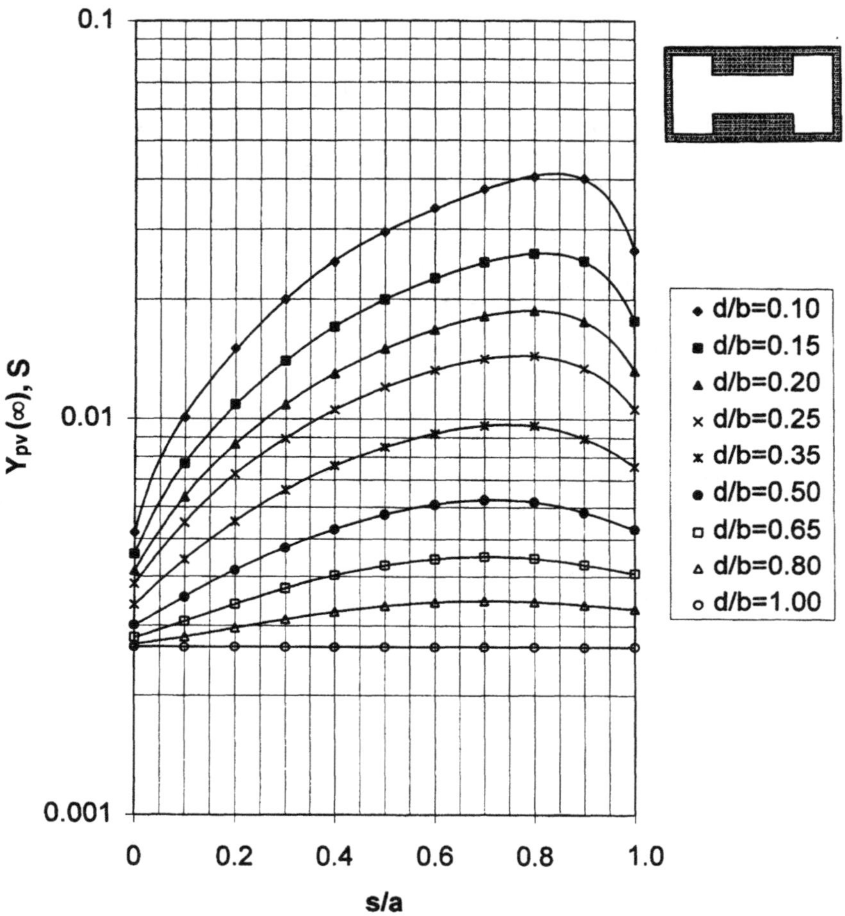

Figure 3.13 Power-voltage admittance of double ridge waveguide ($b/a = 0.50$)
After Hopfer (1955)

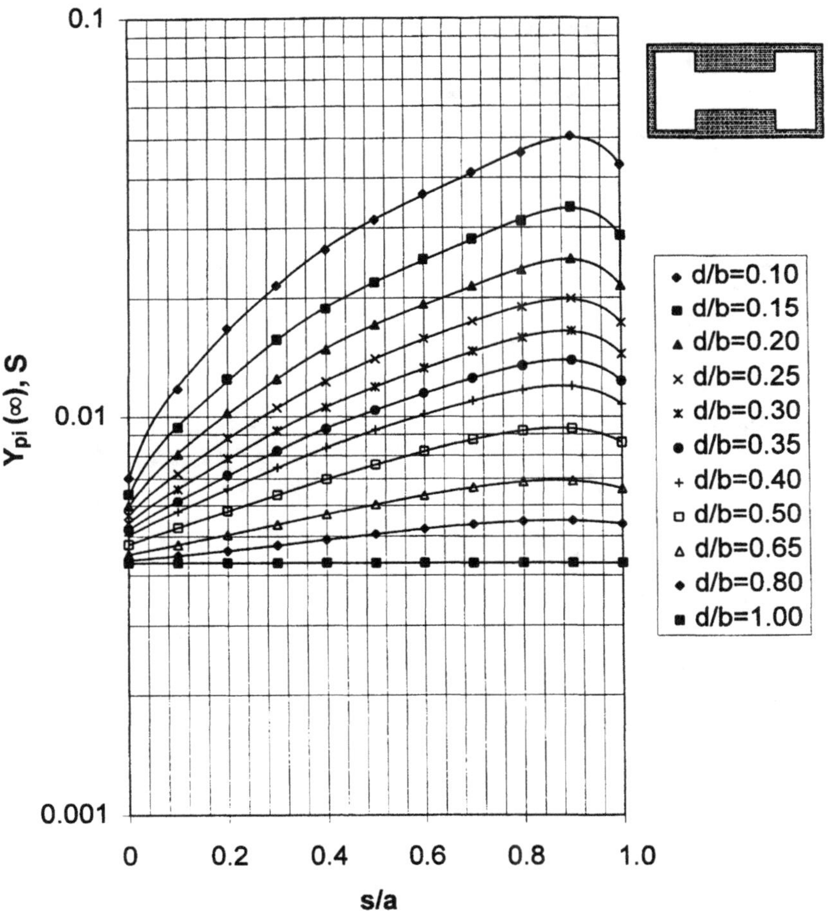

Figure 3.14 Power-current admittance of double ridge waveguide ($b/a = 0.50$)

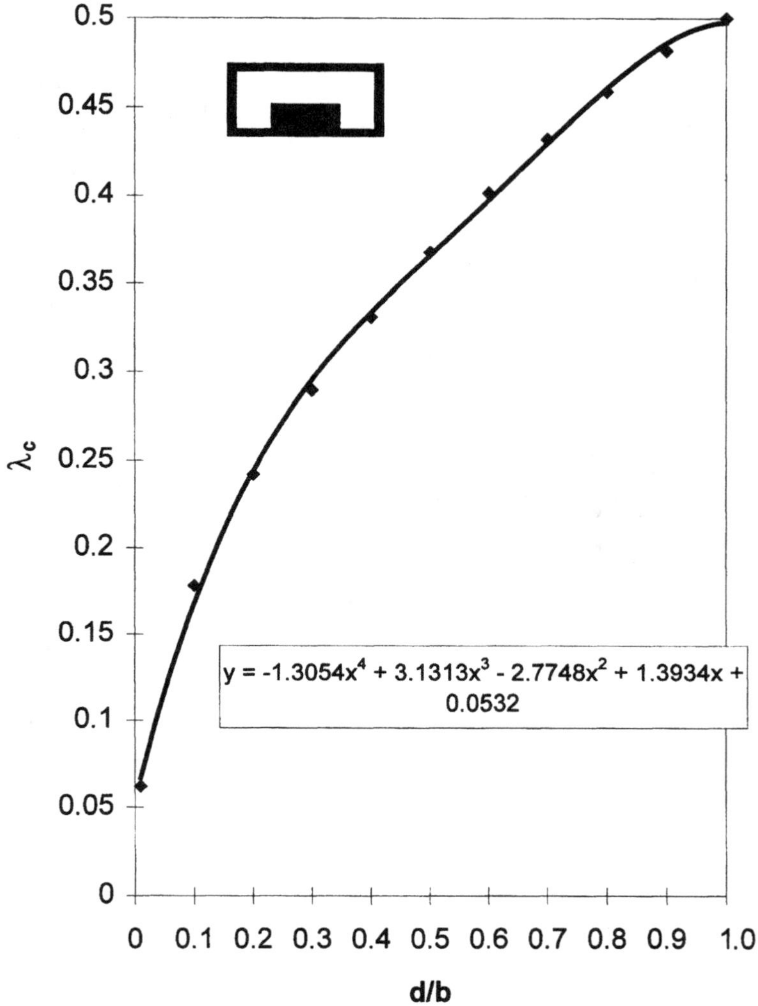

Figure 3.15 a/λ_c *against* d/b *for single ridge waveguide* ($b/a = 0.45$, $s/a = 0.50$)

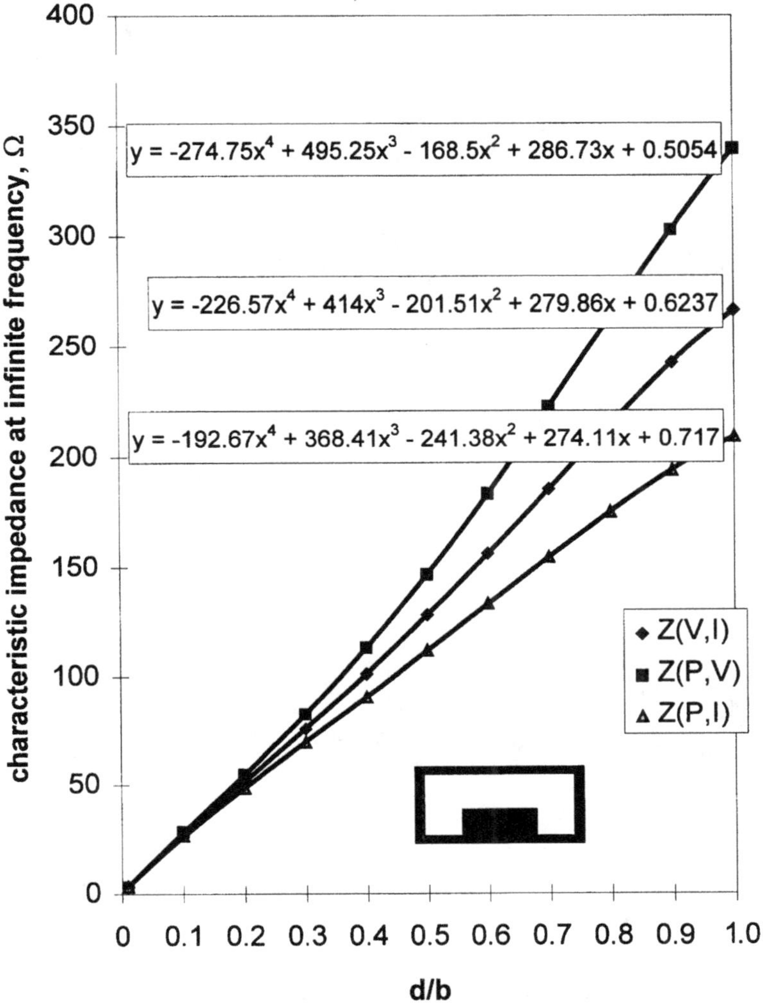

Figure 3.16 Polynomial descriptions of impedance in single ridge waveguide ($b/a = 0.45$, $s/a = 0.50$)

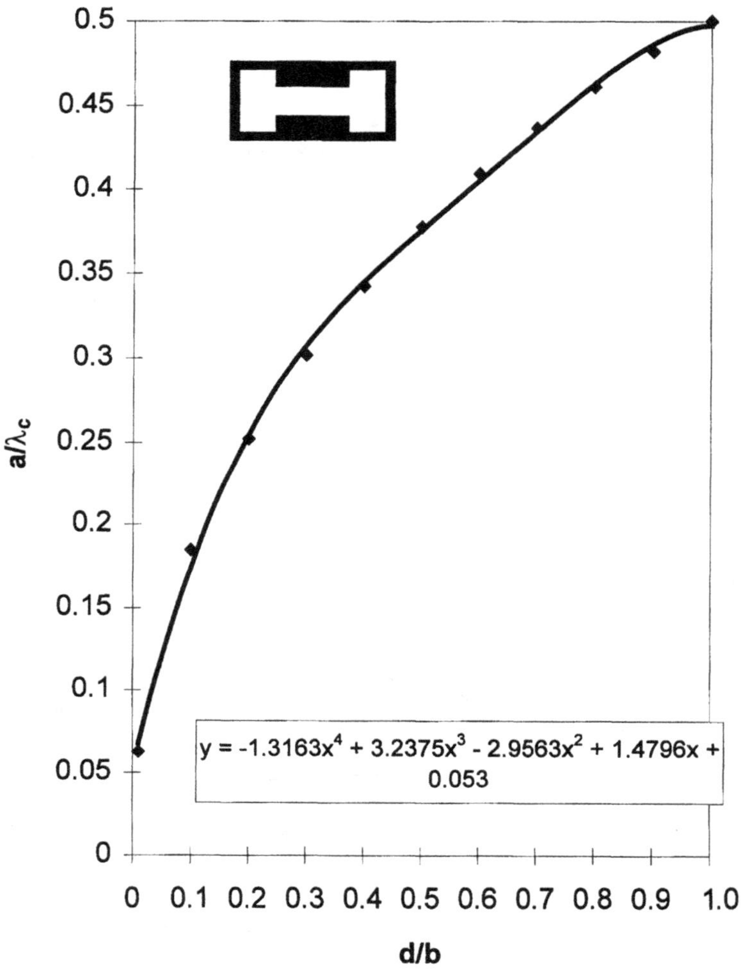

Figure 3.17 a/λ_c *against* d/b *for double ridge waveguide* $(b/a = 0.50,\ s/a = 0.50)$

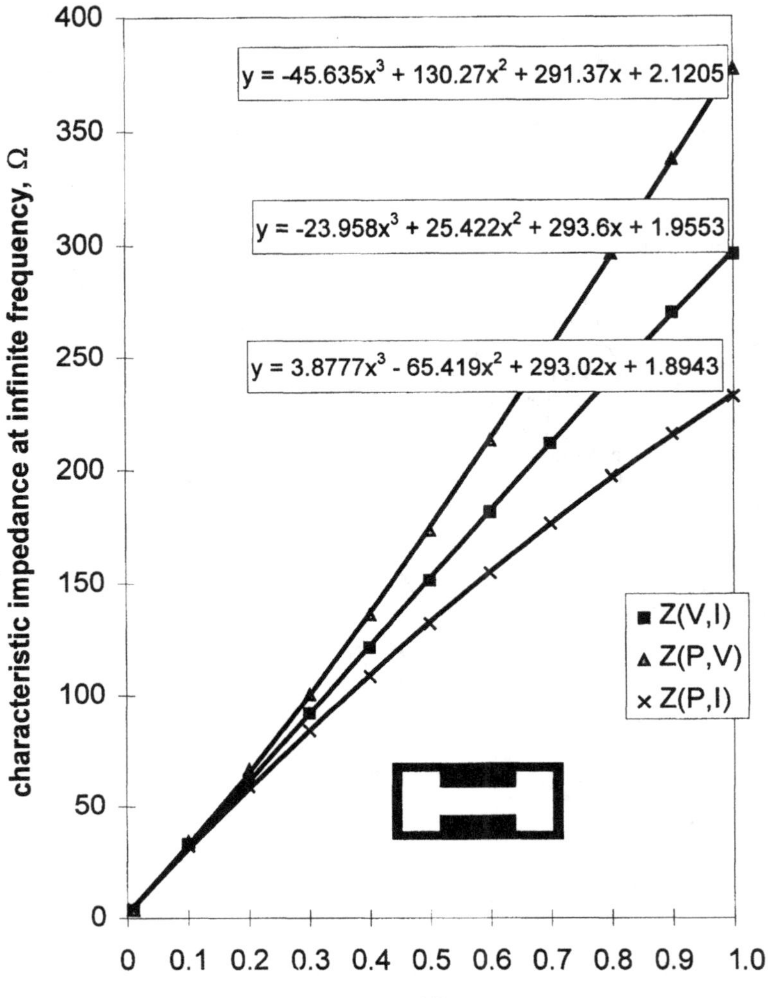

Figure 3.18 Polynomial descriptions of impedance in double ridge waveguide ($b/a = 0.50$, $s/a = 0.50$)

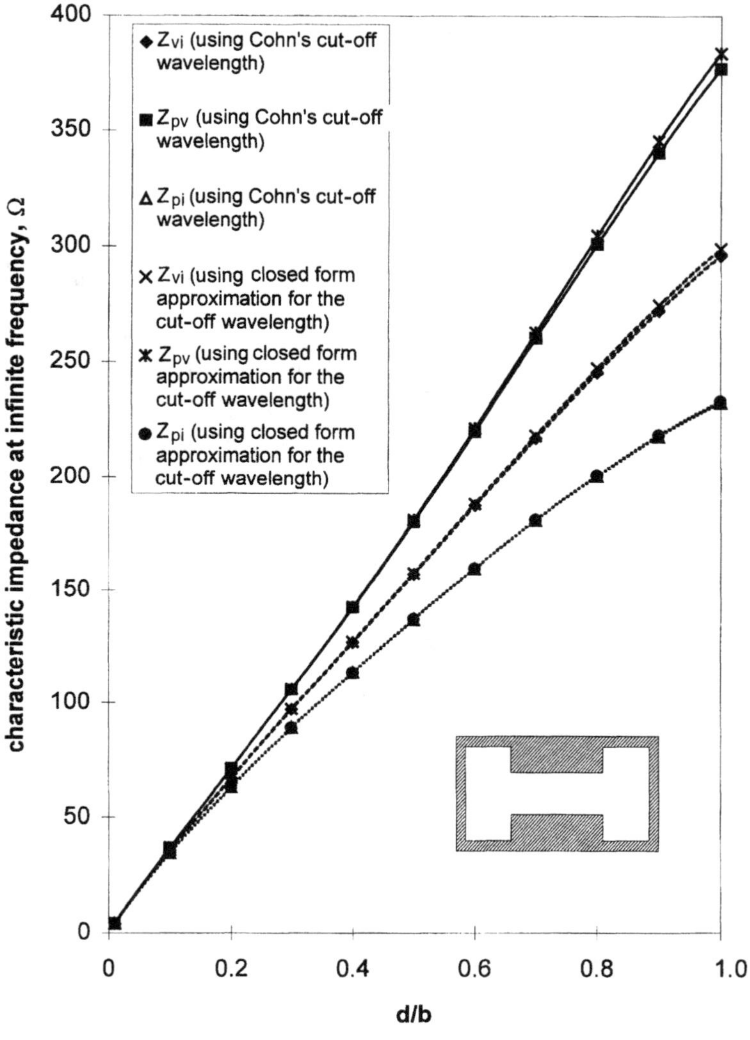

Figure 3.19 Comparison between different definitions of characteristic impedance in double ridge waveguide ($b/a = 0.50$, $s/a = 0.45$)

Chapter 4

Fields, propagation and attenuation in double ridge waveguide

4.1 Introduction

A knowledge of the fields in any waveguide is necessary in order to calculate its power flow, its attenuation and its impedance. The purpose of this chapter is to summarise some approximate closed form relationships for the fields in this type of waveguide and present some calculations based on the finite element method (FEM) procedure. The TE family of solutions is obtained by calculating H_z of the related planar problem region with top and bottom magnetic walls and forming the other components of the field by using Maxwell's equations. The TM family of solutions is obtained by solving the dual planar circuit with top and bottom electric walls. Comparison between the closed form and FEM procedures suggests that the closed form representation is adequate for engineering purposes. Some results on the standing wave solutions of the dominant and higher order modes of this sort of waveguide based on a magnetic field integral equation (MFIE) are also included. The power flow and attenuation are separately summarised.

4.2 Finite element calculation (TE modes)

The only components that can exist at cut-off in a double ridge waveguide in the case of the dominant quasi-TE_{10} mode are those associated with the appropriate planar circuit (E_x, E_y, H_z) and at infinite frequency those associated with TEM propagation (E_x, E_y, H_x, H_y). The calculation of impedance undertaken elsewhere in this text only involves a knowledge of the transverse electric fields at either frequency. The description of the fields in this sort of waveguide starts with a calculation of H_z at cut-off by recognising that it is a solution of the related planar problem region with top and bottom magnetic walls. It continues with the evaluation of the electric fields by employing Maxwell's equations. One way of obtaining H_z is by using an FEM solver. The functional to be minimised in conjunction with this planar problem region is

$$\mathrm{F}(H_z) = \iint_S (-|\nabla_t H_z|^2 + k_0^2 \varepsilon_f |H_z|^2)\, \mathrm{d}s \tag{1}$$

where

$$k_0^2 = \omega_0^2 \varepsilon_0 \mu_0 \tag{2}$$

and ε_f is the relative permittivity of the region. μ_0 and ε_0 are the usual constitutive parameters of free space. The preceding functional automatically satisfies the wave equation and the Neumann boundary condition

$$\frac{\partial H_z}{\partial n} = 0 \tag{3}$$

on the electric side-walls. At a magnetic wall the Dirichlet condition

$$H_z = 0 \tag{4}$$

must be separately enforced.

In the Rayleigh-Ritz approach the true solution is replaced by a trial function which is expanded in terms of a suitable set of real basis or shape functions $\alpha_i(x, y)$ which contains the spatial variation of the problem with complex coefficients u_i. In the finite element problem the complex coefficients represent the fields at the nodes of the elements. This step reduces the problem to a set of simultaneous equations which when extremised produces the required eigenvalue problem:

$$\{[A] - k_a^2[B]\}\bar{U} = 0 \tag{5}$$

$[A]$ and $[B]$ are square matrices given by

$$A_{ij} = \iint_S (\nabla_t \alpha_i) \cdot (\nabla_t \alpha_j)\, \mathrm{d}s \tag{6}$$

and

$$B_{ij} = \iint_S (\alpha_i \alpha_j)\, \mathrm{d}s \tag{7}$$

$\bar{U}$ is a column vector containing the unknowns of the problem.

The transverse components of the electric field of the TE_{10} mode can now be calculated from a knowledge of H_z at cut-off by using Maxwell's equations:

$$H_x = \left(\frac{-\gamma}{k_c^2}\right)\frac{\partial H_z}{\partial x} \tag{8a}$$

$$H_y = \left(\frac{-\gamma}{k_c^2}\right)\frac{\partial H_z}{\partial y} \tag{8b}$$

$$E_x = \left(\frac{j\omega\mu_0}{\gamma}\right)H_y \tag{8c}$$

$$E_y = \left(\frac{-j\omega\mu_0}{\gamma}\right)H_x \tag{8d}$$

$$E_z = 0 \tag{8e}$$

η_0 is the free space wave impedance

$$\eta_0 = \sqrt{\frac{\mu_0}{\varepsilon_0}} \tag{9}$$

The propagation constant γ is determined by the free space and cut-off wave-numbers in the usual way:

$$\gamma^2 = k_0^2 - k_c^2 \tag{10}$$

The waveguide wavelength (λ_g) is separately given in terms of the free space wavelength (λ_0) and the cut-off one (λ_c) by

$$\left(\frac{2\pi}{\lambda_g}\right)^2 = \left(\frac{2\pi}{\lambda_0}\right)^2 - \left(\frac{2\pi}{\lambda_c}\right)^2 \tag{11}$$

where

$$\gamma = j\frac{2\pi}{\lambda_g}$$

$$k_0 = \frac{2\pi}{\lambda_0}$$

$$k_c = \frac{2\pi}{\lambda_c}$$

4.3 Finite element method (TM modes)

In the variational calculation of the TM family of modes the functional to be minimised is

$$\mathrm{F}(E_z) = \iint_s (-|\nabla_t E_z|^2 + k_0^2 \varepsilon_f |E_z|^2)\, \mathrm{d}s \tag{12}$$

with top and bottom electric walls. This functional satisfies the dual boundary conditions met in connection with the problem region with top and bottom magnetic walls. Extremising this functional by using the Rayleigh-Ritz procedure produces a similar eigenvalue equation to that met in the dual problem already dealt with. The complete solution is given in terms of E_z by

$$H_x = \left(\frac{-\gamma}{k_c^2}\right) \frac{\partial E_z}{\partial x} \tag{13a}$$

$$H_z = 0 \tag{13b}$$

$$H_y = \left(\frac{-\gamma}{k_c^2}\right) \frac{\partial E_z}{\partial y} \tag{13c}$$

$$E_x = \left(\frac{j\omega\mu_0}{\gamma}\right) H_y \tag{13d}$$

$$E_y = \left(\frac{-j\omega\mu_0}{\gamma}\right) H_x \tag{13e}$$

4.4 Cut-off space (TE mode)

To calculate the fields at the nodes of the problem region of a typical eigenvector it is first necessary to deduce its eigenvalue. The use of the FEM in solving this kind of eigenvalue problem is a standard result in the literature. The purpose of this section is to outline this solution in the case of the dominant TE mode in a double ridge waveguide. The topology in question is indicated in Figure 4.1. The details of the two discretisations utilised here are illustrated in Figure 4.2. These are fixed by

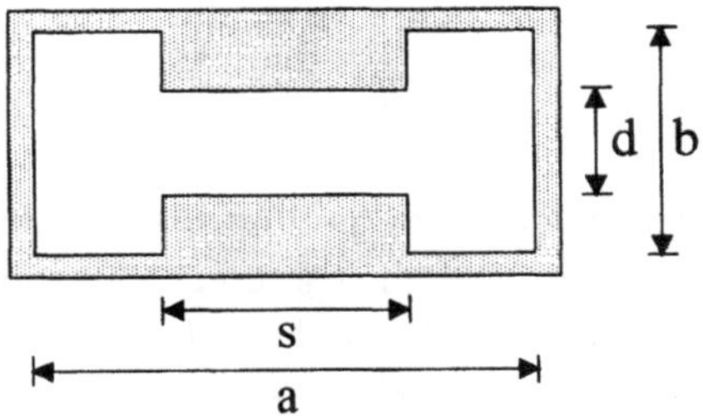

Figure 4.1 Schematic diagram of double ridge waveguide

$$p = 2$$
$$m = 6$$
$$n = 149$$
$$n \times m = 894$$
$$q = 338$$

and

$$p = 2$$
$$m = 6$$
$$n = 168$$

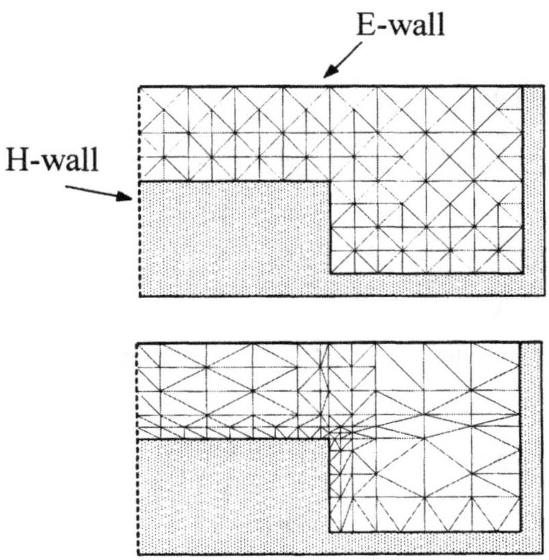

Figure 4.2 Typical discretisations of double ridge waveguide

$$n \times m = 1008$$

$$q = 385$$

respectively.

p is the degree of the interpolation polynomial within each finite element triangle, m is the number of nodes inside each finite element triangle, n is the number of triangles, $n \times m$ is the total number of nodes before assembly of the finite element mesh and q is the number of nodes after assembly of the mesh. The boundary condition at the electric walls are natural boundaries of the functional and are satisfied by definition. The one at the magnetic wall must, however, be separately imposed.

Figure 4.3 indicates one typical result using each mesh arrangement. It also gives a comparison between the FEM employed here and the TRM (transverse resonance method).

4.5 Standing wave solution in double ridge waveguide

The electric field and current density of any mode in this sort of waveguide is readily established once the related eigenvalue is deduced. The FEM is

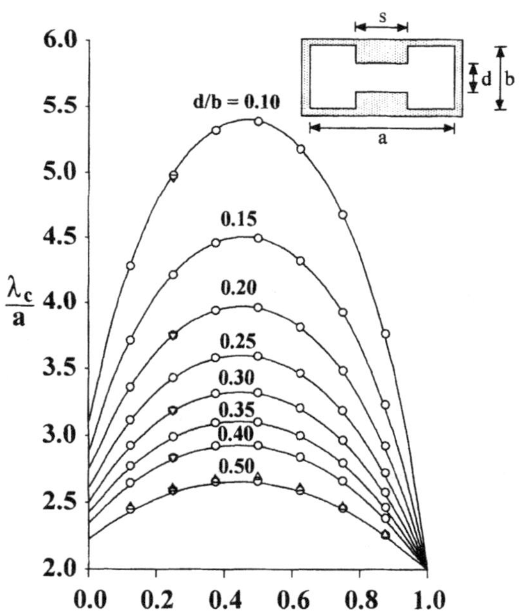

Figure 4.3 Comparison between cut-off spaces of double ridge waveguide using FEM and TRM methods. (Δ, $q = 338$, ○, $q = 385$, $b/a = 0.50$)

Reprinted with permission, Helszajn & McKay (1998, *IEE Proc.*)

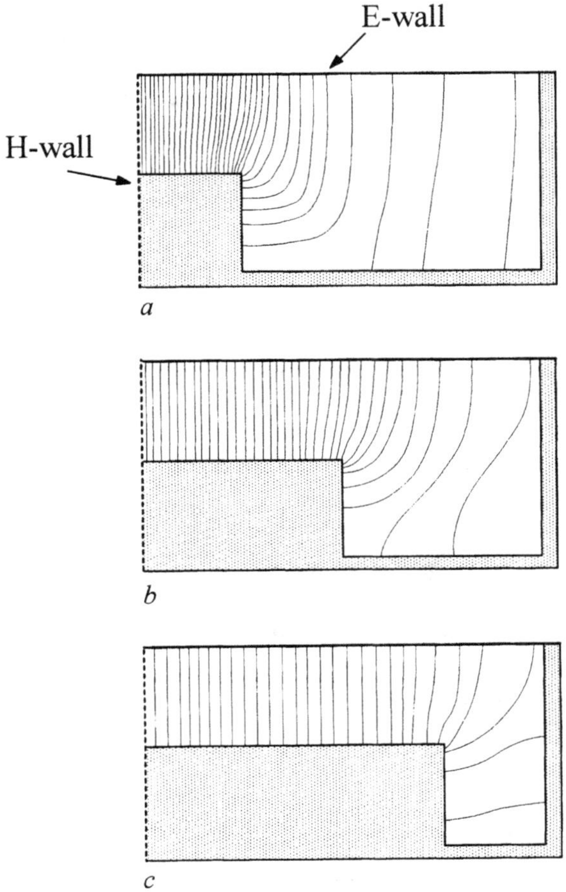

Figure 4.4 Standing wave solutions of dominant TE mode of double ridge waveguide for three different values of s/a (b/a = 0.5, d/b = 0.5)

Reprinted with permission, Helszajn & McKay (1998, *IEE Proc.*)

also admirably suited for this purpose. The governing equation is equation (4). Figure 4.4 illustrates the distribution of the electric field of the dominant mode in a double ridge waveguide for three typical ridge geometries. The density of the lines indicates the relative strength of the field. The details of the discretisations utilised to evaluate these solutions are the same as those employed in the construction of the corresponding cut-off space. Figure 4.5 separately depicts the current distribution in the waveguide. The bunching of the current on either side of the symmetry plane of the ridge is of note. This suggests that the calculation described here respects the singularity in the electric field mentioned elsewhere.

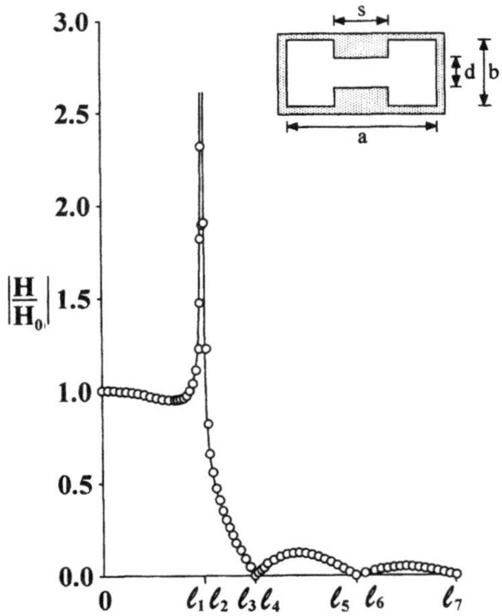

Figure 4.5 Magnetic field at symmetry plane of double ridge waveguide ($b/a = 0.5$, $d/b = 0.5$, $s/a = 0.25$)

Reprinted with permission, Helszajn & McKay (1998, *IEE Proc.*)

4.6 TE fields in double ridge waveguide

While the finite element method provides one means of evaluating the fields in any waveguide a closed form description is often convenient. One such formulation has been given by Getsinger. It is based on retaining a single mode in the gap region whilst expanding that in a typical trough region in an infinite series. This solution is characterised by $E_z = 0$. The fields obtained in this way on matching the two regions are summarised below.

In the gap region (Gestinger, 1962):

$$E_y = \cos k_c\left(\frac{a}{2} - x\right)\exp(-j\beta z) \tag{14a}$$

$$H_x = \left(\frac{-\beta}{\omega\mu_0}\right)\cos k_c\left(\frac{a}{2} - x\right)\exp(-j\beta z) \tag{14b}$$

$$E_z = 0 \tag{14c}$$

$$H_z = \left(\frac{jk_c}{\beta}\right) \sin k_c\left(\frac{a}{2} - x\right) \exp(-j\beta z) \tag{14d}$$

In the trough region:

$$E_x = \sum_{n=0}^{N} A_n\left(\frac{n\pi}{b}\right) \cos(\alpha_n x) \sin\left(\frac{n\pi y}{b}\right) \exp(-j\beta z) \tag{15a}$$

$$E_y = \sum_{n=0}^{N} A_n(-\alpha_n) \sin(\alpha_n x) \cos\left(\frac{n\pi y}{b}\right) \exp(-j\beta z) \tag{15b}$$

$$E_z = 0 \tag{15c}$$

$$H_x = \sum_{n=0}^{N} A_n\left(\frac{\beta\alpha_n}{\omega\mu_0}\right) \sin(\alpha_n x) \cos\left(\frac{n\pi y}{b}\right) \exp(-j\beta z) \tag{15d}$$

$$H_y = \sum_{n=0}^{N} A_n\left(\frac{n\pi\beta}{b\omega\mu_0}\right) \cos(\alpha_n x) \sin\left(\frac{n\pi y}{b}\right) \exp(-j\beta z) \tag{15e}$$

$$H_z = \sum_{n=0}^{N} A_n\left(\frac{-jk_c^2}{\omega\mu_0}\right) \cos(\alpha_n x) \cos\left(\frac{n\pi y}{b}\right) \exp(-j\beta z) \tag{15f}$$

where

$$\beta^2 = k_0^2 - k_c^2$$

$$\alpha_n^2 = k_c^2 - \left(\frac{n\pi}{b}\right)^2$$

and

$$A_n = \frac{-\Gamma_n \cos\left(\frac{k_c s}{2}\right)}{n\pi\alpha_n \sin\left[\alpha_n\left(\frac{a-s}{2}\right)\right]} \left\{ \sin\left[\left(\frac{b+d}{2}\right)\left(\frac{n\pi}{b}\right)\right] - \sin\left[\left(\frac{b-d}{2}\right)\left(\frac{n\pi}{b}\right)\right]\right\}$$

$\Gamma_n = 1$ for $n = 0$ and $\Gamma_n = 2$ for $n = 1, 2, 3, \ldots$.

The fields in the double ridge waveguide summarised here fail at the boundary between the two regions but are adequate everywhere else and provide both a reasonable description of power flow and impedance. The failure at the boundary has its origin in the use of a single mode in the

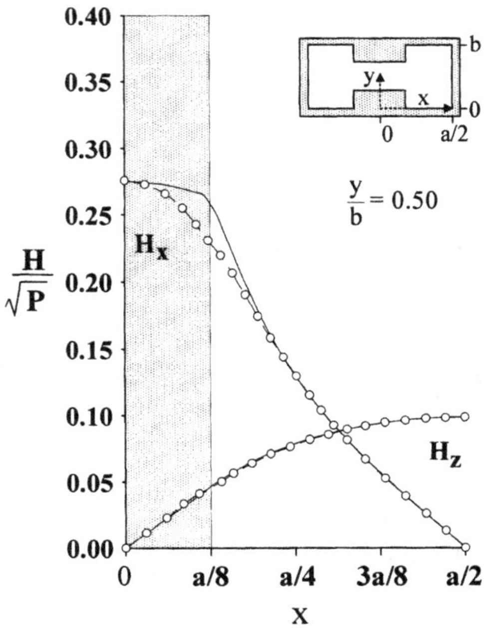

Figure 4.6 Comparison between E_Y and E_Z for TE mode at symmetry plane of double ridge waveguide based on closed form and FEM

Reprinted with permission, Helszajn & McKay (1998, *IEE Proc.*)

ridge gap. The agreement between the finite element method and the closed form solution is indicated in Figure 4.6.

4.7 TM fields in double ridge waveguide

An approximate closed form description of the TM family of solutions in a ridge waveguide has also been described in the literature. This solution is characterised by H_z equal to zero. The fields in the gap region are (Mansour, Tong, MacPhie, 1988):

$$E_x = \gamma \sin \gamma\left(\frac{a}{2} - x\right) \sin\left[\left(\frac{\pi}{d}\right)\left(y + \frac{-b+d}{2}\right)\right] \exp(-j\beta z) \tag{16a}$$

$$E_y = \left(\frac{\pi}{d}\right) \cos \gamma\left(\frac{a}{2} - x\right) \cos\left[\left(\frac{\pi}{d}\right)\left(y + \frac{-b+d}{2}\right)\right] \exp(-j\beta z) \tag{16b}$$

$$E_z = j\left(\frac{k_c^2}{\beta}\right) \cos \gamma\left(\frac{a}{2} - x\right) \sin\left[\left(\frac{\pi}{d}\right)\left(y + \frac{-b+d}{2}\right)\right] \exp(-j\beta z) \tag{16c}$$

and

$$H_x = \left(\frac{-\omega\varepsilon_0}{\beta}\right)\left(\frac{\pi}{d}\right)\cos\gamma\left(\frac{a}{2}-x\right)\cos\left[\left(\frac{\pi}{d}\right)\left(y+\frac{-b+d}{2}\right)\right]\exp(-j\beta z) \tag{16d}$$

$$H_y = \left(\frac{\gamma\omega\varepsilon_0}{\beta}\right)\sin\gamma\left(\frac{a}{2}-x\right)\cos\left[\left(\frac{\pi}{d}\right)\left(y+\frac{-b+d}{2}\right)\right]\exp(-j\beta z) \tag{16e}$$

$$H_z = 0 \tag{16f}$$

where

$$\gamma^2 = k_c^2 - \left(\frac{\pi}{d}\right)^2$$

In the trough region:

$$E_x = \sum_{n=1}^{N}(B_n\alpha_n)\cos(\alpha_n x)\sin\left(\frac{n\pi y}{b}\right)\exp(-j\beta z) \tag{17a}$$

$$E_y = \sum_{n=1}^{N}\left(\frac{B_n n\pi}{b}\right)\sin(\alpha_n x)\cos\left(\frac{n\pi y}{b}\right)\exp(-j\beta z) \tag{17b}$$

$$E_z = \sum_{n=1}^{N}\left(\frac{jB_n k_c^2}{\beta}\right)\sin(\alpha_n x)\sin\left(\frac{n\pi y}{b}\right)\exp(-j\beta z) \tag{17c}$$

$$H_x = \sum_{n=1}^{N}\left(\frac{-B_n\omega\varepsilon_0 n\pi}{\beta b}\right)\sin(\alpha_n x)\cos\left(\frac{n\pi y}{b}\right)\exp(-j\beta z) \tag{17d}$$

$$H_y = \sum_{n=1}^{N}\left(\frac{B_n\omega\varepsilon_0\alpha_n}{\beta}\right)\cos(\alpha_n x)\sin\left(\frac{n\pi y}{b}\right)\exp(-j\beta z) \tag{17e}$$

$$H_z = 0 \tag{17f}$$

where

$$B_n = \frac{2\left(\frac{\pi}{d}\right)\cos\left(\frac{\gamma s}{2}\right)}{b\sin\alpha_n\ell\left[\left(\frac{\pi}{d}\right)^2 - \left(\frac{n\pi}{b}\right)^2\right]}$$

$$\times\left[\sin\left(\frac{n\pi}{b}\right)\left(\frac{b+d}{2}\right) + \sin\left(\frac{n\pi}{b}\right)\left(\frac{b-d}{2}\right)\right]$$

The agreement between this result and the FEM method is indicated in Figure 4.7 for one typical geometry. The calculation of E_y is omitted from Figures 4.6 and 4.7. It is related to H_x by the wave impedance in the usual way,

$$E_y = \left(\frac{\eta_0 k_0}{\beta}\right)H_x \tag{18}$$

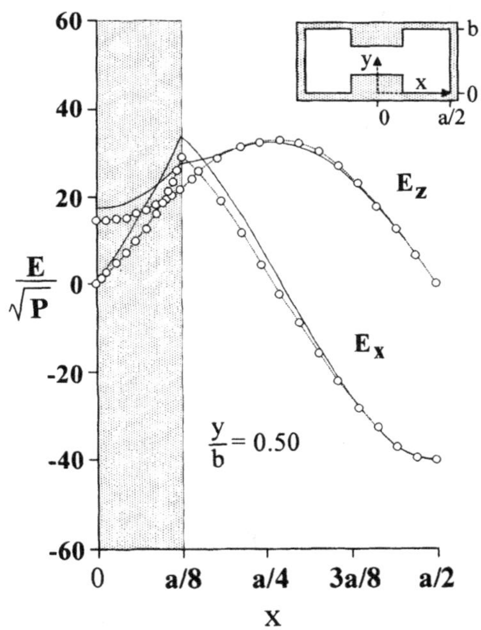

Figure 4.7 Comparison between H_X and H_Z for TM mode at symmetry plane of double ridge waveguide based on closed form and FEM

Reprinted with permission, Helszajn & McKay (1998, *IEE Proc.*)

4.8 MFIE

Another numerical technique that has been employed to investigate the properties of the ridge waveguide is the magnetic field integral equation (MFIE) method. This section summarises some calculations on the electric field distribution of this sort of waveguide. Figures 4.8 and 4.9

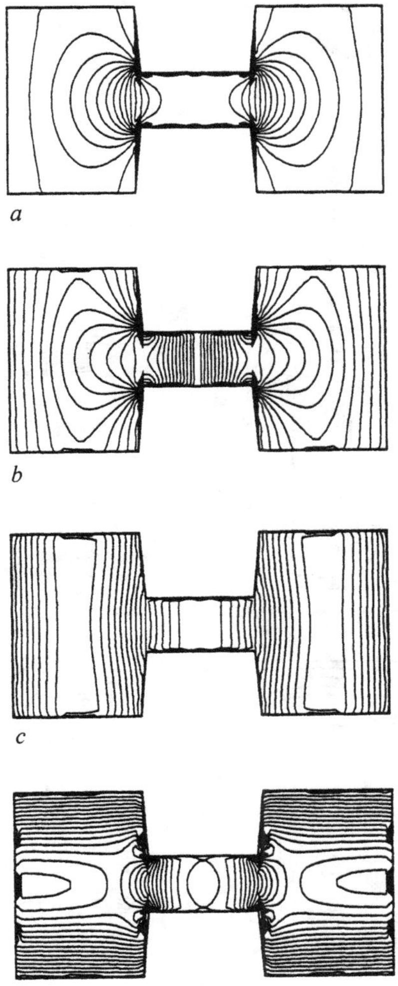

a

b

c

d

Figure 4.8 Electrical field contour plots of a double ridge waveguide (ridge thickness $x/a = 0.3$ and ridge gap $d/a = 0.15$)

a TE_{10}; *b* TE_{20}; *c* TE_{30}; *d* TE_{11} mode

Reprinted with permission, Sun & Balanis (1993)

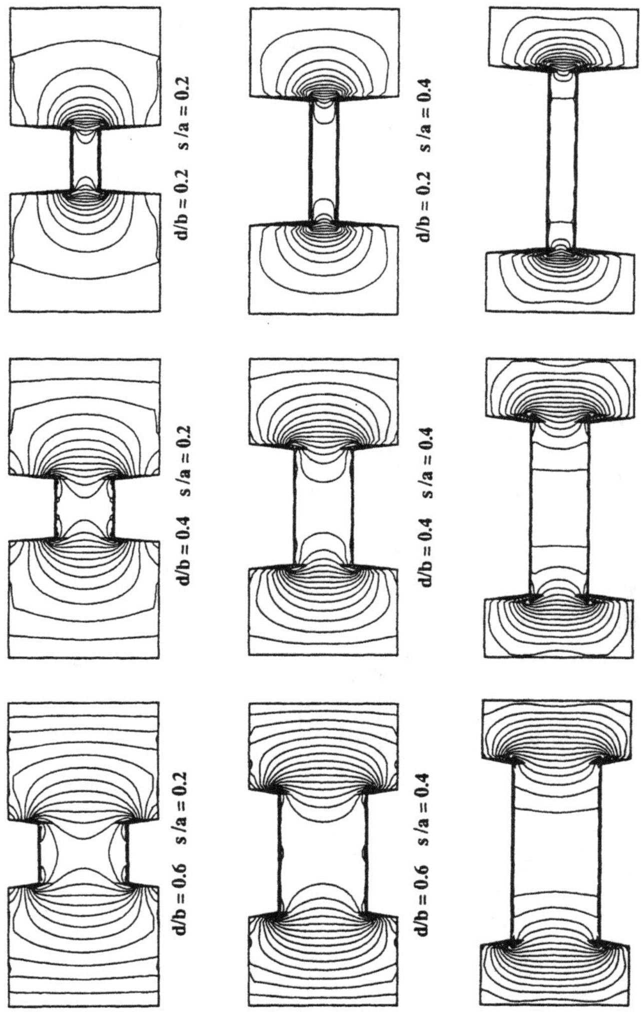

Figure 4.9 Electrical field contour plots of TE_{10} mode in double ridge waveguides with various s/a and d/b ratios

Reprinted with permission, Sun & Balanis (1993)

indicate the first three modes in the waveguide for a number of different aspect ratios of the structure.

4.9 The Poynting vector

The power rating and the power-voltage definition of impedance both require statements of the Poynting vector of the waveguide. The relationship between this quantity at finite and at infinite frequency is given by

$$P(\omega) = P(\infty)\left(\frac{\lambda_0}{\lambda_g}\right) \tag{19}$$

where $P(\infty)$ is

$$P(\infty) = \frac{1}{2}\int_s [\bar{E}_t(\infty) \times \bar{H}_t^*(\infty)]\, ds \tag{20}$$

$\bar{E}_t$ and $\bar{H}_t$ denote the transverse electric and magnetic fields, respectively.

For the purpose of calculation this relationship may be written as

$$\begin{aligned} &\frac{1}{2}\int_s [\bar{E}_t(\infty) \times \bar{H}_t^*(\infty)]\, ds \\ &= A^2\zeta_0 \int_0^d\int_0^{\ell_1} [E_x(\omega_c)E_x^*(\omega_c) + E_y(\omega_c)E_y^*(\omega_c)]\, dx\, dy \\ &+ A^2\zeta_0 \int_0^b\int_{\ell_4}^{\ell_5} [E_x(\omega_c)E_x^*(\omega_c) + E_y(\omega_c)E_y^*(\omega_c)]\, dx\, dy \end{aligned} \tag{21}$$

The factor A may be deduced by equating the Poynting vector to 1 W.

4.10 Attenuation in waveguides

An important quantity in the description of any waveguide is its dissipation per unit length. When this quantity is small, the time average power (P_t) transmitted through the waveguide may be written approximately as

$$P_t \to P_t \exp(-2\alpha z) \tag{22}$$

The power loss per unit length (P_ℓ) is then given by

$$P_\ell = -\frac{\partial P_t}{\partial z} = 2\alpha P_t \tag{23}$$

The attenuation per unit length (α) is therefore described by

$$\alpha = \frac{P_\ell}{2P_t} \tag{24}$$

Since P_t has been evaluated in the preceding section, it only remains to deduce P_ℓ. This may be done by determining the dissipation in each wall of the waveguide. This may be done by forming the following quantity:

$$P_\ell = \frac{1}{2} Z_s \int_A |J|^2 \, dA \tag{25}$$

Z_s is the skin resistance of the metal enclosure:

$$Z_s = \sqrt{\frac{\omega\mu_0}{2\sigma}} \tag{26}$$

J is the current density and σ is the resistivity of the waveguide wall.

Figure 4.10 indicates the attenuation in a double ridge waveguide based on the MFIE method.

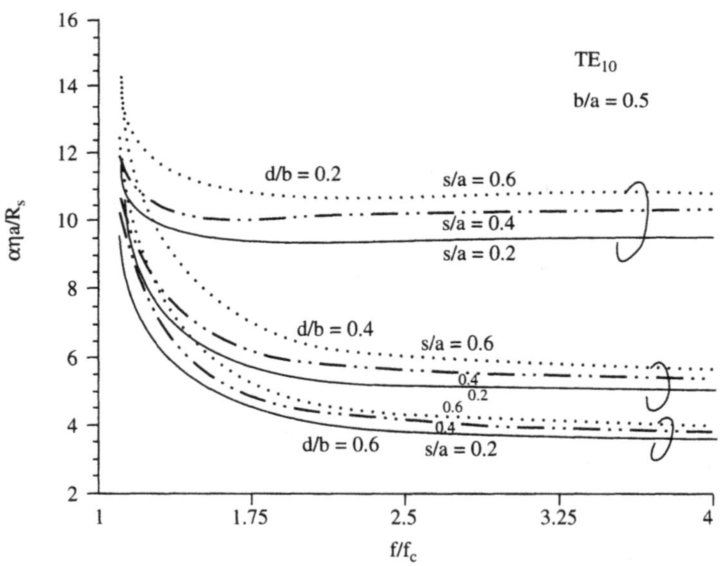

Figure 4.10 Normalised attenuation constant of double ridge waveguide (f_c = cut-off frequency of TE_{10} mode)

Reprinted with permission, Sun & Balanis (1993)

Chapter 5

Impedance of double ridge waveguide using the finite element method

J. Helszajn and M. McKay

5.1 Introduction

One means of calculating the cut-off space and impedance of a ridge waveguide which avoids the need to describe the fringing fields at the edges of the ridges is the finite element method (FEM). One way of obtaining any of the definitions of impedance at any frequency in such a waveguide is to make use of the relationship between that at finite and infinite frequencies. The impedance at infinite frequency is readily deduced by recognising that the electric field distribution is invariant at cut-off and at infinite frequency and that the magnetic field at infinite frequency is simply related to the electric field there by the impedance of free space. This suggests that a knowledge of the electric field at infinite frequency or at the cut-off space is in practice sufficient for calculation. One simple means of solving this problem at the cut-off frequency is to recognise that these quantities correspond to the eigenvectors and eigenvalues of the related planar circuit with electric side-walls and top and bottom magnetic ones. Since this eigenvalue problem is readily solved by using the FEM it provides one classic solution to this sort of problem region. The FEM employed here is based on standard nodal triangular elements and thus mesh refinement is necessary in the neighbourhood of any singularities to ensure convergence. The resulting number of nodes can be reduced by supplementing the basis functions with singular trial functions or by implementing scalar singular elements to emulate the behaviour of the field distributions at metal discontinuities. While the main endeavour of this chapter is the evaluation of the voltage-current definition of impedance in a ridge waveguide the power-voltage and power-current ones are also calculated for completeness' sake. Each of these two calculations may also be reduced to the ratio of two suitable integrals in the electric field at cut-off.

One advantage of numerical methods is that it is not necessary to introduce the notion of fringing effects at the side-walls of the ridge structure as is the case with some early calculations based on the transverse resonance method (TRM). Some calculations on the impedance of waveguides with trapezoidal ridges are included for completeness.

5.2 Voltage-current definition of impedance

One means of calculating the impedance of a homogeneous waveguide is to form this quantity at infinite frequency prior to recovering it at finite frequency by introducing the dispersive factor λ_g/λ_0. The topology under consideration is indicated in Figure 5.1. The calculation of the impedance in a ridge waveguide at infinite frequency relies on the fact that the distribution of the electric field in a homogeneous waveguide is the same at all frequencies, including cut-off and infinite frequency, and that its wave impedance at infinite frequency is that of free space. Since the electric field at cut-off is readily evaluated using the FEM it provides one attractive means of tackling this problem. One possible definition of impedance in a homogeneous waveguide at finite frequency is the voltage-current one. It is given by

$$Z_{VI}(\omega) = Z_{VI}(\infty)\left(\frac{\lambda_g}{\lambda_0}\right) \tag{1}$$

where

$$Z_{VI}(\infty) = \frac{V(\infty)}{I(\infty)} \tag{2}$$

$V(\infty)$ and $I(\infty)$ are the voltage at the centre of the waveguide and the total longitudinal current in either the top or bottom half of the waveguide at infinite frequency, respectively. λ_g is the usual waveguide wavelength,

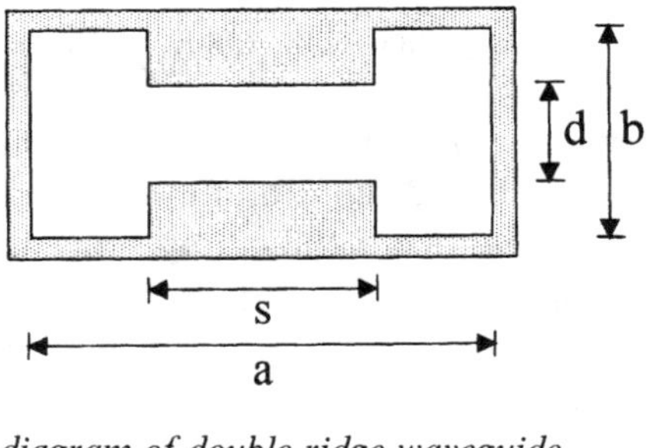

Figure 5.1 Schematic diagram of double ridge waveguide

$$\left(\frac{2\pi}{\lambda_g}\right)^2 = \left(\frac{2\pi}{\lambda_0}\right)^2 - \left(\frac{2\pi}{\lambda_c}\right)^2 \tag{3}$$

Making use of the observation that the distributions of the electric fields are identical at both cut-off and infinite frequency gives

$$\frac{V(\infty)}{I(\infty)} = \frac{V(\omega_c)}{I(\omega_c)} \tag{4}$$

In terms of the original variables,

$$V(\infty) = \int_0^d E_0(\omega_c)\,\mathrm{d}\ell \tag{5}$$

$$\begin{aligned} I(\infty) = 2\zeta_0 \int_0^{\ell_1} E(\omega_c)_{\text{wall}}\,\mathrm{d}\ell \\ + 2\zeta_0 \int_{\ell_2}^{\ell_3} E(\omega_c)_{\text{wall}}\,\mathrm{d}\ell \\ + 2\zeta_0 \int_{\ell_4}^{\ell_5} E(\omega_c)_{\text{wall}}\,\mathrm{d}\ell \\ + 2\zeta_0 \int_{\ell_6}^{\ell_7} E(\omega_c)_{\text{wall}}\,\mathrm{d}\ell \end{aligned} \tag{6}$$

provided that

$$H(\infty) = \zeta_0 E(\infty)_{\text{wall}} \tag{7}$$

$E(\omega_c)_{\text{wall}}$ is the electric field at the wall of the waveguide at cut-off. ζ_0 is the free space wave admittance, and $\ell_1, \ell_2, \ldots \ell_7$ are the integration paths. The latter quantities are specified by

$$\ell_1 = \frac{s}{2}$$

$$\ell_2 = \frac{(b-d)}{2}$$

$$\ell_3 = 0$$

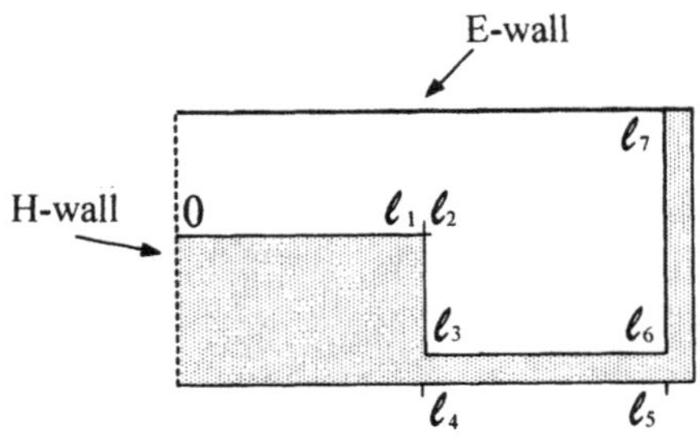

Figure 5.2 Path of current integral

$$\ell_4 = \frac{s}{2}$$

$$\ell_5 = \frac{a}{2}$$

$$\ell_6 = 0$$

$$\ell_7 = \frac{b}{2}$$

A scrutiny of the symmetry attached to the problem region of the double ridge waveguide suggests that, insofar as the dominant TE_{10} mode is concerned, it is sufficient to solve one-quarter of the original topology. The electrical paths met in connection with this problem region are indicated in Figure 5.2. The solution is complete once the electric fields on the interior metal boundaries are established.

5.3 Calculation of voltage-current definition of impedance

The main task of this chapter is the calculation of the voltage-current definition of impedance at infinite frequency in a ridge waveguide by using the FEM. This is done by calculating $V(\infty)$ and $I(\infty)$ in equations (5) and (6) for each of the two different finite element meshes illustrated in Figure 5.3.

The required transverse components of the electric field can now be calculated from a knowledge of H_z at cut-off by using Maxwell's equations,

$$E_x = \frac{-\eta_0}{k_c}\left(\frac{\partial H_z}{\partial y}\right) \tag{8a}$$

$$E_y = \frac{H\eta_0}{k_c}\left(\frac{\partial H_z}{\partial x}\right) \tag{8b}$$

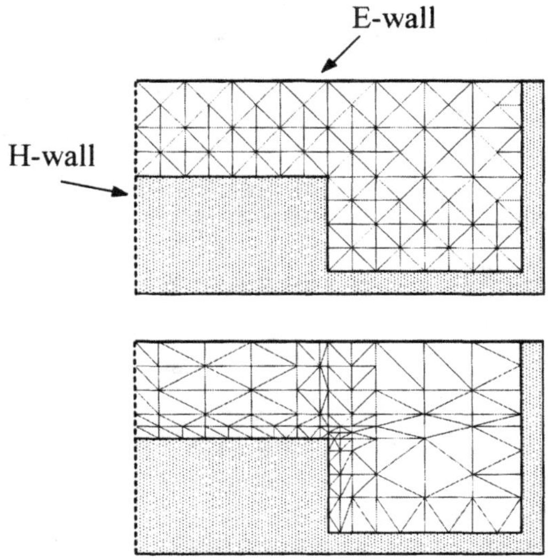

Figure 5.3 Discretisations of double ridge waveguide

The other components of the TEM wave are related at infinite frequency by

$$E_x = \eta_0 H_y \tag{9a}$$

$$E_y = -\eta_0 H_x \tag{9b}$$

η_0 is the free-space wave impedance

$$\eta_0 = \sqrt{\frac{\mu_0}{\varepsilon_0}} \tag{10}$$

Figure 5.4 compares the solutions obtained in this way for one range of ridge geometry. A comparison between the voltage-current definition of impedance based on the finite element method and the closed-form expression developed in Chapter 3 is separately indicated in Figure 5.5.

5.4 Power-current and power-voltage definitions of impedance

The two other definitions of impedance met in this sort of waveguide besides that of the voltage-current are the power-current and power-voltage ones. The power-voltage definition of impedance based on the FEM has also been dealt with in the literature but has so far been restricted to one value of the s/a ratio. The required definitions are

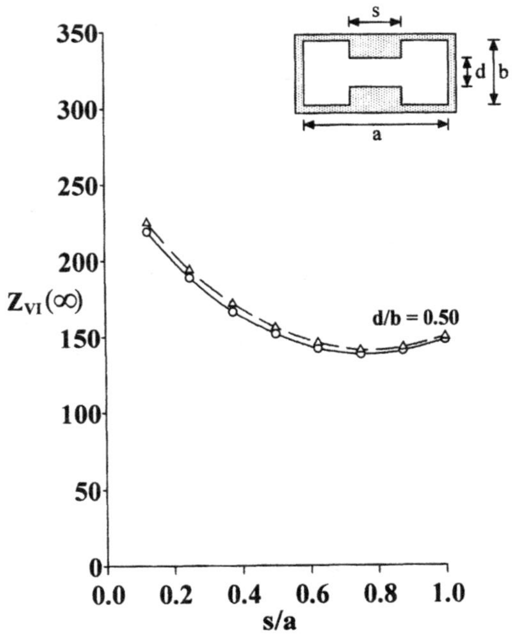

Figure 5.4 Voltage-current definition of impedance of double ridge waveguide using two possible discretisations (Δ, $q = 338$, $\bigcirc$, $q = 385$, $b/a = 0.50$)

Reprinted with permission, Helszajn & McKay (1998, *IEE Proc.*)

$$Z_{PV}(\omega) = Z_{PV}(\infty)\left(\frac{\lambda_g}{\lambda_0}\right) \tag{11}$$

and

$$Z_{PI}(\omega) = Z_{PI}(\infty)\left(\frac{\lambda_g}{\lambda_0}\right) \tag{12}$$

where

$$Z_{PV}(\infty) = \frac{V(\infty)V^*(\infty)}{2P_t(\infty)} \tag{13}$$

and

$$Z_{PI}(\infty) = \frac{2P_t(\infty)}{I(\infty)I^*(\infty)} \tag{14}$$

Since the three classic definitions of impedance are related, a knowledge of any two is sufficient to describe the third. The approach utilised here is to calculate the power-current one in terms of the voltage-current and

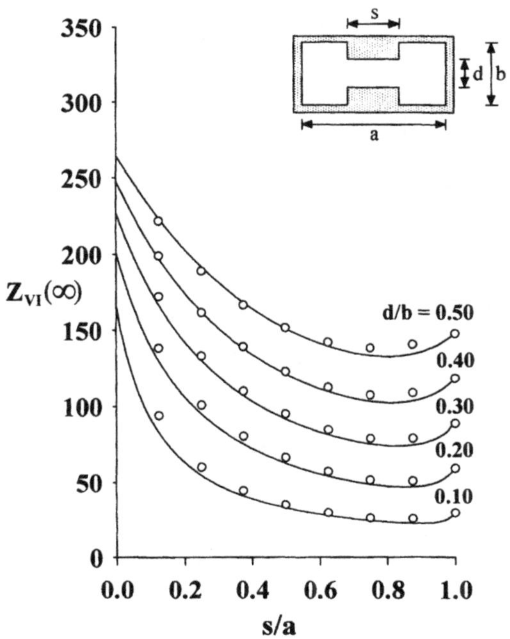

Figure 5.5 Comparison between voltage-current definition of impedance of double ridge waveguide using Sharma & Hoefer (1983) definition and FEM ($b/a = 0.50$)

Reprinted with permission, Helszajn & McKay (1998, *IEE Proc.*)

power-voltage definitions rather than directly. This is done by noting the classic relationship between the three quantities,

$$Z_{VI}^2(\infty) = Z_{PV}(\infty)Z_{PI}(\infty) \tag{15}$$

Making use, again, of the fact that the field distributions at cut-off and infinite frequency are identical gives

$$\frac{V(\infty)V^*(\infty)}{2P_t(\infty)} = \frac{V(\omega_c)V^*(\omega_c)}{2P_t(\omega_c)} \tag{16}$$

This quantity is again independent of the absolute value of the electric field. It can therefore be readily evaluated using the FEM without having to be concerned with the absolute value of the fields met in any calculation. For computational purposes equation (16) may be expressed as

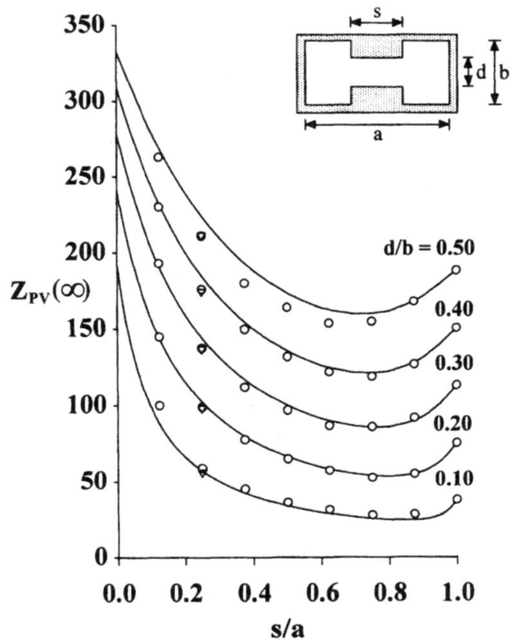

Figure 5.6 Comparison between power-voltage definition of impedance of double ridge waveguide using Hopfer (1955) and FEM definitions (▽ Garb & Kastner (1995), ○, q = 385, b/a = 0.50)

Reprinted with permission, Helszajn & McKay (1998, *IEE Proc.*)

$$Z_{\mathrm{PV}}(\infty) = \frac{\int_0^d [E_y(\omega_c)E_y^*(\omega_c)]\,\mathrm{d}x}{2\zeta_0\left\{\int_0^d\int_0^{\ell_1}[E_x(\omega_c)E_x^*(\omega_c) + E_y(\omega_c)E_y^*(\omega_c)]\,\mathrm{d}x\,\mathrm{d}y + \int_0^b\int_{\ell_4}^{\ell_5}[E_x(\omega_c)E_x^*(\omega_c) + E_y(\omega_c)E_y^*(\omega_c)]\,\mathrm{d}x\,\mathrm{d}y\right\}} \tag{17}$$

Figure 5.6 compares the results obtained with the FEM in Garb and Kastner (1995) and the closed form expression based on the TRM described in Chapter 3.

Figure 5.7 summarises the power-current definition of impedance in this sort of waveguide.

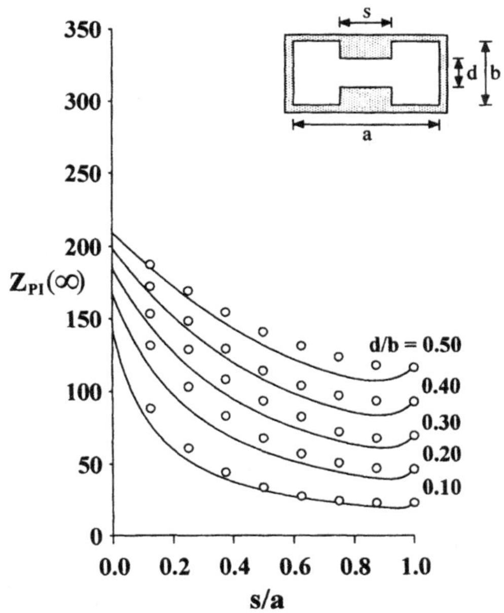

Figure 5.7 Power-current definition of impedance of double ridge waveguide (○, q = 385, b/a = 0.50)

Reprinted with permission, Helszajn & McKay (1998, *IEE Proc.*)

5.5 Impedance of ridge waveguide using trapezoidal ridges

The impedance of rectangular waveguides with trapezoidal ridges is also of some interest. Figure 5.8 depicts the geometry under consideration. It is fixed by the ratio s_1/s. The purpose of this section is to briefly summarise some calculations on its voltage-current definition of impedance at infinite frequency. This is done for s_1/s equal to 1.0, 0.80 and 0.60 with $b/a = 0.50$, $s/a = 0.50$ and $d/b = 0.50$. The results obtained here are summarised by $s_1/s = 1.0$, $\lambda_c/a = 2.69$, $Z_{VI}(\infty) = 151.91\Omega$, $s_1/s = 0.80$, $\lambda_c/a = 2.70$, $Z_{VI} = 150.75\Omega$ and $s_1/s = 0.60$, $\lambda_c/a = 2.71$ and $Z_{VI} = 149.51\Omega$. These calculations suggest, at first sight, that varying these quantities has no notable effect on the impedance of this sort of waveguide. The relative magnitude of the electric field is indicated graphically in Figure 5.9.

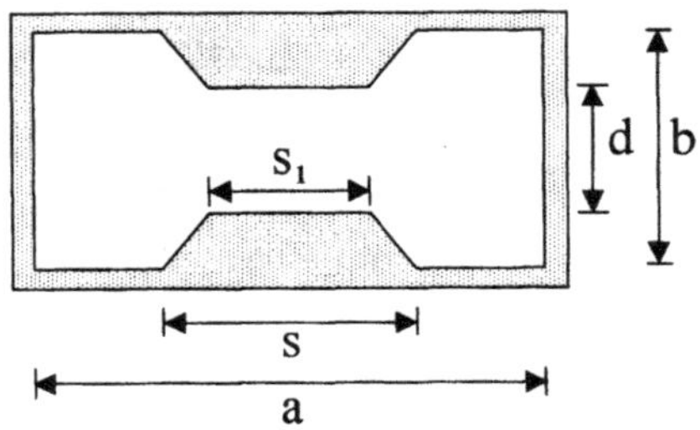

Figure 5.8 Schematic diagram of ridge waveguide using trapezoidal ridges

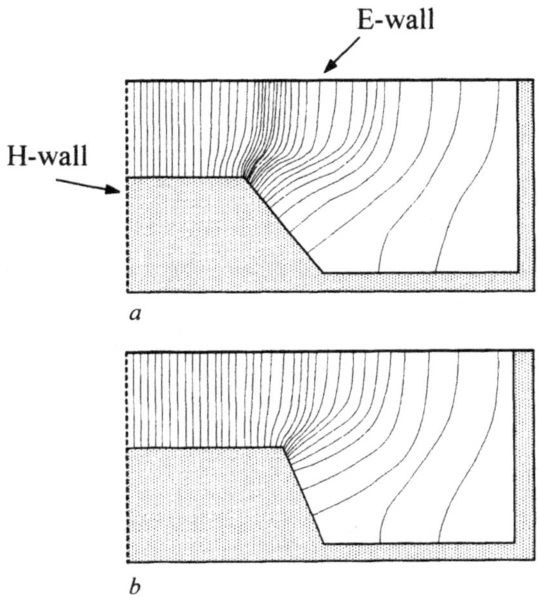

Figure 5.9 Standing wave solution in ridge waveguide using trapezoidal ridges

Chapter 6

Characterisation of single ridge waveguide using the finite element method

M. McKay and J. Helszajn

6.1 Introduction

The various definitions of impedance in a single ridge waveguide are also of some interest. The purpose of this chapter is to summarise some calculations. The description of this problem region follows closely that introduced in connection with the double ridge arrangement except that the limits of the integral used to define the current in the waveguide walls are in this instance different. It corresponds, instead, with the position at which the polarity of the electric field on the boundary contour on either side of the symmetric plane reverses. This condition ensures that the magnitude of the total currents flowing in the equivalent top and bottom walls of the waveguide are equal. A careful scrutiny of this problem suggests that the interior contour of the waveguide breaks up in every instance into one interval consisting of the top and side-walls of the structure and a second interval comprising the remaining walls. The calculation undertaken here suggests that the error resulting in approximating the discontinuity at a typical ridge edge by a shunt capacitance and retaining the contributions from the top and side-walls of the ridge and bottom waveguide wall is negligible. This observation may be understood by recognising that the field is concentrated between the ridge and upper wall of the waveguide. This difficulty does not arise in the power-voltage definition of impedance. The power-current definition is simply obtained by using the relationship between the three classic definitions of impedance. The analysis of the field distribution and power-voltage definition of impedance in various nonsymmetrical ridge structures has included mode matching techniques, a variational method and a surface integral approach. The work outlined here is based on the nodal FEM.

6.2 Cut-off space of single ridge waveguide

The calculation of impedance at finite frequency involves a knowledge of the cut-off space. The single ridge structure differs from the double ridge geometry in that it only supports one plane of symmetry about its wide dimension. The cut-off space of this structure is the same as that of the double ridge structure with twice the height and twice the gap opening. The evaluation of the cut-off space of various ridge topologies by the FEM is a standard result in the literature. The geometry in question is depicted in Figure 6.1. A comparison of the results obtained in this work and that based on the transverse resonance condition is shown in Figure 6.2. The discretisation used to obtain these data is defined by

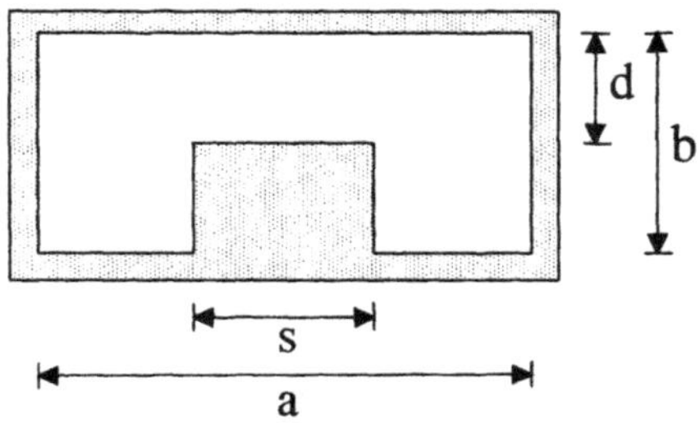

Figure 6.1 Schematic diagram of single ridge waveguide

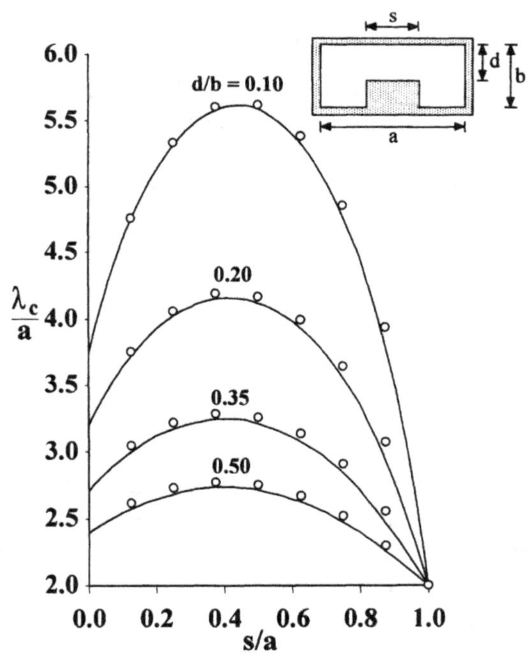

Figure 6.2 Cut-off space of single ridge waveguide ($b/a = 0.45$)

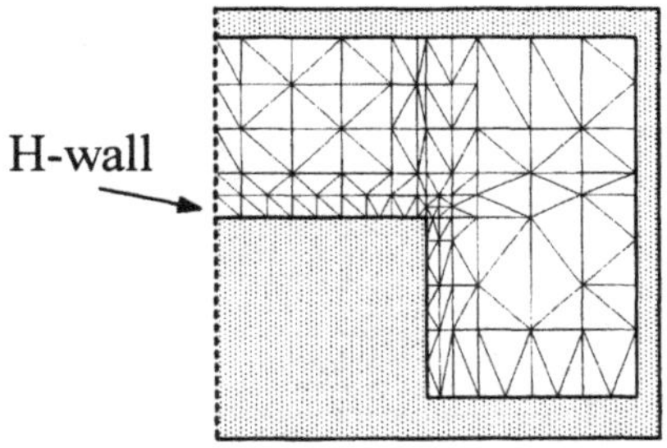

Figure 6.3 Discretisation of single ridge waveguide

$$p = 2$$

$$m = 6$$

$$n = 168$$

$$n \times m = 1008$$

$$q = 385$$

p is the degree of the polynomial approximation inside each element, m is the number of nodes, n is the number of elements, $n \times m$ is the number of nodes prior to assembly of the mesh and q is the number of nodes after assembly of the mesh. The mesh arrangement is indicated in Figure 6.3.

6.3 Fields in single ridge waveguide

A scrutiny of the symmetry attached to the problem region of the single ridge waveguide suggests that, insofar as the dominant TE_{10} mode is concerned, it is again sufficient to solve one-half of the original topology. The electrical paths of the problem region are indicated in Figure 6.4:

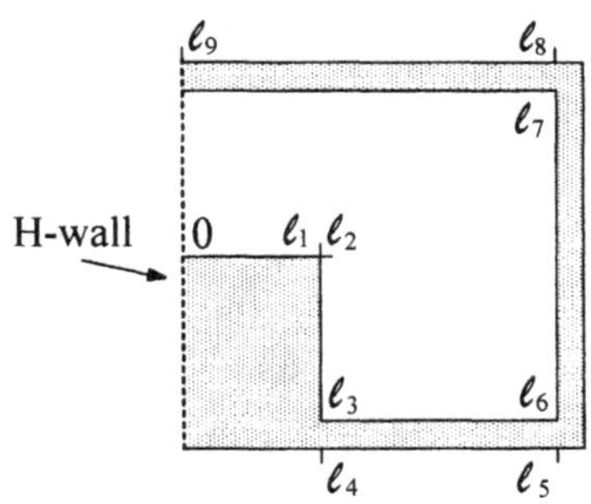

Figure 6.4 Path of current integral

$$\ell_1 = \frac{s}{2} \qquad \ell_2 = (b - d)$$

$$\ell_3 = 0 \qquad \ell_4 = \frac{s}{2}$$

$$\ell_5 = \frac{a}{2} \qquad \ell_6 = 0$$

$$\ell_7 = b \qquad \ell_8 = \frac{a}{2}$$

$$\ell_9 = 0$$

Figure 6.5 shows a typical electric field distribution around the interior contour of the single ridge waveguide for two typical geometries. The distribution of a typical electric field indicated in this illustration is compatible with the magnetic field of the partitioned double ridge waveguide at the same walls dealt with in Chapter 3. It indicates that a unique voltage-current definition of impedance is obtained by taking either the current in the top

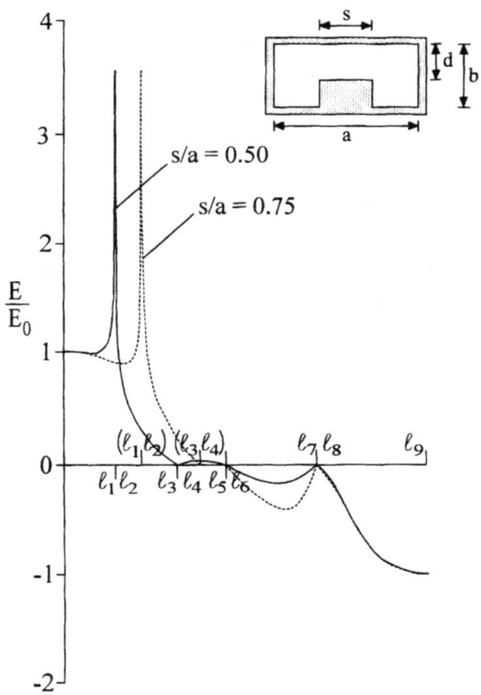

Figure 6.5 Electric field distribution on interior waveguide wall ($b/a = 0.45$, $d/b = 0.35$)

and side walls or that in the ridge and bottom walls for the purpose of calculation. It is also observed that the electric field is zero at all 90 degree corners of the structure.

Figure 6.6 separately depicts the distribution of the electric fields of the dominant mode in a single ridge waveguide for three different geometries.

Approximate closed form expressions for the fields in a single ridge waveguide are also of interest. These may be constructed from those in the double ridge waveguide summarised in Chapter 4 without difficulty. The fields in a single ridge waveguide are given from those in the double ridge by replacing d by $2d$ and b by $2b$ in the latter arrangement prior to introducing an electric wall at the symmetry plane of the problem region. The dimensions s and a are unaltered by this operation, as is readily understood.

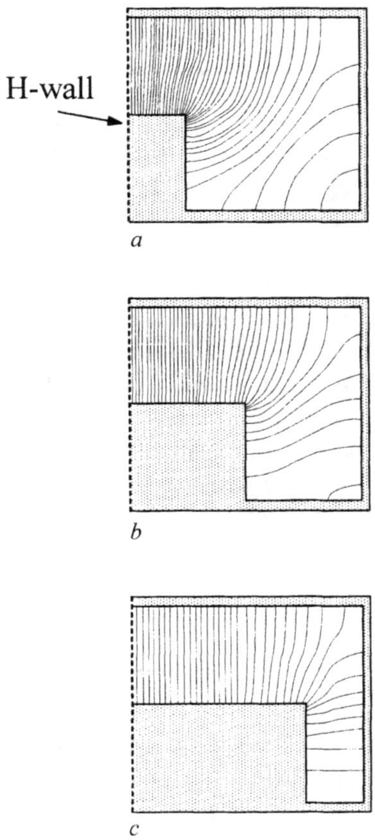

Figure 6.6 Standing wave solutions of single ridge waveguide ($b/a = 0.45$, $d/b = 0.50$)
(a) $s/a = 0.25$; (b) $s/a = 0.50$; (c) $s/a = 0.75$

6.4 Impedance of single ridge waveguide

The voltage-current definition of impedance in a homogeneous waveguide at finite frequency is again given in terms of the calculated one at infinite frequency by

$$Z_{\mathrm{VI}}(\omega) = Z_{\mathrm{VI}}(\infty)\left(\frac{\lambda_g}{\lambda_0}\right) \tag{1}$$

where

$$Z_{\mathrm{VI}}(\infty) = \frac{V(\infty)}{I(\infty)} \tag{2}$$

and λ_g is the usual guide wavelength

$$\left(\frac{2\pi}{\lambda_g}\right)^2 = \left(\frac{2\pi}{\lambda_0}\right)^2 - \left(\frac{2\pi}{\lambda_c}\right)^2 \tag{3}$$

The calculation of the voltage between the ridge and upper waveguide wall proceeds as in the case of the double ridge geometry but the absence of a symmetry plane means that the integration limits associated with the two possible current paths in the top and bottom walls of the single ridge structure must be given separate consideration. One way to overcome this difficulty is to recognise that the polarisation of the electric field reverses around the boundary contour of the waveguide. The limits on the integrals met in connection with the calculation of the total current may therefore be taken to coincide with the two locations where the electric field passes through zero on either side of the symmetry plane.

Figure 6.7 compares the voltage-current definition of impedance obtained here with a closed form expression.

The power-voltage definition of impedance at finite frequency is given in terms of that at infinite frequency by

$$Z_{\mathrm{PV}}(\omega) = Z_{\mathrm{PV}}(\infty)\left(\frac{\lambda_g}{\lambda_0}\right) \tag{4}$$

The absence of any current integral in this definition avoids the difficulty encountered in the calculation of Z_{VI}. The calculation proceeds as for the double ridge waveguide (Helszajn & McKay, 1998, *IEE Proc.*). Figure 6.8 indicates the agreement between the FEM result and some previous calculations.

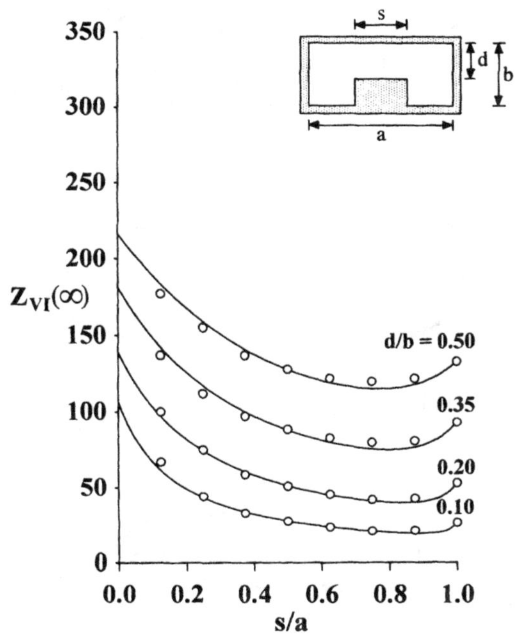

Figure 6.7 Comparison between voltage-current definition of impedance of single ridge waveguide using FEM (○, q = 385) and Sharma Hoefer (1983) definition (b/a = 0.45)

The power-current definition of impedance is obtained easily from the classic relationship

$$Z_{VI}^2(\infty) = Z_{PV}(\infty) Z_{PI}(\infty) \tag{5}$$

It is depicted in Figure 6.9.

6.5 Insertion loss in single ridge waveguide

The insertion loss in a single ridge waveguide is also, of course, of some interest. It is defined in a like manner to that given in connection with the double ridge arrangement,

$$\alpha = \frac{P_\ell}{2P_t} \text{ (dB)}$$

Figure 6.10 indicates some normalised results for the dominant TE_{10} mode based on some MFIE calculations. They are carried out on three typical values of gap ratios (d/b) for parametric values of s/a. The aspect ratio of the waveguide is $b/a = 0.45$.

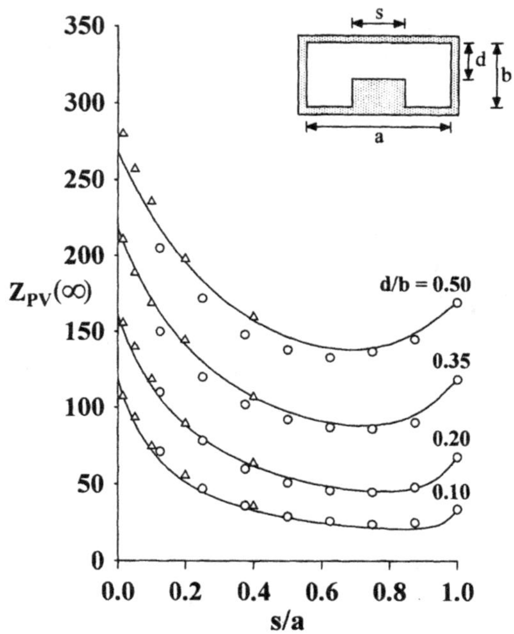

Figure 6.8 Comparison between power-voltage definition of impedance of single ridge waveguide using FEM (○) and Hopfer, 1955 and Utsumi, 1985 (Δ), (b/a=0.45)

6.6 Higher order modes

The bandwidth of any waveguide is fixed by the onset of the first higher order mode of the structure. This problem has already been tackled in connection with the double ridge waveguide and is briefly dealt with here in the case of the single ridge arrangement. It is defined as the ratio of the cut-off wavelengths of the fundamental and the first higher order mode. The first higher order mode in this waveguide has odd symmetry about the problem region. The transverse resonance condition from which its cut-off number may be obtained is

$$-\cot\theta_1 - \left(\frac{Y_{02}}{Y_{01}}\right)\cot\theta_2 + \frac{B}{Y_{01}} = 0$$

The original variables entering into this equation are available in Chapter 3.

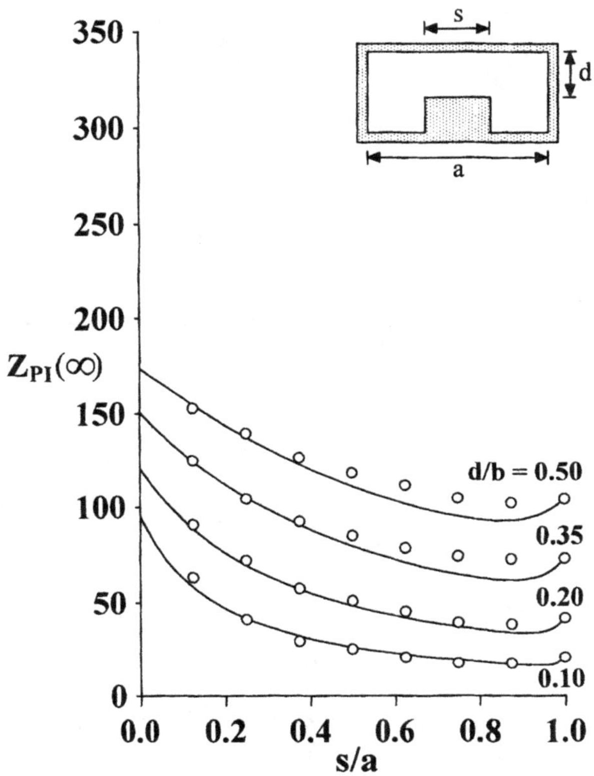

Figure 6.9 Power-current definition of impedance of single ridge waveguide using FEM (○, q = 385), (b/a=0.45)

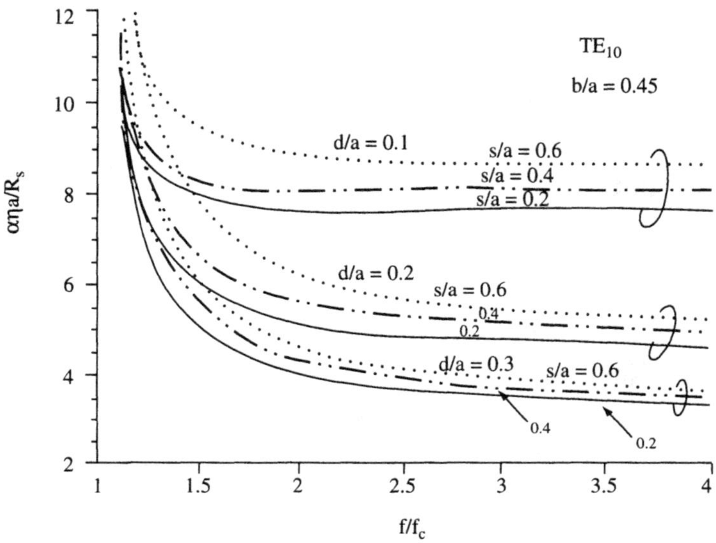

Figure 6.10 Normalised attenuation constant of single ridge waveguide (f_c = cut-off frequency of TE_{10} mode)

Chapter 7

Propagation constant and impedance of dielectric loaded ridge waveguide using a hybrid finite element solver

M. McKay and J. Helszajn

7.1 Introduction

The finite element method (FEM) has been widely utilised in the analysis of the cut-off space and propagation constant of dielectric loaded waveguides containing isotropic and anisotropic media. It includes variational expressions based on the two-component E_z/H_z hybrid notation and the three-component (magnetic $\bar{H}$ or electric $\bar{E}$) vector formulation. The purpose of this chapter is to give some calculations on the cut-off space, the propagation constant and the impedance of the ridge waveguide with a dielectric filler between its ridges based on a hybrid E_z/H_z finite element calculation. Modes in this sort of waveguide may in general be described as quasi-LSE_{mn} or quasi-LSM_{mn}, depending on whether E_x or H_x is equal to zero as the dielectric loaded ridge waveguide is reduced to a rectangular one. The first three modes may be denoted as quasi-LSE_{10}, -LSE_{20}, -LSE_{11} or -LSM_{01}. One feature of this type of waveguide is that the separation between the cut-off frequencies of the dominant and first order quasi-LSE_{mo} modes is increased. The exact details of the cut-off space are dependent on the dielectric constant (ε_r) of the insert and the geometry of the ridge waveguide. Some existing experimental data obtained on the propagation constant of a square waveguide are in engineering agreement with the theory. A feature of any dielectric loaded waveguide is the existence of planes of quasi-circular polarisation at the boundary between the two dielectrics. It has been utilised historically in the design of a number of nonreciprocal 2-port devices. A calculation of the ellipticity in this sort of waveguide is dealt with in a separate chapter. The spurious modes encountered in the numerical solutions of this sort of waveguide may be suppressed by either enforcing the divergence free constraint in association with tangential and normal boundary conditions or by employing vector elements to ensure only tangential continuity of the field components across element boundaries. The voltage-current definition of impedance for various ridge topologies is separately investigated.

7.2 Hybrid functional

The problem region tackled in this chapter consists of a ridge waveguide with a dielectric filler between its two ridges. Figure 7.1 indicates the topology considered here. In general this structure supports hybrid modes, except in the case of the pure TE_{mo} ones which can exist when the dielectric extends across the full waveguide. The solution to this type of problem involves the cut-off space, the propagation constant and one or more definitions of impedance. In an inhomogeneous geometry the latter two quantities have to be calculated at each and every frequency.

One solution to this sort of problem is to use a variational technique. This involves constructing an energy functional which when extremised by using the Ralyeigh-Ritz method produces a solution which satisfies the original wave equation. The functional formulation for this sort of problem region may be developed in terms of the longitudinal field components (E_z/H_z) or the three component magnetic (**H**) or electric (**E**) field vectors. The former choice is adopted here. The required construction starts by noting the vector form of the Helmholtz equation

$$\mathbf{L}\boldsymbol{\phi} = 0 \tag{1}$$

where $\bar{\mathrm{L}}$ is the matrix of the form

$$\mathbf{L} = \begin{bmatrix} L_{\mathrm{ee}} & L_{\mathrm{eh}} \\ L_{\mathrm{he}} & L_{\mathrm{hh}} \end{bmatrix} \tag{2}$$

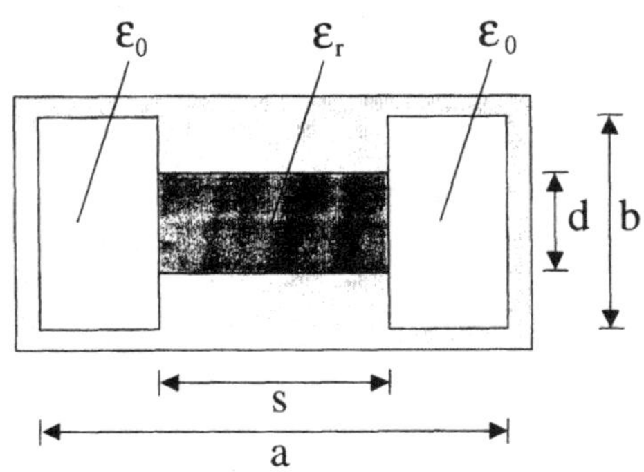

Figure 7.1 Schematic diagram of dielectric loaded ridge waveguide

and the hybrid vector $\boldsymbol{\phi}$ represents the field variables,

$$\boldsymbol{\phi} = \begin{bmatrix} E_z \\ H_z \end{bmatrix} \tag{3}$$

For an isotropic dielectric region the wave equations associated with E_z and H_z are uncoupled and the entries of the matrix of the form reduce to the standard scalar Helmholtz equations. The entries of the form are separately given by

$$L_{\mathrm{ee}} = \varepsilon_0 \varepsilon_{\mathrm{r}} (\nabla_{\mathrm{t}}^2 + k_{\mathrm{cr}}^2) \tag{4a}$$

$$L_{\mathrm{hh}} = \mu_0 (\nabla_{\mathrm{t}}^2 + k_{\mathrm{cr}}^2) \tag{4b}$$

$$L_{\mathrm{eh}} = 0 \tag{4c}$$

$$L_{\mathrm{he}} = 0 \tag{4d}$$

where

$$k_{\mathrm{cr}}^2 = k_0^2 \varepsilon_{\mathrm{r}} - \beta^2 \tag{5}$$

and

$$k_0 = \omega \sqrt{\varepsilon_0 \mu_0} \tag{6}$$

k_{cr}, ε_{r} and β represent the cut-off wavenumber, the relative permittivity and the propagation constant of a typical region, respectively. The construction of the associated functional proceeds by premultiplying the vector form of the Helmholtz equation in a typical region by the transpose of the conjugate vector field, integrating the ensuing quadratic equation over the cross-

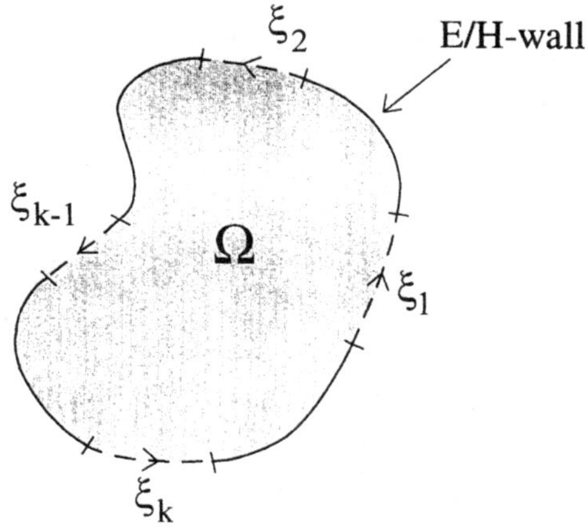

Figure 7.2 Homogeneous problem region with inhomogeneous boundary condition

section of the region in question and utilising Green's theorem in a plane to map a surface integral into a contour one. Application of the appropriate boundary conditions on the contour ξ of a homogeneous region of cross-sectional area Ω produces the functional of a typical region,

$$\begin{aligned} F_r(E_z, H_z) = {} & \varepsilon_r \iint_{\Omega} [-(\nabla_t E_z) \cdot (\nabla_t E_z^*) + k_{cr}^2 |E_z|^2] \, d\Omega \\ & + \eta_0^2 \iint_{\Omega} [-(\nabla_t H_z) \cdot (\nabla_t H_z^*) + k_{cr}^2 |H_z|^2] \, d\Omega \\ & + \eta_0 \left(\frac{\beta}{k_0}\right) \sum_1^k \int_{\xi_k} \left[H_z^* \left(\frac{\partial E_z}{\partial t}\right) - E_z^* \left(\frac{\partial H_z}{\partial t}\right) \right] dt \\ & + j\eta_0 \left(\frac{k_{cr}^2}{k_0}\right) \sum_1^k \int_{\xi_k} [(E_z^* H_t) - (H_z^* E_t)] \, dt \end{aligned} \tag{7}$$

t is an anticlockwise directed unit vector tangential to the boundary ξ, and k denotes the number of media interfaces met on the boundary.

η_0 is the free-space wave impedance

$$\eta_0 = \sqrt{\frac{\mu_0}{\varepsilon_0}} \tag{8}$$

Figure 7.2 summarises this situation.

The first surface integral term is the energy functional associated with the planar problem region with top and bottom electric walls. It automatically satisfies the Neumann boundary condition

$$\frac{\partial H_z}{\partial n} = 0 \tag{9}$$

at any magnetic side-wall. The Dirichlet boundary condition

$$E_z = 0 \tag{10}$$

must, however, be separately enforced at any electric side-walls. The second surface integral corresponds to the energy of the dual problem with top and bottom magnetic walls for which the natural boundary condition is the Neumann one at electric side-walls and the Dirichlet condition must be enforced at magnetic side-walls. The two contour integral terms also produce real parts and ensure conservation of energy across dielectric boundaries.

The energy functional for the total inhomogeneous problem region can therefore be expanded in terms of a typical region

$$\mathrm{F}(E_z, H_z) = \sum_{1}^{r} \mathrm{F_r}(E_z, H_z) \tag{11}$$

It is of note that β has the same value in each region and that the second contour integral in equation (7) makes no contribution to the overall functional in equation (11) since each common contour ξ_k is traversed along opposite directions in the connected media and may be discarded.

The Rayleigh-Ritz procedure approximates the field solutions (E_z, H_z) in terms of a set of linearly independent shape functions $\alpha(x, y)$ prior to extremising the functional to produce a matrix eigenvalue problem. Noting that a priori fields exist at electric, magnetic and resistive walls, E_z and H_z in a typical region may be expanded in terms of the free components in its interior and the free and prescribed components on its boundaries:

$$E_z \approx \sum_{1}^{f_\mathrm{e}} e_{\mathrm{rf}}\alpha_{\mathrm{f}}(x, y) + \sum_{1}^{p_\mathrm{e}} e_{\mathrm{rp}}\alpha_{\mathrm{p}}(x, y) \tag{12a}$$

and

$$H_z \approx \sum_{1}^{f_\mathrm{h}} h_{\mathrm{rf}}\alpha_{\mathrm{f}}(x, y) + \sum_{1}^{p_\mathrm{h}} h_{\mathrm{rp}}\alpha_{\mathrm{p}}(x, y) \tag{12b}$$

The following work is only concerned with electric and magnetic walls and the prescribed nodes are hereafter set to zero. Introducing the preceding approximations into equation (11) and extremising with respect to the unconstrained variables (e_{rf}, h_{rf}) produces the eigenvalue problem for the inhomogeneous region,

$$\sum_{1}^{r} \frac{1}{\varepsilon_\mathrm{r} - \left(\dfrac{\beta}{k_0}\right)^2} \begin{bmatrix} \dfrac{\varepsilon_\mathrm{r}}{\eta_0^2}[S]_{(1)} & \dfrac{1}{2\eta_0}\left(\dfrac{\beta}{k_0}\right)[U] \\ \dfrac{1}{2\eta_0}\left(\dfrac{\beta}{k_0}\right)[U]^\mathrm{T} & [S]_{(2)} \end{bmatrix} \begin{bmatrix} [e_{\mathrm{rf}}] \\ [h_{\mathrm{rf}}] \end{bmatrix}$$

$$= \sum_{1}^{r} k_0^2 \begin{bmatrix} \dfrac{\varepsilon_\mathrm{r}}{\eta_0^2}[T]_{(1)} & 0 \\ 0 & [T]_{(2)} \end{bmatrix} \begin{bmatrix} [e_{\mathrm{rf}}] \\ [h_{\mathrm{rf}}] \end{bmatrix} \tag{13}$$

The order of the symmetric matrices $[S]_{(1)}$ and $[S]_{(2)}$ corresponds to the degrees of freedom (r) of the free nodal electric (f_e) and magnetic (f_m) fields of the region, respectively. A similar definition applies to the $[T]$

matrices. The entries of the **S** and **T** matrices are the classic ones met in connection with a planar circuit with either top and bottom magnetic or electric walls

$$S_{ij} = \int_{\Omega} (\nabla_t \alpha_i) \cdot (\nabla_t \alpha_j) \, d\Omega \tag{14a}$$

$$T_{ij} = \int_{\Omega} \alpha_i \alpha_j \, d\Omega \tag{14b}$$

$[U]$ is a matrix of order $(f_e \times f_m)$ and represents the physical coupling between the electric and magnetic fields at the dielectric boundaries of the region. $[U]_{(T)}$ is the transpose of $[U]$. The entries of this matrix are given by

$$U_{ij} = \sum_{1}^{k} \int_{\xi_k} \left(\alpha_i \frac{\partial \alpha_j}{\partial t} - \alpha_j \frac{\partial \alpha_i}{\partial t} \right) dt \tag{14c}$$

At cut-off, e_{rf} and h_{rf} are decoupled.

7.3 Cut-off space of dielectric loaded rectangular ridge waveguide

The first task of this chapter is the evaluation of the cut-off space of the dielectric loaded double ridge waveguide. The aspect ratio of the waveguide (b/a) is 0.50 and the ridges are described by parametric values of d/b and s/a, respectively. At cut-off, β is zero and the functional in equation (7) reduces to the sum of the two independent functionals associated with the planar problem region with top and bottom electric and magnetic walls, respectively. Imposing appropriate symmetry planes and extremising the functional in conjunction with the FEM yields the quasi-LSE and -LSM modes of the problem region. The purpose of this section is to summarise some results. The segmentation employed in this work is depicted in Figure 7.3. The discretisations in the dielectric and air regions are defined by

$$p = 2$$

$$m = 6$$

$$n = 43$$

$$n \times m = 258$$

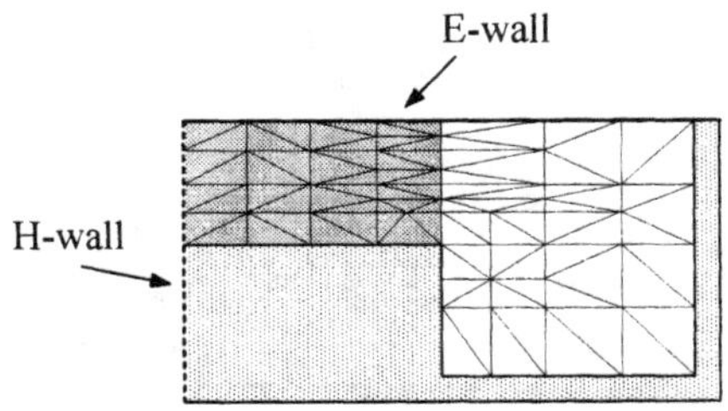

Figure 7.3 Discretisation of dielectric loaded ridge waveguide

and

$$p = 2$$

$$m = 6$$

$$n = 49$$

$$n \times m = 294$$

respectively.

p is the degree of the interpolation polynomial within each finite element triangle, m is the number of nodes inside each finite element triangle, n is the number of triangles and $n \times m$ is the total number of nodes before assembly of the finite element mesh.

The corresponding free (f) and prescribed (p) electric and magnetic nodes in the individual regions are

$$f_h = 97$$

$$f_e = 88$$

$$p_h = 9$$

$$p_e = 18$$

and

$$f_h = 120$$

$$f_e = 91$$

$$p_h = 0$$

$$p_e = 29$$

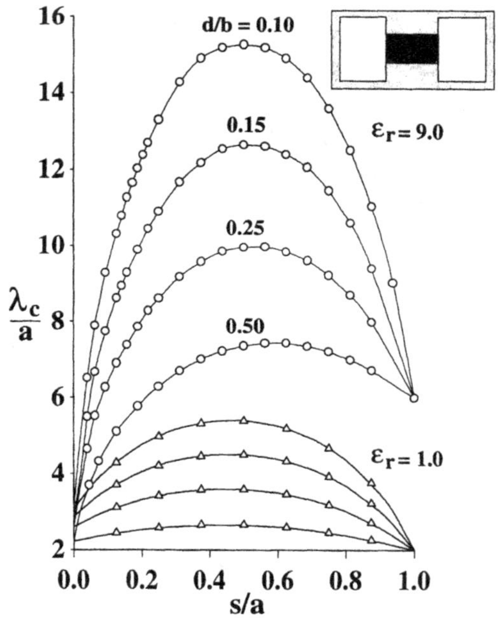

Figure 7.4 Cut-off space of quasi-LSE$_{10}$ mode in dielectric loaded ridge waveguide for $\varepsilon_r = 1$ and $\varepsilon_r = 9.0$

Reprinted with permission, McKay & Helszajn (2000)

Figure 7.4 indicates the variation of the normalised cut-off wavelength of the dominant quasi-LSE$_{10}$ mode when a dielectric ($\varepsilon_r = 9$) is inserted between the ridges. Figure 7.5 indicates the separation between the quasi-LSE$_{10}$ and -LSE$_{20}$ modes.

Although the emphasis of this section has been on the calculation of the quasi-LSE$_{10}$ and -LSE$_{20}$ modes, in general, the cut-off space of ridge and dielectric loaded waveguides is more complicated. Figure 7.6 indicates one typical range of topologies for which the quasi-LSE$_{11}$ or quasi-LSM$_{01}$ mode is actually the first higher order one.

Figure 7.7 depicts the electric and magnetic field patterns of the quasi-LSE$_{10}$ and -LSE$_{11}$ modes for one typical geometry.

7.4 Propagation constant in dielectric loaded rectangular ridge waveguide

The purpose of this section is to summarise some calculations on the propagation constant of the dominant quasi-LSE$_{10}$ mode in a double ridge

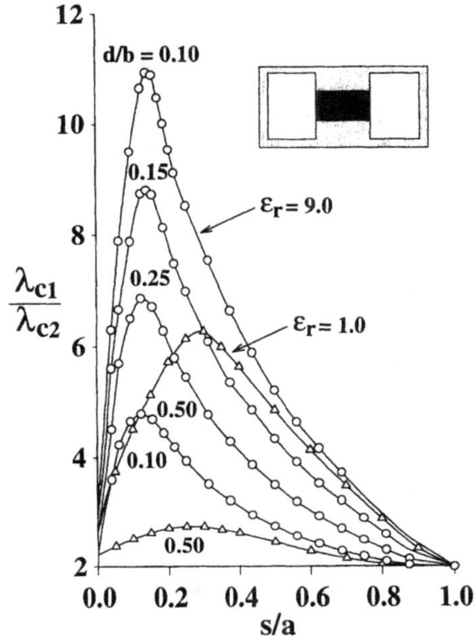

Figure 7.5 Ratio of cut-off numbers of quasi-LSE_{10} and -LSE_{20} modes for $\varepsilon_r = 1$ and $\varepsilon_r = 9.0$

Reprinted with permission, McKay & Helszajn (2000)

waveguide with a dielectric filler between the ridges. This is done for one waveguide topology and two different materials. Figure 7.8 indicates the agreement between some calculations on a standard rectangular waveguide with a dielectric insert and some calculations using the transverse resonance method ($d/b = 1.0$). Figure 7.9 shows the main result obtained here.

Some calculations on a ridge waveguide with an inhomogeneous dielectric filler have also been carried out for completeness' sake. The topology considered here is illustrated in Figure 7.10 and some typical data are given in Figure 7.11.

7.5 Propagation constant in dielectric loaded square waveguide

Figure 7.12 indicates some calculations on the normalised propagation constant of a square ridge waveguide with a dielectric filler between its ridges. Some measurements in the open literature are separately superimposed on this result. The measurements are obtained by using a 1-port reflection

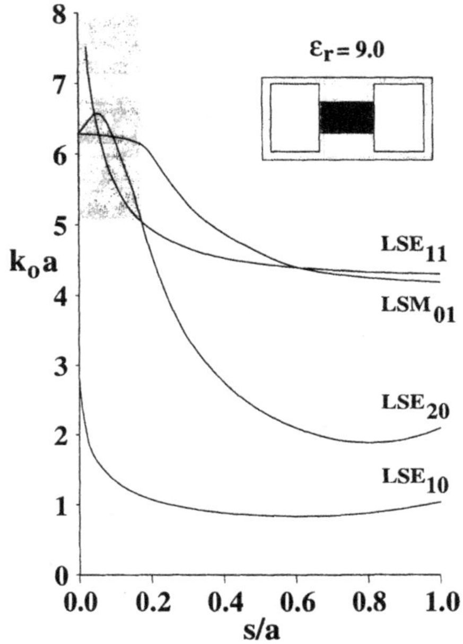

Figure 7.6 Higher order modes in dielectric loaded ridge waveguide $\varepsilon_r = 9.0$ $(b/a = 0.50,\ d/b = 0.50)$

Reprinted with permission, McKay & Helszajn (2000)

cavity. The segmentation utilised here is the same as that employed in the rectangular geometry.

7.6 Voltage-current definition of impedance

The impedance of a dielectric loaded double ridge waveguide is also of some interest in microwave engineering. The purpose of this section is to give some calculations on its voltage-current definition. The voltage is obtained in the usual way by evaluating the electric field along the magnetic symmetry plane extending between the two ridges,

$$E_y = \frac{jk_0\eta_0}{(k_0^2\varepsilon_r - \beta^2)}\left(\frac{\partial H_z}{\partial x}\right) - \left(\frac{j\beta}{k_0^2\varepsilon_r - \beta^2}\right)\left(\frac{\partial E_z}{\partial y}\right) \tag{15}$$

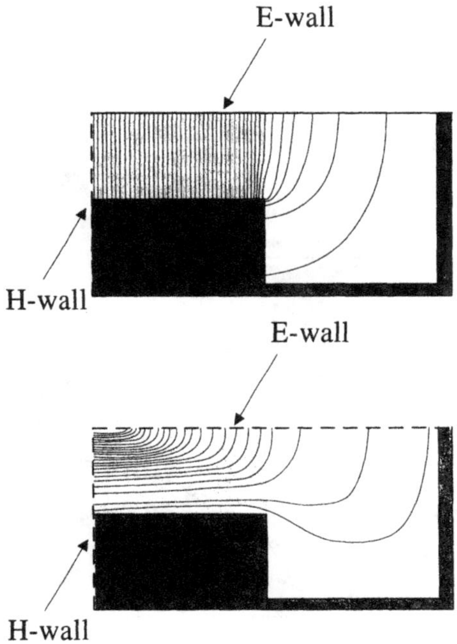

Figure 7.7 Electric and magnetic field distribution of quasi-LSE_{10} and -LSE_{11} modes at cut-off ($\varepsilon_r = 9.0$, $b/a = 0.50$, $s/a = 0.50$, $d/b = 0.50$)

Reprinted with permission, McKay & Helszajn (2000)

The current is separately obtained by integrating the magnetic field along the contour defined by the electric symmetry plane. This calculation may be undertaken by using H_x and H_y as follows:

$$H_x = \frac{-j\beta}{(k_0^2\varepsilon_r - \beta^2)}\left(\frac{\partial H_z}{\partial x}\right) + \frac{jk_0\varepsilon_r}{\eta_0(k_0^2\varepsilon_r - \beta^2)}\left(\frac{\partial E_z}{\partial y}\right)$$

$$H_y = \frac{-j\beta}{(k_0^2\varepsilon_r - \beta^2)}\left(\frac{\partial H_z}{\partial y}\right) - \frac{jk_0\varepsilon_r}{\eta_0(k_0^2\varepsilon_r - \beta^2)}\left(\frac{\partial E_z}{\partial x}\right) \qquad (16)$$

Figure 7.13 indicates the result in the case of a dielectric constant of 9.0. Unlike the homogeneous problem region, for which it is sufficient to calculate the impedance at infinite frequency and use the dispersion relationship to obtain that at the actual frequency, it is necessary in this instance to make the calculation at each individual frequency. Each curve is truncated at the cut-off frequency of the quasi-LSE_{20} mode in the respective topologies.

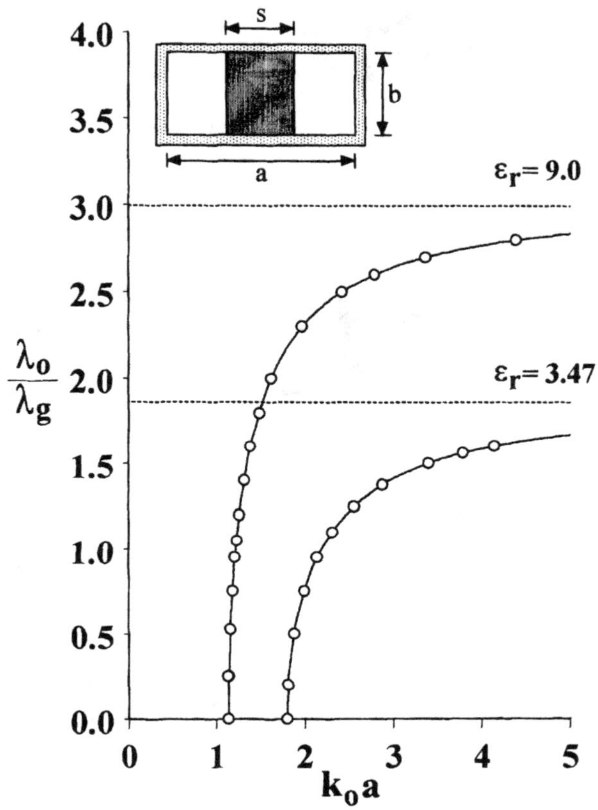

Figure 7.8 Comparison between propagation constants of dielectric loaded rectangular waveguide using TRM and FEM techniques (b/a = 0.50)

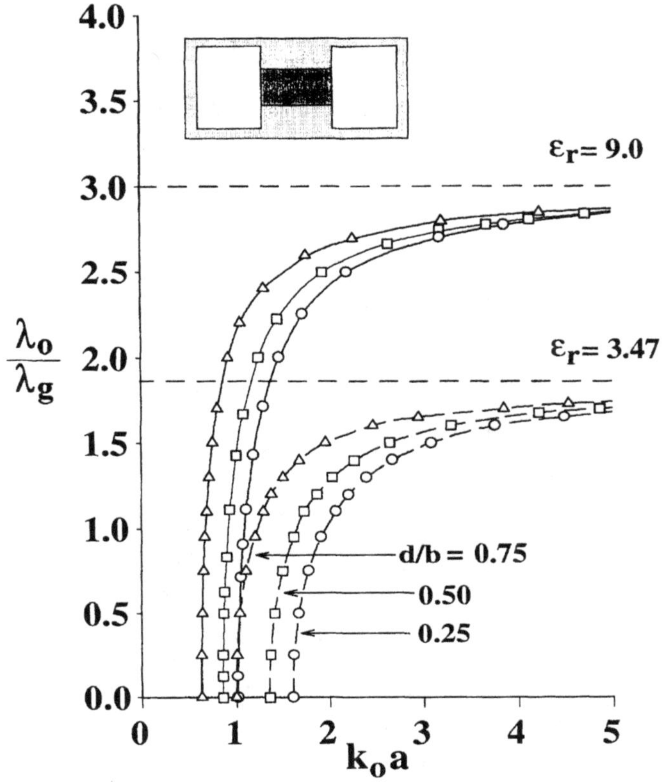

Figure 7.9 Propagation constant of dielectric loaded ridge waveguide ($\varepsilon_r = 3.47$ and 9.0, $b/a = 0.50$)

Reprinted with permission, McKay & Helszajn (2000)

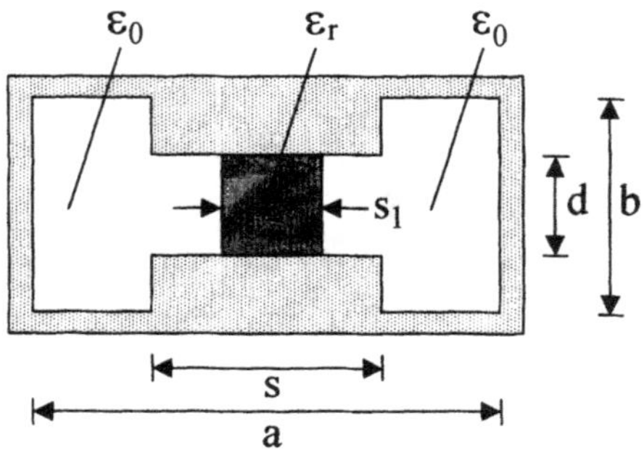

Figure 7.10 Schematic diagram of partially dielectric loaded ridge waveguide

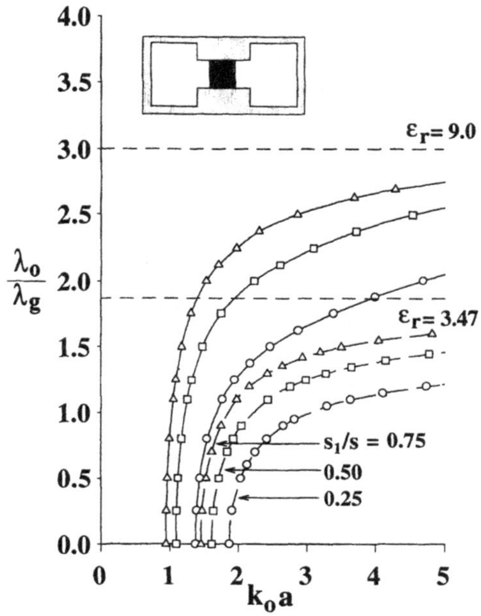

Figure 7.11 Propagation constant of partially dielectric loaded ridge waveguide ($\varepsilon_r = 3.47$ and 9.0, $b/a = 0.50$, $s/a = 0.50$)

Reprinted with permission, McKay & Helszajn (2000)

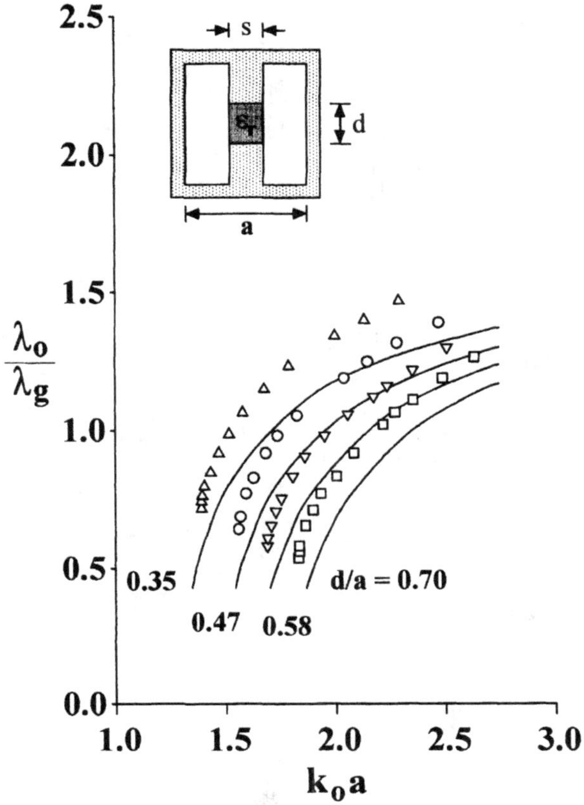

Figure 7.12 Comparison of theoretical and experimental propagation constants in square ridge waveguide FEM; experiment: □, ▽, △, ○ *($\varepsilon_r = 3.47$, $s/a = 0.25$, $a = 8.55$ mm)*

Reprinted with permission, McKay & Helszajn (2000)

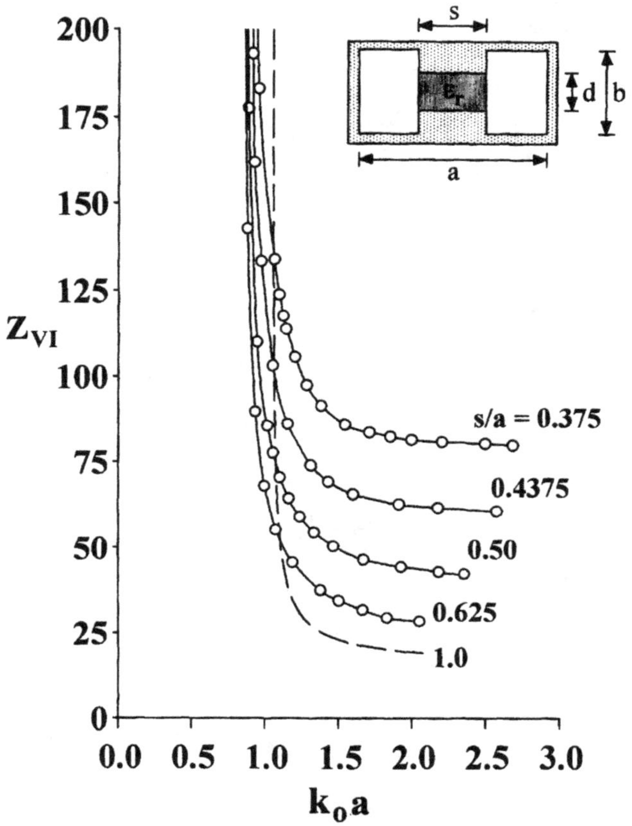

Figure 7.13 Voltage-current definition of impedance in dielectric loaded ridge waveguide, $\varepsilon_r = 9.0$ ($b/a = 0.50$, $d/b = 0.50$)

Reprinted with permission, McKay & Helszajn (2000)

Chapter 8

Circular polarisation in ridge and dielectric loaded ridge waveguides

8.1 Introduction

An important concept in microwave engineering is that of circular polarisation. The two possibilities correspond to two equal vectors in space quadrature with one or the other advanced or retarded in time quadrature. Counter-rotating magnetic fields occur naturally on either side of the symmetry plane of an ordinary rectangular waveguide propagating the dominant TE mode; and at the interface and everywhere outside two different dielectric regions. Furthermore, in each instance, the hand of rotation is interchanged if the direction of propagation is reversed. Situations in which the rotation of these waves are different in the two directions of propagation are, of course, of special interest. One interesting aspect of the single or double ridge waveguides is that neither display planes of circular polarisation in its trough regions. The single ridge waveguide, however, supports such planes on its flat wall and the double ridge one on the symmetry plane defined by its electric wall. The exact nature of the polarisation in a ridge waveguide with a dielectric filler between its ridges based on the FEM method is separately summarised. Since either half-space of the more simple problem of a dielectric rib between two parallel plates has many of the properties of the half-space revealed by a dielectric brick in a ridge waveguide it is investigated prior to tackling the exact problem. One feature of the parallel plate waveguide is the fact that the alternating magnetic field is quasi-circularly polarised with counter-rotating hands at each interface between the two dielectric regions and everywhere outside it.

8.2 Circular polarisation

An alternating radio frequency wave is either vertically, horizontally or clockwise and anticlockwise elliptically or circularly polarised. Since circular polarisation plays such an important role in the operation of ferrite and other devices it is helpful to recall its definition in some detail. The two possibilities correspond to counter-rotating waves and are defined in Cartesian co-ordinates by

$$\mathbf{E}^-(\omega t, \beta z) = (\mathbf{a}_x E_0 + j\mathbf{a}_y E_0)\exp j(\omega t - \beta z)$$

$$\mathbf{E}^+(\omega t, \beta z) = (\mathbf{a}_x E_0 - j\mathbf{a}_y E_0)\exp j(\omega t - \beta z)$$

Whether $\mathbf{E}^+$ or $\mathbf{E}^-$ represents a clockwise or anticlockwise circularly polarised wave may be readily established by constructing the real parts of these quantities and evaluating the same at $\omega t = 0$, $\beta z = 0, \frac{\pi}{2}, \pi, \frac{3\pi}{2}$, etc. or at $\beta z = 0$, $\omega t = 0, \frac{\pi}{2}, \frac{3\pi}{2}$, etc. This gives

$$\mathbf{E}^-(\omega t, \beta z) = \mathbf{a}_x E_0 \cos(\omega t - \beta z) + \mathbf{a}_y E_0 \sin(\omega t - \beta z)$$

$$\mathbf{E}^+(\omega t, \beta z) = \mathbf{a}_x E_0 \cos(\omega t - \beta z) - \mathbf{a}_y E_0 \sin(\omega t - \beta z)$$

Taking $\omega t = 0$ by way of an example gives

$$\mathbf{E}^-(0, \beta z)\pi = \mathbf{a}_x E_0 \cos(\beta z) - \mathbf{a}_y E_0 \sin(\beta z)$$

$$\mathbf{E}^+(0, \beta z) = \mathbf{a}_x E_0 \cos(\beta z) + \mathbf{a}_y E_0 \sin(\beta z)$$

A scrutiny of the preceding equations also indicates that

$$\mathbf{E}^+(\omega t, \beta z) + \mathbf{E}^-(\omega t, \beta z) = \mathbf{a}_x E_0 \cos(\omega t - \beta z)$$

This result suggests that a linearly polarised wave can always be decomposed into a linear combination of counter-rotating circularly polarised waves. Figures 8.1*a* and *b* show pictorial displays of the two magnetic field patterns at $\omega t = 0$ in free space entering into the description of this problem region. Figure 8.2 indicates the corresponding solution for the electric field.

8.3 Open half-space of asymmetrically dielectric loaded ridge waveguide

The open half-space between the ridges of an asymmetrically dielectric loaded ridge waveguide has at first sight some similarity to the open regions of a single dielectric rib between two parallel plates. The topology of the

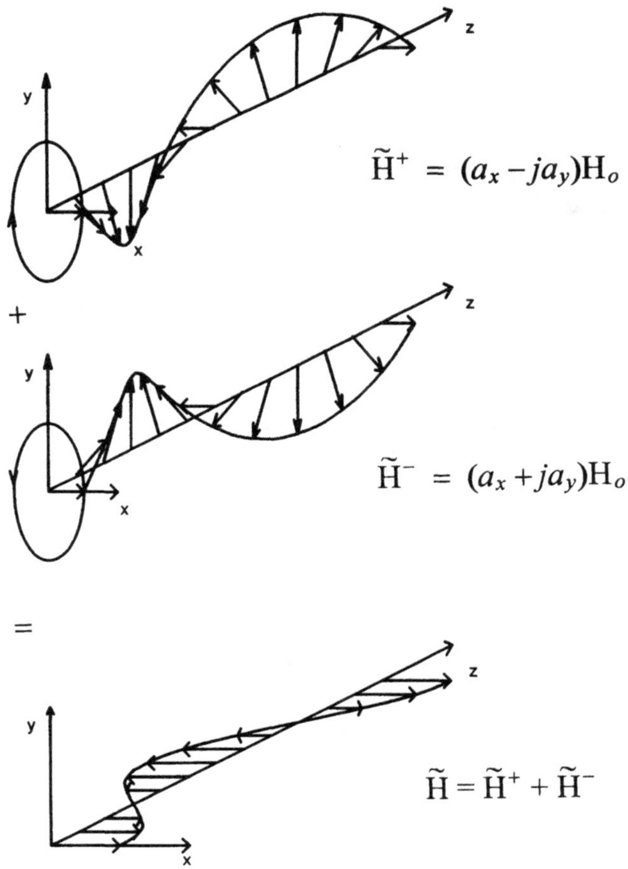

Figure 8.1 Linear polarisation, clockwise, counter-clockwise circular polarisation (magnetic field)

parallel plate waveguide may therefore be utilised to establish the nature of the half-space revealed by a ridge waveguide with the other half-space filled by a suitable dielectric medium. The solution to this problem is a classic one in the literature. The topology considered here is illustrated in Figure 8.3. One property of the parallel plate waveguide is that it displays counter-rotating elliptically polarised magnetic fields at each boundary between the two dielectric regions and everywhere outside. Another property is that the hands of polarisations are interchanged when the direction of propagation is reversed. Such a waveguide may therefore be employed, in conjunction with suitably magnetised ferrite plates, in the construction of nonreciprocal isolators and phase shifters. It may also be used to provide some insight into the operation of nonreciprocal ridge components.

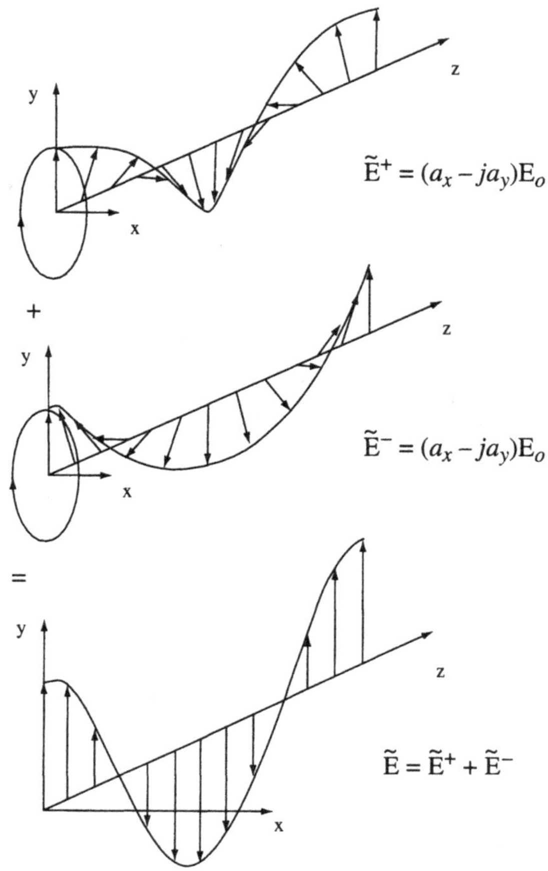

Figure 8.2 Linear polarisation, clockwise, counter-clockwise circular polarisation (electric field)

8.4 Circular polarisation in dielectric-loaded parallel plate waveguides with open side-walls

Dielectric-loaded parallel plate waveguides with open side-walls or with electric or magnetic side-walls all support, under appropriate boundary conditions, planes of circular polarisation at the boundaries between dielectric and air regions. The configuration treated in this section is the open side-wall arrangement in Figure 8.3. A solution of this problem region indicates that the fields decay exponentially outside the dielectric region and that the ratio of the transverse to longitudinal components of the RF magnetic field

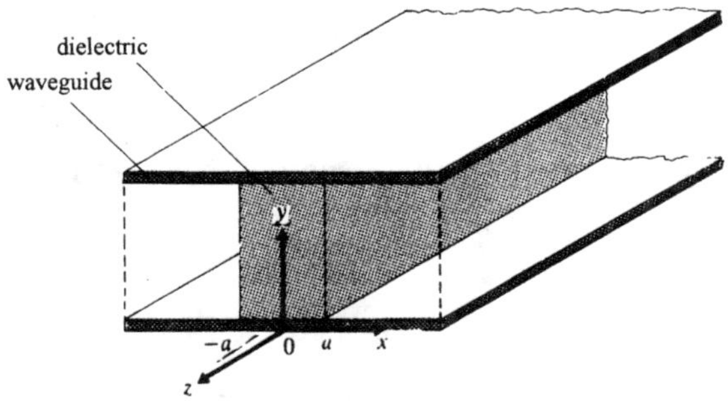

Figure 8.3 Schematic diagram of dielectric-loaded parallel plate waveguide with open side-walls

at the interface between the two dielectric regions and everywhere outside it is given by

$$\frac{H_x}{H_z} = \frac{-j}{\sqrt{1-(k_0/\beta)^2}} \tag{1}$$

where k_0 is the free space propagation constant and β is the phase constant along the structure.

To maintain the ellipticity below 1.05 (say) it is necessary to have

$$\frac{\beta}{k_0} \geqslant 3 \tag{2}$$

The derivation of this important result starts by establishing the field components of the TE family of modes in the three-region dielectric-loaded parallel plate waveguide with open side-walls. The solutions are labelled even or odd according to whether an electric or magnetic wall can be introduced along the plane of symmetry at $x = 0$. The dominant mode in such a waveguide is the so-called even one with no low frequency cut-off condition. One solution in the dielectric region is

$$H_z = A\sin(k_1 x)\exp(-j\beta z) \tag{3}$$

It satisfies the wave equation with

$$-k_1^2 + (-\beta^2 + \omega^2\mu_0\varepsilon_0\varepsilon_1) = 0 \tag{4}$$

and the magnetic wall boundary condition at the symmetry plane of the dielectric region. The other field components in region 1 are now readily constructed in terms of H_z as

$$H_x = \frac{-j\beta}{k_1} A\cos(k_1 x)\exp(-j\beta z) \tag{5}$$

$$E_y = \frac{j\mu_0\omega}{k_1} A\cos(k_1 x)\exp(-j\beta z) \tag{6}$$

A suitable decaying solution in region 2 in keeping with the open wall boundary condition adopted for this region at $x = -\infty$ is

$$H_z = B\exp[k_2(a+x) - j\beta z] \tag{7}$$

It satisfies the wave equation with

$$k_2^2 + (-\beta^2 + \omega^2\mu_0\varepsilon_0) = 0 \tag{8}$$

The constant B is determined by noting that H_z must be continuous at the boundary between the two regions

$$B = -A\sin(k_1 a) \tag{9}$$

The complete solution in region 2 is therefore described by the dispersion relationship in equation (8) and by

$$H_z = -A\sin(k_1 a)\exp[k_2(a+x) - j\beta z] \tag{10}$$

$$H_x = \frac{-j\beta}{k_2} A\sin(k_1 a)\exp[k_2(a+x) - j\beta z] \tag{11}$$

$$E_y = \frac{j\mu_0\omega}{k_2} A\sin(k_1 a)\exp[k_2(a+x) - j\beta z] \tag{12}$$

The solution in region 3 has the same form as that in 2 but with the sign of H_z reversed:

$$H_z = A\sin(k_1 a)\exp[k_2(a-x) - j\beta z] \tag{13}$$

$$H_x = \frac{-j\beta}{k_2} A\sin(k_1 a)\exp[k_2(a-x) - j\beta z] \tag{14}$$

$$E_y = \frac{j\mu_0\omega}{k_2} A\sin(k_1 a)\exp[k_2(a-x) - j\beta z] \tag{15}$$

and satisfies the same wave equation as that in region 2.

The magnetic field is therefore elliptically polarised with one hand of rotation everywhere in region 2:

$$\frac{H_x}{H_z} = \frac{j\beta}{k_2} \tag{16}$$

That in region 3 is elliptically polarised with the other hand of rotation as asserted:

$$\frac{H_x}{H_z} = \frac{-j\beta}{k_2} \tag{17}$$

For completion, it is now necessary to evaluate β and the field patterns of the structure. This may be done by first noting that the propagation constant β must be the same in each region:

$$k_2^2 + \omega^2 \mu_0 \varepsilon_0 = -k_1^2 + \omega^2 \mu_0 \varepsilon_0 \varepsilon_1 \tag{18}$$

and furthermore noting that the electric field E_y must be continuous across the two regions. Applying this boundary condition gives

$$k_2 = k_1 \tan(k_1 a) \tag{19}$$

The preceding two relationships may now be employed to evaluate $k_1 a$ and $k_2 a$ for parametric values of ε_1 and $\omega^2 \mu_0 \varepsilon_0$, and either separation constant may be employed with the appropriate wave equation to determine β. A knowledge of these three parameters is sufficient to construct the field patterns of the waveguide. Figure 8.4 depicts one result. The relationship between β/k_2 and $(ka)\sqrt{\varepsilon_1}$ is separately illustrated in Figure 8.5.

The derivation of the second family of TE solutions for which the introduction of an electric wall at the plane of symmetry leaves the solution unperturbed is outside the remit of this text. It is actually the next higher order mode of this class of waveguide. It is indicated in Figure 8.6 for completeness.

8.5 Circular polarisation in dielectric loaded ridge waveguide

A scrutiny of the field pattern of a dielectric loaded ridge waveguide indicates that it also supports planes of elliptical or circular polarisation at the boundary between the two dielectric regions and everywhere outside. One way to solve this problem region is to use the two-component E_z/H_z hybrid FEM outlined in Chapter 7. One solution which displays a quasi-circularly polarised solution at the symmetry plane of the problem region is the dominant quasi-LSE$_{10}$ mode of the structure. One feature of this result is that the alternating magnetic field is not only circularly polarised at the boundary between the two dielectrics but also everywhere outside. This result is obtained by calculating H_x from a knowledge of E_z and H_z:

$$H_x = \frac{-j\beta}{(-\beta^2 + k_0^2 \varepsilon_r)} \left(\frac{\partial H_z}{\partial x} \right) + \frac{jk_0 \varepsilon_r}{\eta_0(-\beta^2 + k_0^2 \varepsilon_r)} \left(\frac{\partial E_z}{\partial y} \right)$$

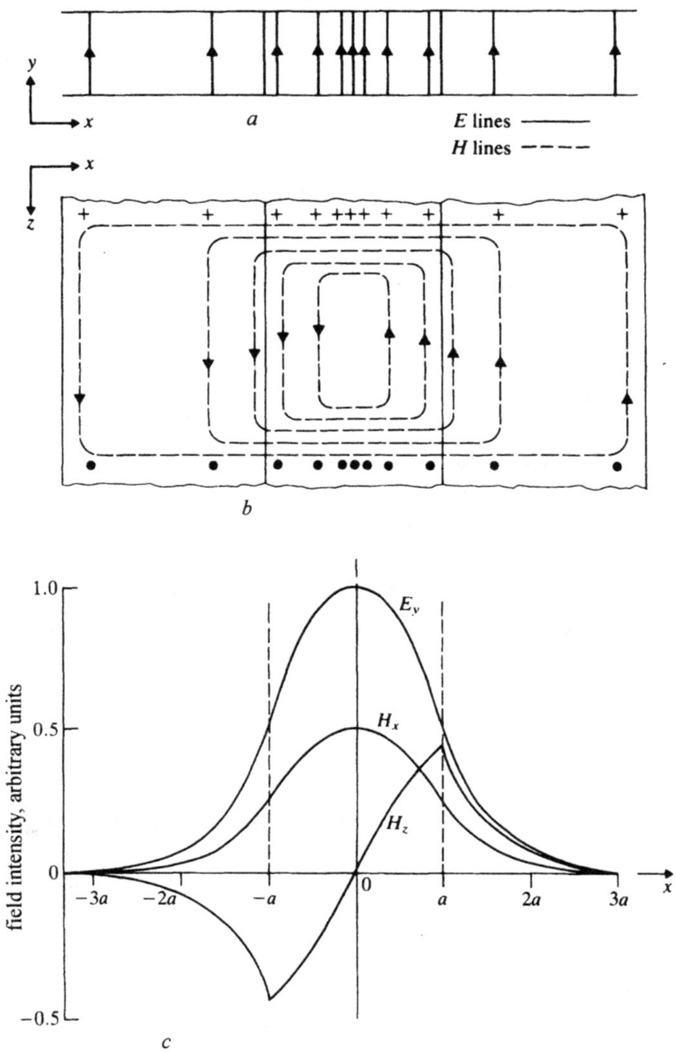

Figure 8.4 Electric and magnetic fields of dielectric-loaded parallel plate waveguide with open side-walls (dominant even mode solution)

Source: Cohn (1959)

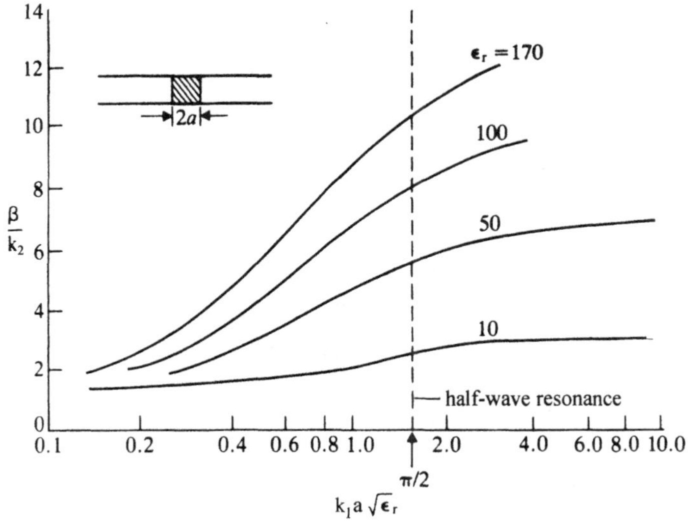

Figure 8.5 Relationship between β/k_2 and $(k_I a)\sqrt{\varepsilon_1}$

Source: Anderson & Hines (1960)

A typical result at the symmetry plane of the waveguide is indicated in Figure 8.7. It applies for a normalised phase constant of

$$\frac{\beta}{k_0} = 2.5$$

The relationship between the normalised frequency and the ellipticity at the dielectric interface is separately illustrated in Figure 8.8. The truncation of each curve corresponds to the onset of the quasi-LSE_{20} mode in the waveguide. The finite element mesh employed in obtaining these results is indicated in Figure 8.9.

8.6 Circular polarisation in homogeneous ridge waveguide

A property of a ridge is that is does not have planes of circular polarisation on the floors of the trough regions. It does, however, display such polarisations at the plane of the electric wall defined by the symmetry of the waveguide. The failure of a single ridge waveguide to exhibit this sort of

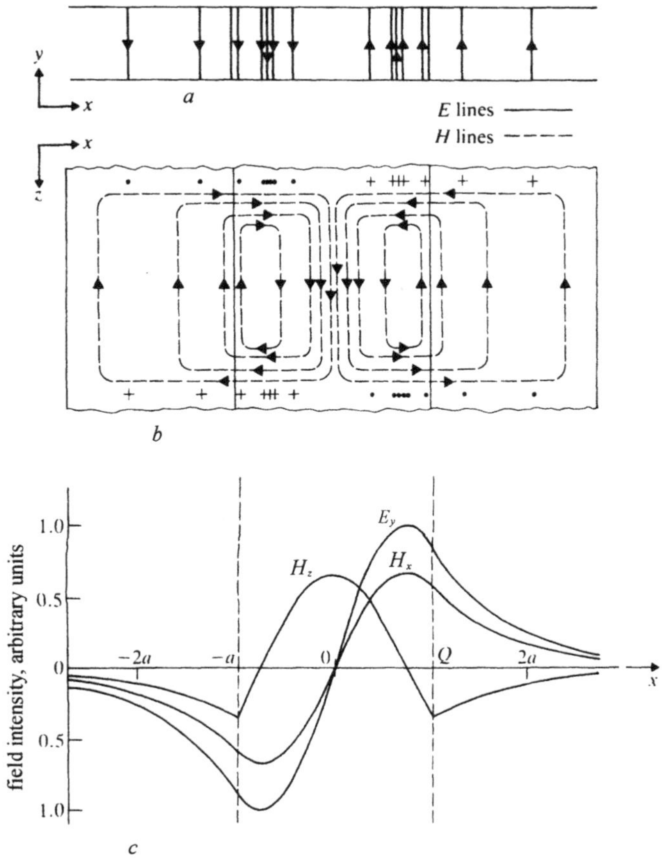

Figure 8.6 Electric and magnetic fields of dielectric-loaded parallel plate waveguide with open side-walls (dominant odd mode solution)

Source: Cohn (1959)

polarisation in the trough regions may be understood by mapping a rectangular waveguide into a ridge arrangement with equivalent path length around its periphery. This operation is illustrated in Figure 8.10. It highlights how the side-walls of the rectangular waveguide, which do not support planes of circular polarisation, are translated into the bottom walls of the ridge channels.

Since the closed form descriptions of the fields in a ridge waveguide are in good agreement with those produced by using the finite element method these may be used for the purpose of calculation. Figure 8.11 shows one typical result at six different planes along the thickness of the waveguide

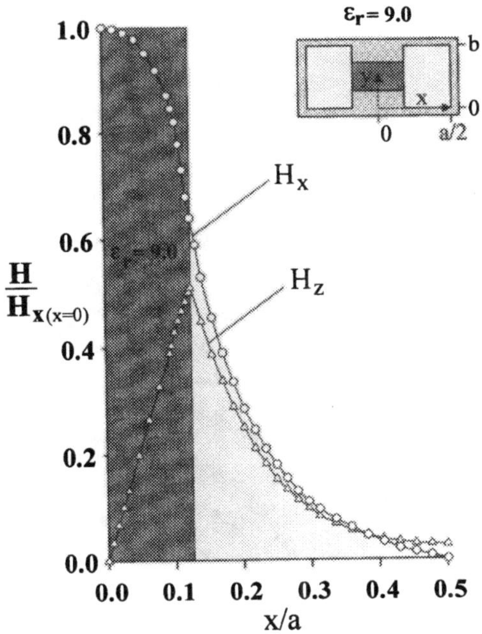

Figure 8.7 Circular polarisation at symmetry plane in dielectric loaded ridge waveguide ($s/a = 0.25$, $d/b = 0.50$, $b/a = 0.50$)

Reprinted with permission, Helszajn & McKay (2000)

inside the half-space formed by placing an electric wall at its symmetry wall. Figure 8.12 illustrates a similar result for another geometry. Figures 8.13 and 8.14 compare some closed form and finite element calculations in the same two geometries.

The closed form relationships for the fields in this sort of waveguide may also be readily employed to calculate the positions of circular polarisation at the plane of the electric wall for different values of the gap spacing d/b. This result is indicated in Figure 8.15. It is obtained by retaining the first 15 modes in the trough and ridge regions.

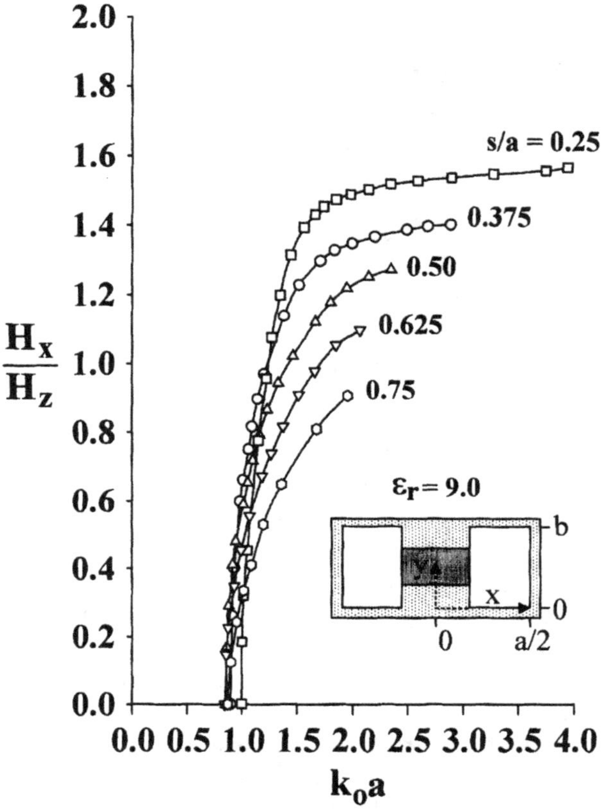

Figure 8.8 Relationship between ellipticity at electric symmetry plane and normalised frequency in dielectric loaded ridge waveguide for parametric values of s/a (d/b = 0.50)

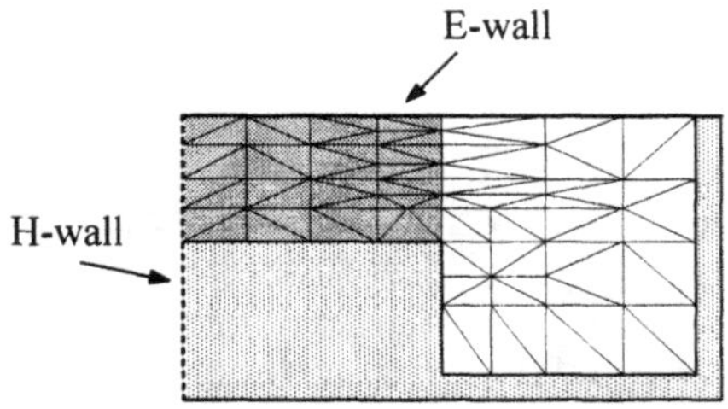

Figure 8.9 Finite element discretisation of dielectric loaded ridge waveguide

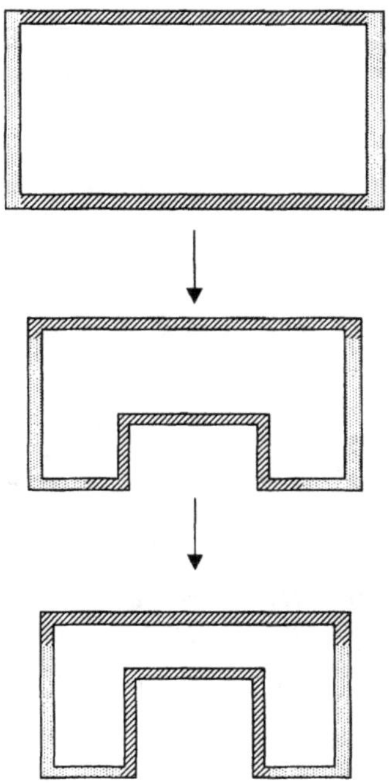

Figure 8.10 Mapping between rectangular and single ridge waveguides

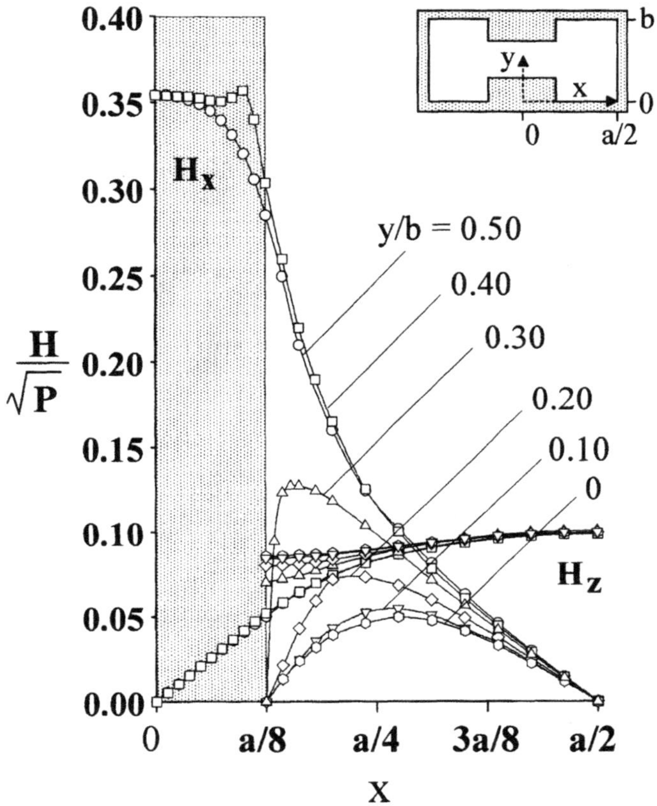

Figure 8.11 FEM calculations of H_x and H_z in layered planes parallel to the electric symmetry wall of the waveguide ($s/a = 0.25$, $b/a = 0.50$, $d/b = 0.35$, $k_0/k_c = 2.0$)

Reprinted with permission, Helszajn & McKay (2000)

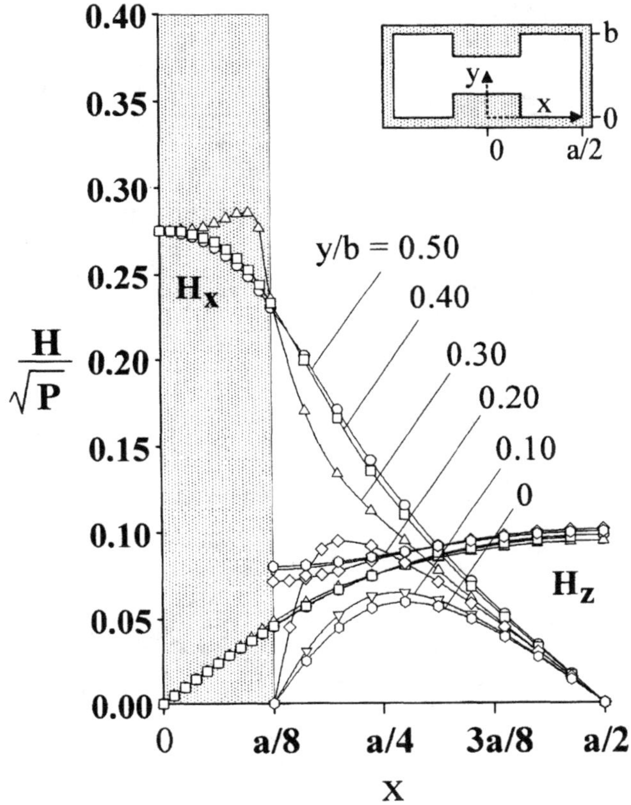

Figure 8.12 FEM calculations of H_x and H_z in layered planes parallel to the electric symmetry wall of the waveguide ($s/a = 0.25$, $b/a = 0.50$, $d/b = 0.50$, $k_0/k_c = 2.0$)

Reprinted with permission, Helszajn & McKay (2000)

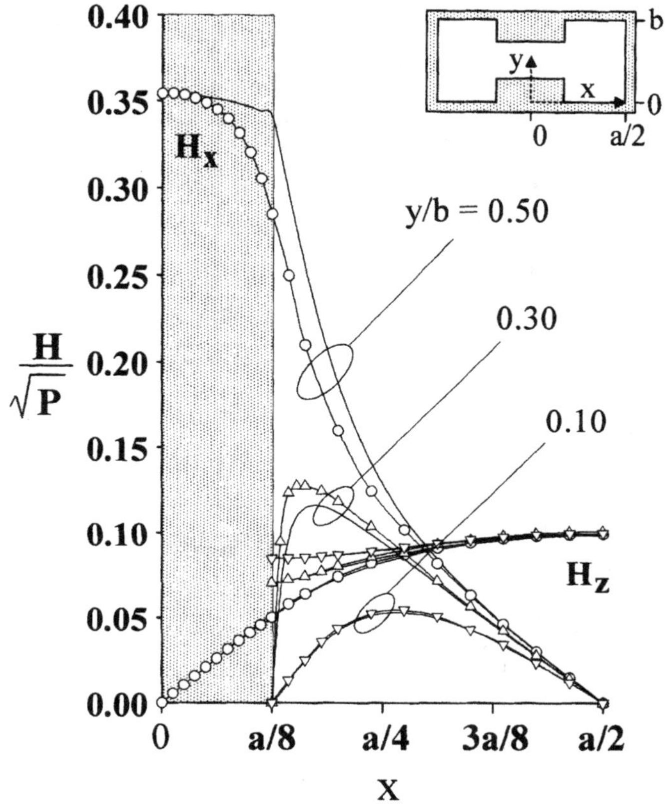

Figure 8.13 Comparison between FEM and closed form calculations of H_x and H_z in layered planes parallel to the electric symmetry wall of the waveguide ($s/a = 0.25$, $b/a = 0.50$, $d/b = 0.35$, $k_0/k_c = 2.0$)

Reprinted with permission, Helszajn & McKay (2000)

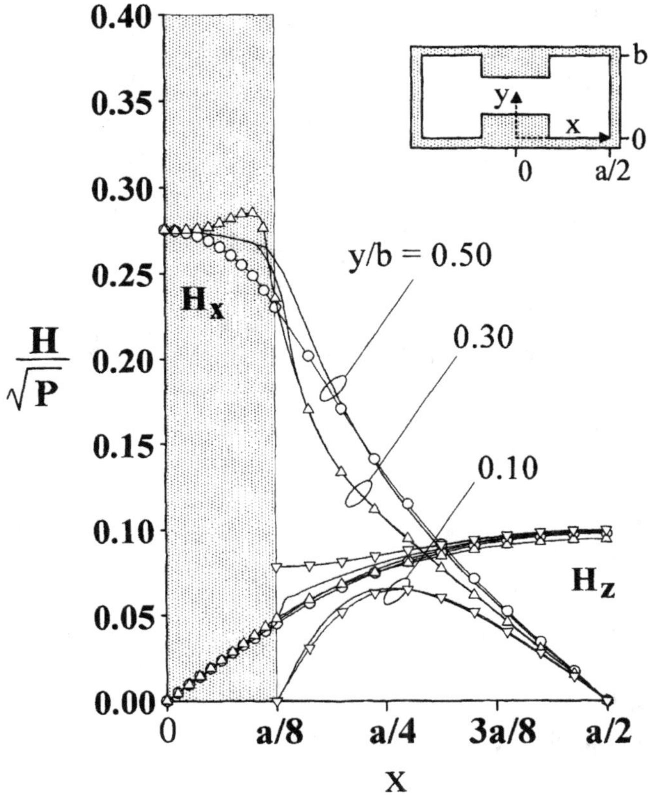

Figure 8.14 Comparison between FEM and closed form calculations of H_x and H_z in layered planes parallel to the electric symmetry wall of the waveguide ($s/a = 0.25$, $b/a = 0.50$, $d/b = 0.50$, $k_0/k_c = 2.0$)

Reprinted with permission, Helszajn & McKay (2000)

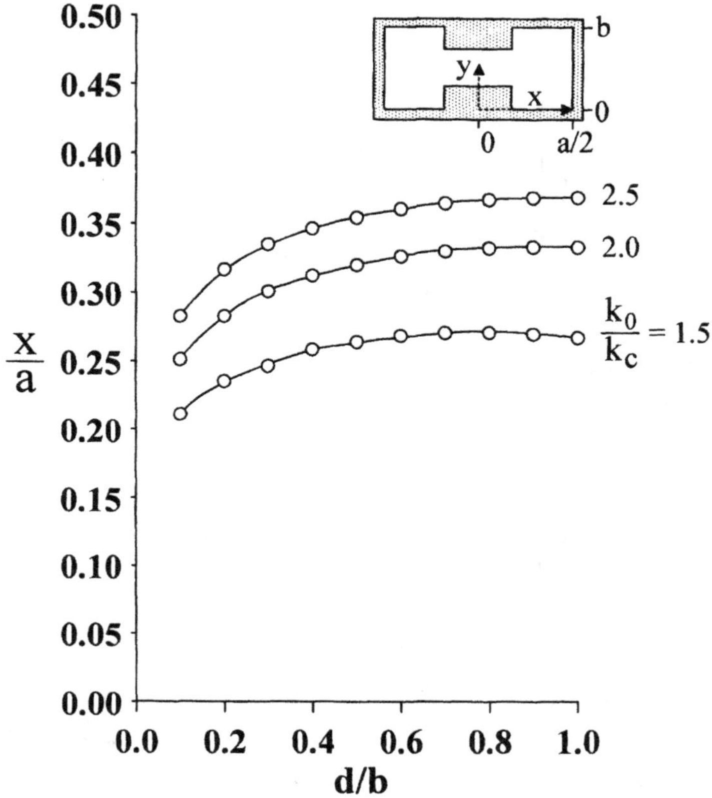

Figure 8.15 Position of circular polarisation at plane of electric symmetry wall in double ridge waveguide ($s/a = 0.25$, $b/a = 0.50$)

Reprinted with permission, Helszajn & McKay (2000)

Chapter 9

Quadruple ridge waveguide

9.1 Introduction

The ridge waveguide is not in practice restricted to the rectangular waveguide with one or two ridges. One or more ridges have by now been introduced into circular, square and triangular waveguides. A property of the round or square waveguide symmetrically loaded by four ridges is that its dominant mode can be decomposed into counter-rotating circular polarised waves on its axis. This sort of ridge waveguide supports Faraday rotation provided it is perturbed by a gyromagnetic material along its axis. Since the dominant mode solution of this structure has two-fold symmetry it is sufficient, insofar as it is concerned, to investigate the problem regions revealed by introducing suitable orthogonal magnetic and electric walls in all combinations. Its mode nomenclature coincides with that of the round or square waveguide obtained by removing the ridges. The nodal finite element method (FEM) again provides one means of investigating this sort of isotropic waveguide. The effect of depositing dielectric tiles on the ridges is also given some attention. Its description necessitates the use of a hybrid (E_z/H_z) or vector H_x, H_y, H_z functional. A quarter-wave plate can be readily realised in this sort of waveguide by employing unequal orthogonal pairs of ridges.

9.2 Quadruple ridge waveguide

One quadruple ridge waveguide consists of a square waveguide symmetrically loaded by four ridges in the manner indicated in Figure 9.1*a*. Its details are again completely described by a normalised gap (d/a) and a normalised ridge width (s/a). The normalised cut-off number (λ_c/a) completes its definition. The field patterns and cut-off of this type of waveguide space may be

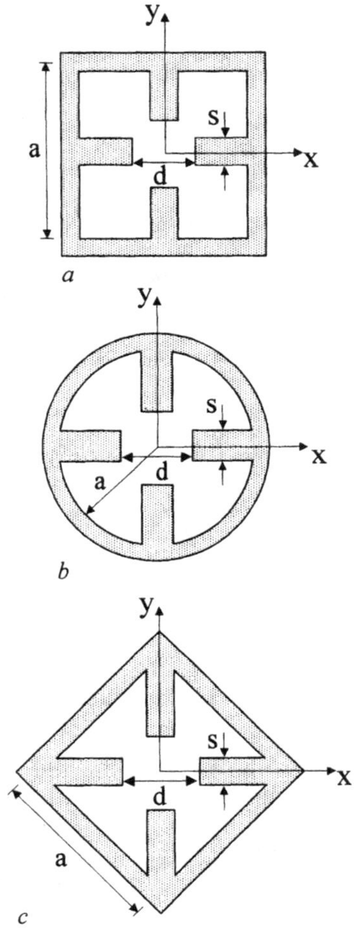

Figure 9.1 Schematic diagrams of square and round ridge waveguides

evaluated without ado by once more using the FEM or other numerical techniques. The geometry of the related circular waveguide is separately illustrated in Figure 9.1*b*. Its topology is fixed by the same variables as those employed to specify the square arrangement except that the side dimension represents, in this instance, the diameter of the waveguide. A square quadruple ridge waveguide with diagonal ridges is depicted in Figure 9.1*c*.

The cut-off spaces of these sorts of waveguides have been calculated using the FEM, the mode matching method (MMM), the magnetic field integral equation (MFIE) and other numerical procedures. The cut-off spaces of the TE family of modes of these waveguides correspond to those of planar circuits with top and bottom magnetic walls and an electric side-wall. Its TM mode spectrum corresponds to that of planar circuits with top and bottom electric walls and a magnetic side-wall. The field patterns of either problem region are obtained from the eigenvectors of the problem region in question in conjunction with Maxwell's equations.

9.3 Cut-off space in quadruple ridge waveguide using MFIE method

A comprehensive investigation of the cut-off spaces of various quadruple ridge waveguides has been undertaken based on the MFIE method. The cut-off spaces obtained in this way are separately depicted in Figures 9.2–9.4. The TE_{21U} and TE_{21L} solutions are obtained by introducing orthogonal electric and magnetic walls at the symmetric planes of the problem region, respectively. A property of these mode charts is the splitting of the degenerate $TE_{2,1}$ modes by the introduction of the ridges. This feature may be understood by recalling that an inward deformation of an electric wall of a cavity resonator in the vicinity of a pure electric field produces an increase in the resonant frequency whereas the same deformation in the vicinity of a pure magnetic field has the opposite effect. The field patterns of the split $TE_{2,1}$ modes are indicated in Figure 9.5.

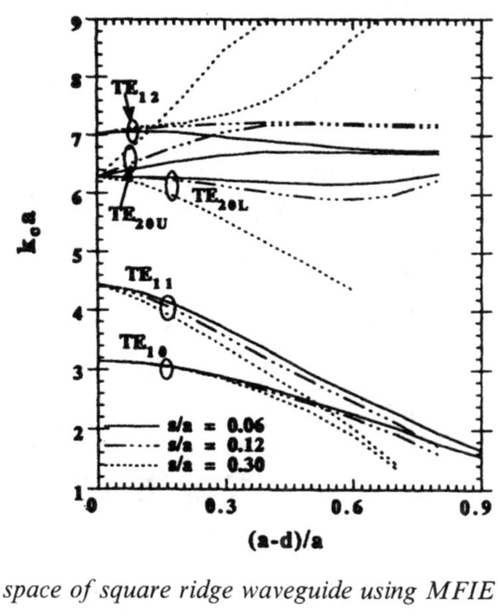

Figure 9.2 Cut-off space of square ridge waveguide using MFIE

Reprinted with permission, Sun & Balanis (1994)

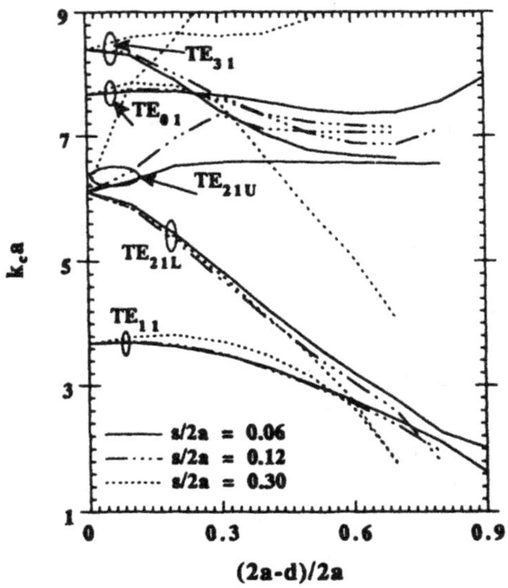

Figure 9.3 Cut-off space of round ridge waveguide using MFIE

Reprinted with permission, Sun & Balanis (1994)

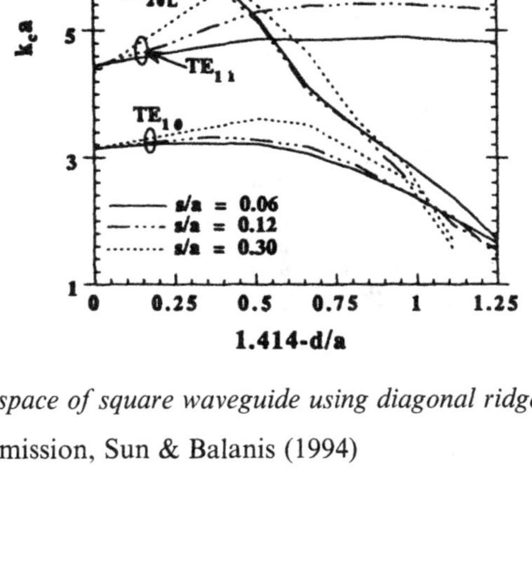

Figure 9.4 Cut-off space of square waveguide using diagonal ridges using MFIE

Reprinted with permission, Sun & Balanis (1994)

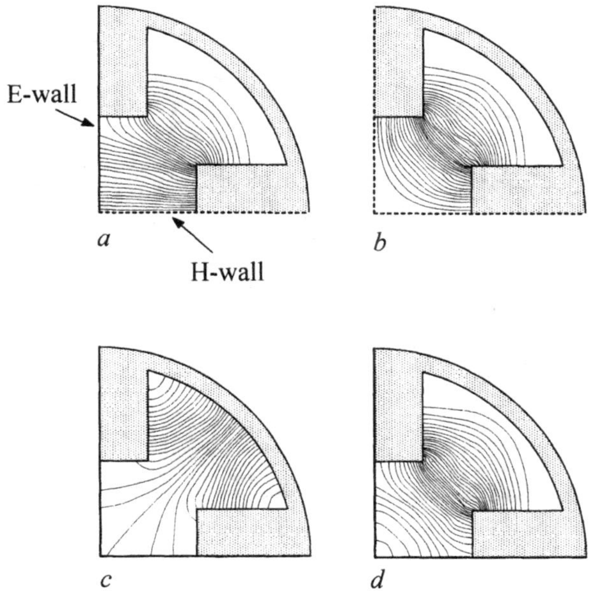

Figure 9.5 Field patterns of quadruple ridge circular waveguide using FEM
a TE_{11}; *b* TE_{21L}; *c* TE_{21U}; *d* TE_{01}

9.4 Cut-off space of ridge waveguide using MMM

A mode matching procedure has also been utilised to establish the cut-off space of a ridge waveguide in the special case of one or more conical ridges. Figure 9.6 depicts the details of a typical ridge. The use of conical instead of rectangular ridges permits closed form radial variables to be used throughout. Figures 9.7–9.10 illustrate the cut-off spaces of these sorts of structures. The details of the radial MMM encountered in this problem region are described in the original literature, as are a number of related filter structures.

9.5 Cut-off space of quadruple ridge waveguide using FEM

The cut-off space of the TE family of solutions in a quadruple ridge waveguide may also be again readily established by using the FEM. It coincides with the eigenvalues of the related planar geometry with top and bottom magnetic walls, and an electric side-wall. The functional met with this

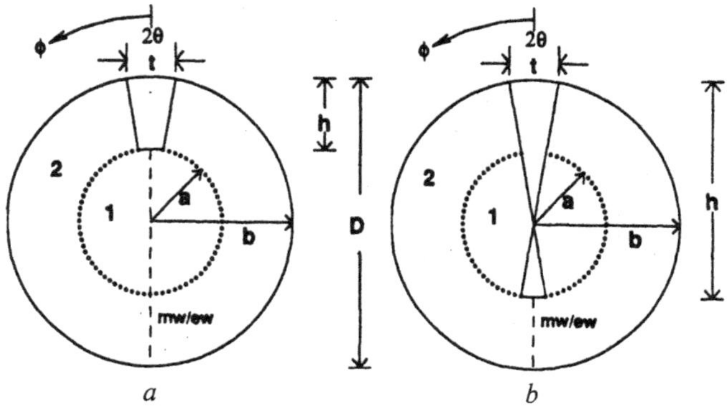

Figure 9.6 Geometry of a single ridge circular waveguide

a Ridge depth < radius; *b* ridge depth > radius

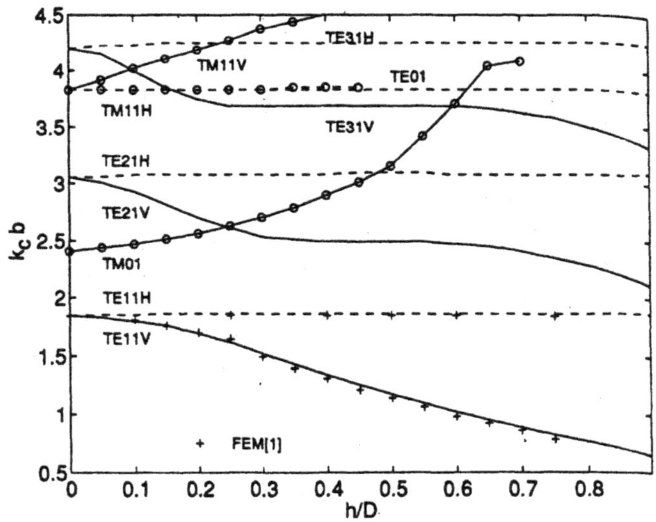

Figure 9.7 Cut-off space of circular waveguide with single conical ridges using the MMM ($t/D = 0.04$)

Reprinted with permission, Balaji & Vahldieck (1996)

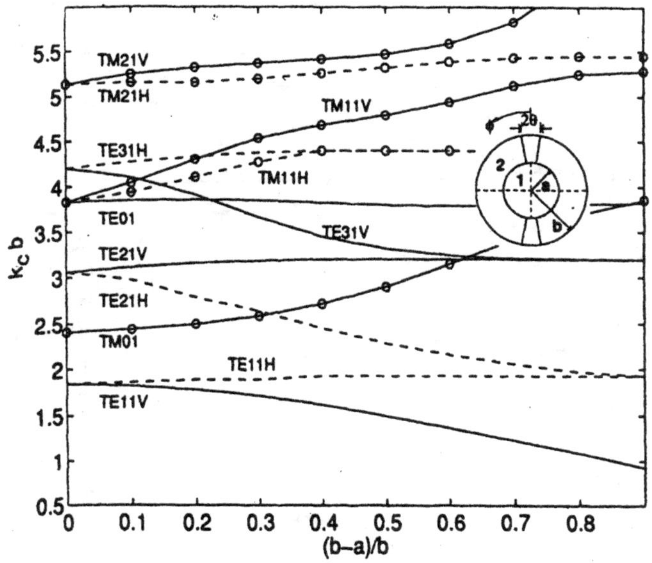

Figure 9.8 Cut-off space of circular waveguide with two conical ridges using the MMM ($2\theta = 10°$)

Reprinted with permission, Balaji & Vahldieck (1996)

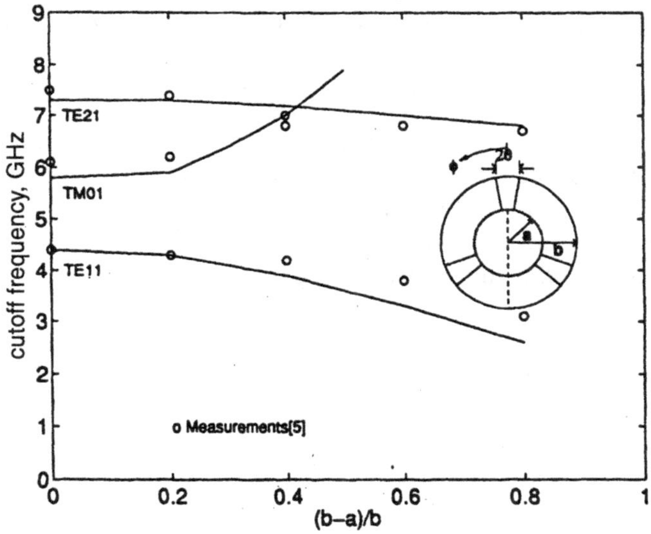

Figure 9.9 Cut-off space of circular waveguide with three conical ridges using the MMM ($2\theta = 10°$)

Reprinted with permission, Balaji & Vahldieck (1996)

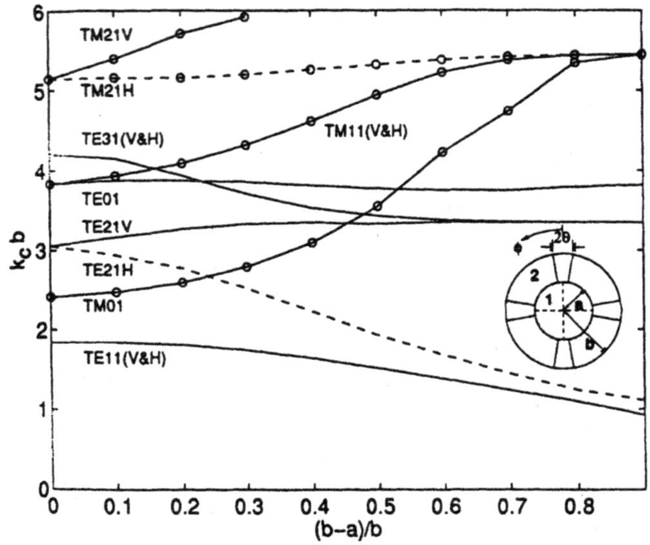

Figure 9.10 Cut-off space of circular waveguide with four conical ridges using the MMM ($b = 20mm$, $2\theta = 20°$)

Reprinted with permission, Balaji & Vahldieck (1996)

topology is that encountered in connection with the single or double ridge geometry. The matrix equation produced by extremising this functional is

$$\{[A] - k_{\mathrm{a}}^2[B]\}H_z = 0 \tag{1}$$

The square matrices $[A]$ and $[B]$ have the meaning met in connection with the single and double ridge problem regions. k_{a} and H_z represent a typical eigenvalue and a typical eigenvector.

Since this structure has four fold symmetry it is sufficient to solve the problem region obtained by partitioning it into four equal regions by introducing orthogonal electrical and magnetic walls in all combinations. One possible discretisation in the case of the circular configuration is specified by

$$p = 2$$

$$m = 6$$

$$n = 71$$

$$n \times m = 426$$

$$q = 172$$

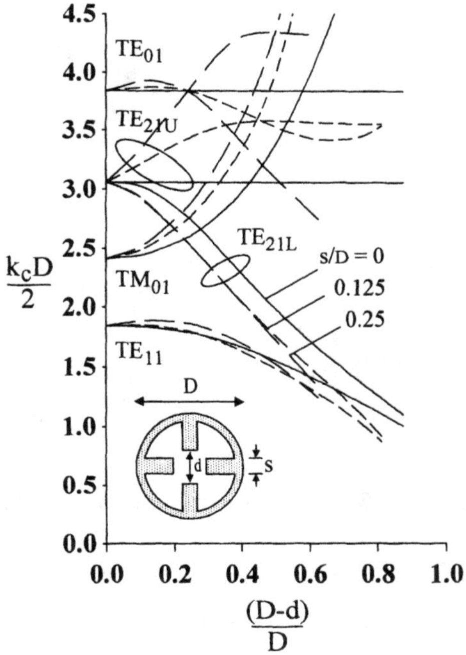

Figure 9.11 Cut-off space of quadruple ridge circular waveguide using FEM

p is the degree of the polynomial approximation inside each element, m is the number of nodes, n is the number of elements, $n \times m$ is the number of nodes prior to assembly of the mesh and q is the number of nodes after assembly of the mesh. Figure 9.11 summarises the required cut-off space.

The cut-off space of the TM eigensolution is obtained by replacing the top and bottom magnetic walls by electric walls and again introducing orthogonal electric and magnetic walls in all combinations. The functional in this instance involves E_z instead of H_z:

$$\{[A] - k_a^2[B]\}E_z = 0 \quad (2)$$

The cut-off space of the TM eigensolution is separately superimposed on Figure 9.11. The first symmetric TM_{01} mode.

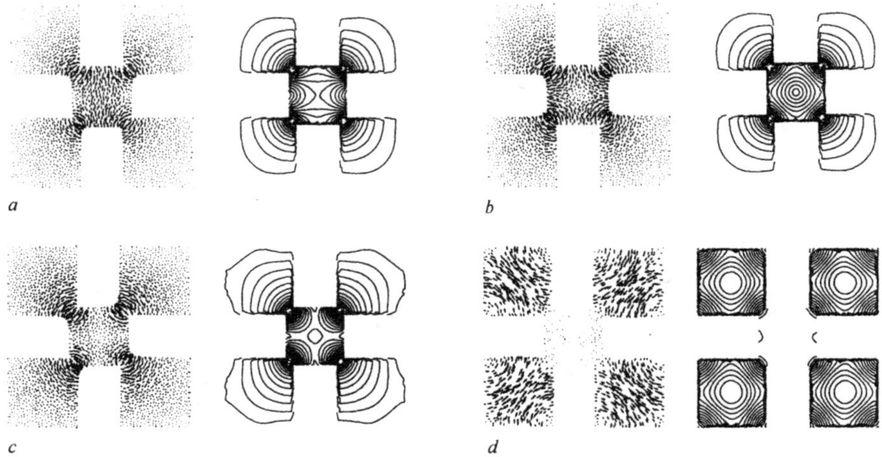

Figure 9.12 Modal electrical field distributions in a square quadruple-ridge waveguide using MFIE

a TE_{10} mode; *b* TE_{11} mode; *c* TE_{20L} mode; *d* TE_{20U} mode

Reprinted with permission, Sun & Balanis (1994)

9.6 Fields in quadruple ridge waveguide

The fields in the round ridge waveguide can be established easily by recalling that the eigenvectors of the problem region coincide with the magnetic field H_z or E_z at the nodes of the two possible planar problem regions. The other components of the field are then deduced by using Maxwell's equations. Figures 9.12–9.14 reproduce, however, some results on the dominant and higher order TE modes in a round waveguide based on an MFIE calculation. The mode nomenclature met in the description of the round waveguide is that obtained by withdrawing the ridges. The nature of the fields in the square waveguide can be inferred from those of the round geometry without difficulty. The field patterns of the TE family of solutions correspond to the planar problem region with top and bottom magnetic walls and an electric side-wall. Those of the TM eigensolutions are obtained by replacing the top and bottom magnetic walls with electric walls.

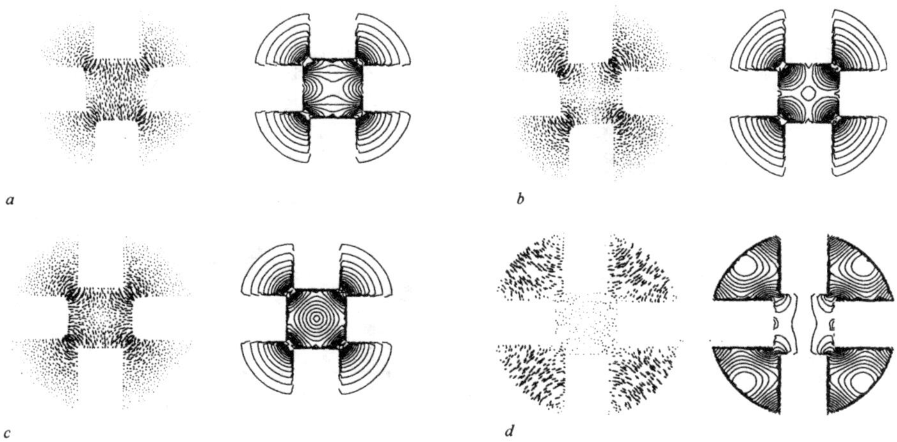

Figure 9.13 Modal electrical field distributions in a circular quadruple-ridge waveguide using MFIE

a TE_{11} mode; *b* TE_{01} mode; *c* TE_{21L} mode; *d* TE_{21U} mode using MFIE

Reprinted with permission, Sun & Balanis (1994)

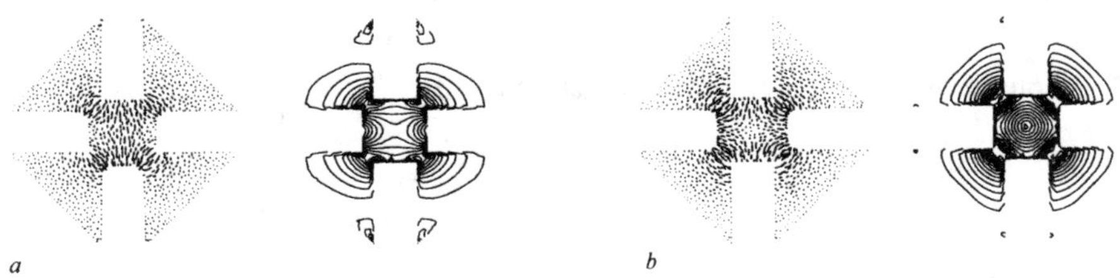

Figure 9.14 Modal electrical field distributions in a diagonal quadruple-ridge waveguide using MFIE

a TE_{10} mode; *b* TE_{20L} mode using MFIE

Reprinted with permission, Sun & Balanis (1994)

9.7 Cut-off space of dielectric loaded quadruple ridge waveguide

The quadruple ridge waveguide can also be dielectric loaded in various ways. Figure 9.15 illustrates some possibilities. Since such a typical structure supports hybrid modes, an E_z/H_z or a three-component vector formulation is necessary for its description. The hybrid E_z/H_z formulation is adopted here. It reduces to the set of simultaneous equations reproduced below:

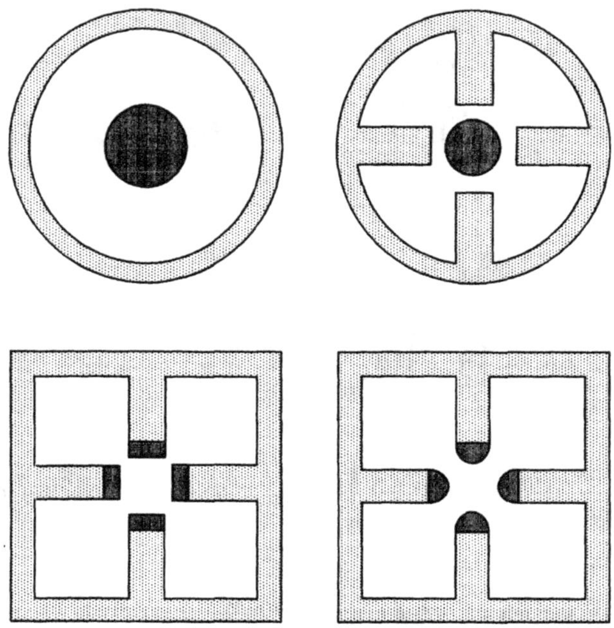

Figure 9.15 Schematic diagrams of round and square ridge waveguide loaded by dielectric tiles and rods

Reprinted from Helszajn & Shrimpton (1996)

$$\sum_{1}^{r} \frac{1}{\varepsilon_r - \left(\frac{\beta}{k_0}\right)^2} \begin{bmatrix} \frac{\varepsilon_r}{\eta_0^2}[S]_{(1)} & \frac{1}{2\eta_0}\left(\frac{\beta}{k_0}\right)[U] \\ \frac{1}{2\eta_0}\left(\frac{\beta}{k_0}\right)[U]_{(1)} & [S]_{(2)} \end{bmatrix} \begin{bmatrix} [e_{rf}] \\ [h_{rf}] \end{bmatrix}$$

$$= \sum_{1}^{r} k_0^2 \begin{bmatrix} \frac{\varepsilon_r}{\eta_0^2}[T]_{(1)} & 0 \\ 0 & [T]_{(2)} \end{bmatrix} \begin{bmatrix} [e_{rf}] \\ [h_{rf}] \end{bmatrix} \quad (3)$$

The notation entering in this result is in keeping with that met in Chapter 8.

Since there is at first sight no standard size tabulated for the quadruple ridge waveguide the dimensions chosen here are based on the corresponding opening of a double ridge rectangular waveguide. Adopting this convention gives

$$\frac{s}{d} = 0.25$$

The gap between the ridges is again described by

$$\frac{d}{D}$$

Once the waveguide size is settled it is necessary to fix the aspect ratio (S/H) of the ferrite tile. This quantity is bracketed in this work by

$$2 \leqslant \frac{s}{h} \leqslant 4$$

The definition of the geometry is complete once the gap between the ridges is selected. This parameter is defined in terms of a filling factor

$$k = \frac{2h}{d}$$

Its maximum value, to prevent the orthogonal pairs of ridges touching, is

$$k_{\max} = 1 - \left(\frac{ks}{2h}\right)$$

The range of filling factors investigated here is bracketed by

$$\frac{k_{\max}}{2} \leqslant k \leqslant \frac{3k_{\max}}{4}$$

The value of the relative dielectric constant (ε_r) completes the physical description of the waveguide structure in question.

The cut-off space for one gyromagnetic arrangement is depicted in Figure 9.16. The segmentation employed in obtaining this result is indicated in Figure 9.17.

The field patterns in this sort of waveguide can be deduced without difficulty once the cut-off space and propagation constant at any frequency are available. Figure 9.18 illustrates some results on a square ridge waveguide with four dielectric tiles.

Since the problem region under consideration is inhomogeneous the propagation constant cannot be deduced from a knowledge of the cut-off space but must be calculated at each and every frequency.

Figure 9.19 summarises some experimental work on the propagation constant of one square ridge waveguide using a single pair of ridges in the 7.0–15 GHz band for different values of d/a with $s/a = 0.25$. The side of the waveguide is 8.55 mm. The thickness of the tiles was 11 mm and the width was 2.14 mm. The relative dielectric constant of the dielectric tiles was 15.0.

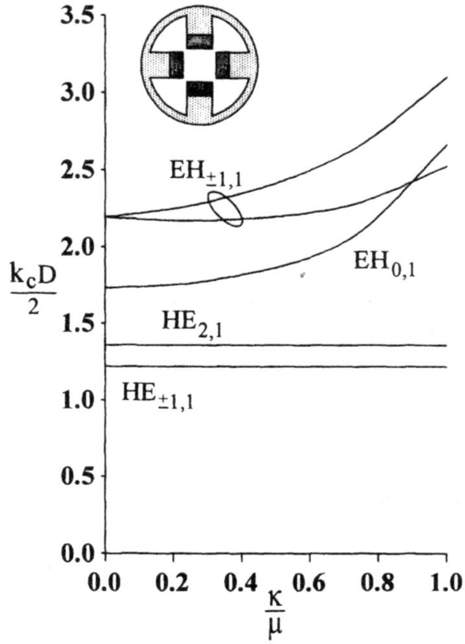

Figure 9.16 Cut-off space of quadruple ridge waveguide loaded by dielectric tiles

Reprinted with permission, McKay & Helszajn (2000)

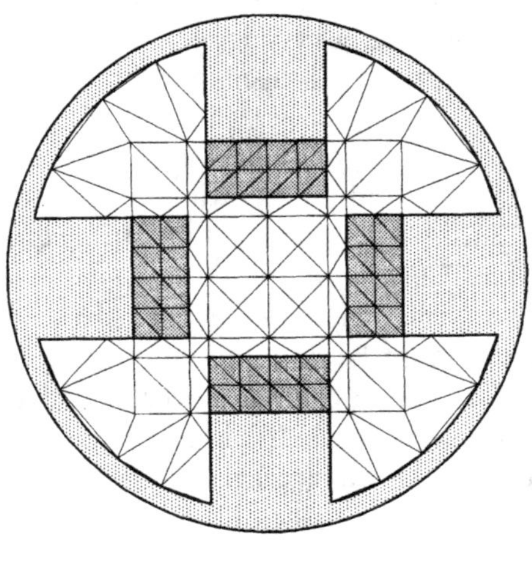

Figure 9.17 Discretisation of quadruple ridge waveguide loaded by dielectric tiles

Reprinted with permission, McKay & Helszajn (2000)

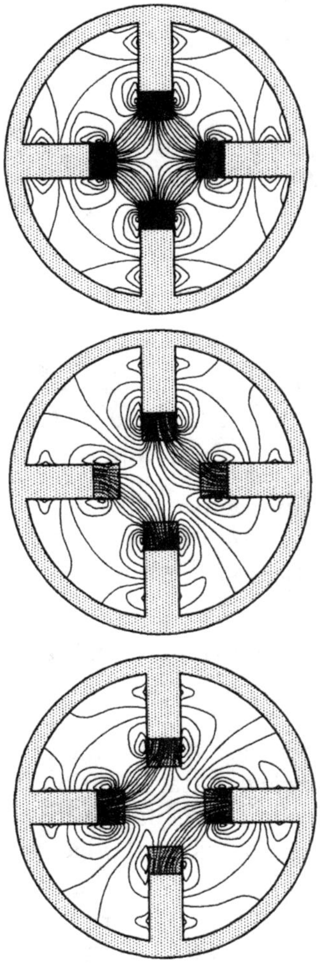

Figure 9.18 Standing wave solutions of quadruple ridge waveguide loaded by dielectric tiles

Reprinted with permission, McKay & Helszajn (2000)

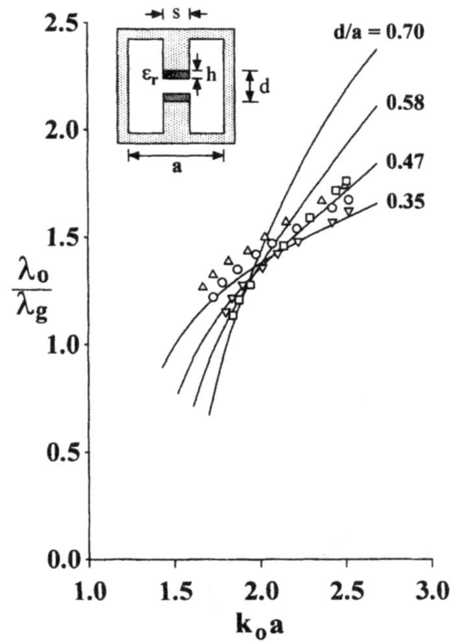

Figure 9.19 Experimental phase constant of square waveguide ($s/a = 0.15$, $s/a = 0.25$) using single pair of dielectric loaded ridges

Reprinted with permission, Helszajn & Shrimpton (1996)

9.8 Impedance in quadruple ridge circular waveguide using conical ridges

A specification of the impedance of the quadruple ridge circular waveguide is also essential for its complete characterisation. The purpose of this section is to summarise some calculations on its power-voltage definition in the case of a circular waveguide with four conical ridges. This quantity is given in the usual way by

$$Z_{\mathrm{PV}}(\omega) = \frac{U^2}{2P} \tag{4}$$

P is the power carried along the waveguide, and U is the voltage across a typical gap.

Figure 9.20 indicates the relationship between the impedance at infinite frequency against the gap factor d/b for two values of s/a based on some FEM calculations. A scrutiny of these data suggests that the impedance of this sort of waveguide is proportional to the gap factor d/b. This is a general result. The relationship between the impedances at finite and infinite frequencies is

$$Z_{PV}(\omega) = Z_{PV}(\infty)\left(\frac{\lambda_g}{\lambda_0}\right) \tag{5}$$

Figure 9.21 compares the impedances of double and quadruple circular ridge waveguides for one typical situation.

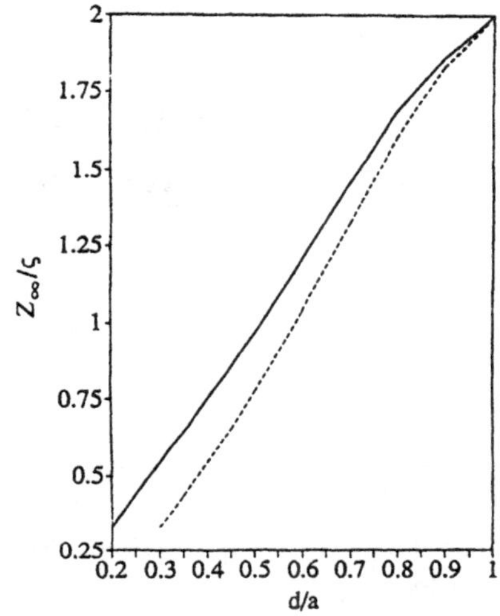

Figure 9.20 Power-voltage definition of impedance in round waveguide

Reprinted with permission, Balaji & Vahldieck (1996)

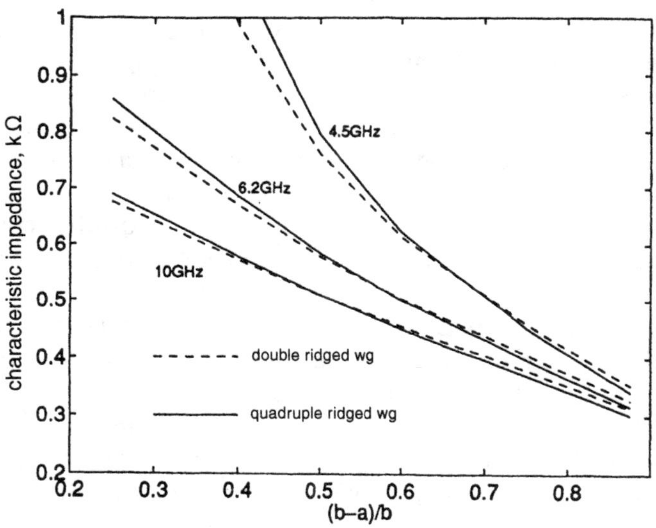

Figure 9.21 Characteristic impedance of quadruple and double ridge circular waveguide against ridge depth, b = 2 cm, ridge thickness (2θ) = 10°

Reprinted with permission, Balaji & Vahldieck (1996)

Chapter 10

Faraday rotation in gyromagnetic quadruple ridge waveguide

10.1 Introduction

A feature of either the round or square quadruple ridge waveguide is the existence of counter-rotating degenerate magnetic fields on its axis. Such a degeneracy can be split by a suitably magnetised gyromagnetic insulator, thereby producing so-called Faraday rotation of the polarisation of the field pattern along the direction of propagation. This situation may be understood by recalling that such a medium will display one value of scalar permeability if the alternating radio frequency magnetic field rotates in the same direction as the electron spin and another value if it rotates in the opposite direction. One unique property of Faraday rotation is that it is nonreciprocal. This means that a wave propagating in one direction which is rotated by an angle θ does not rotate back to its original position when it is returned to its starting point. It is therefore a suitable structure for the construction of nonreciprocal Faraday rotation devices such as circulators, phase shifters and isolators. It is also an appropriate prototype for a host of reciprocal power dividers and other components. Propagation in a dual mode triple ridge gyromagnetic waveguide is handled separately.

One model of a Faraday rotation bit is a nonreciprocal 4-port directional coupler. The chapter includes the scattering matrix of the arrangement. A three-component magnetic field formulation of the sort of functional met in the Rayleigh-Ritz calculation of propagation in this type of waveguide is included separately. It is employed in the calculations of propagation of a number of the gyromagnetic waveguides addressed in this chapter. The descriptions of some typical nonreciprocal components are also given.

10.2 Faraday rotation section

A circular quadruple waveguide symmetrically loaded along its axis by a suitably magnetised gyromagnetic material supports Faraday rotation along the direction of propagation. Figure 10.1 indicates four structures which have the symmetry in question. A physical appreciation of Faraday rotation may be obtained by decomposing a linearly polarised wave into a pair of degenerate counter-rotating ones which are then split by the gyrotropy. If the linearly polarised wave is now reconstituted in terms of the split counter-rotating ones then its polarisation is rotated as it propagates along the waveguide. This situation is illustrated in Figure 10.2 in the case of the magnetic field and in Figure 10.3 in the case of the electric one. The angle of rotation is defined by the difference between the split phase constants $\beta_\pm$,

$$\theta = \left(\frac{\beta_+ - \beta_-}{2}\right)\ell \tag{1}$$

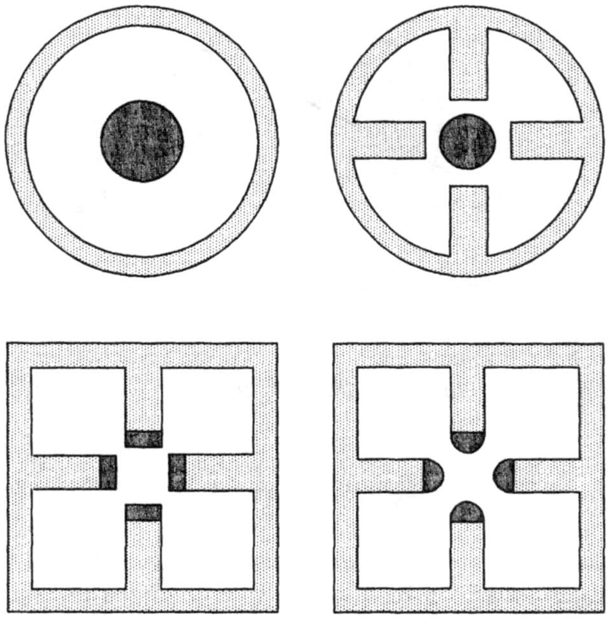

Figure 10.1 Schematic diagrams of round and square ridge waveguides loaded with ferrite regions

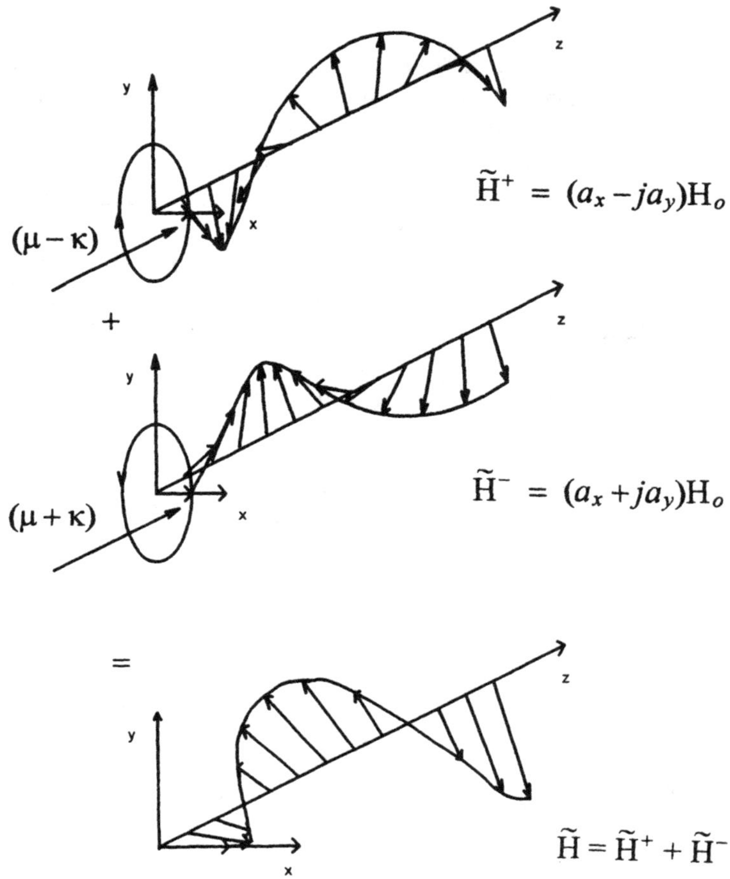

Figure 10.2 Faraday rotation (magnetic field)

The phase constant of the rotated wave is related to the sum of the propagation constants,

$$\beta_0 = \left(\frac{\beta_+ + \beta_-}{2}\right) \tag{2}$$

This result is obtained easily by decomposing a linearly polarised wave at the origin into a linear combination of two counter-rotating waves with different propagation constants,

$$\begin{bmatrix} E_x \\ E_y \end{bmatrix} = \frac{E_0}{2}\begin{bmatrix} 1 \\ -j \end{bmatrix}\exp(j\beta_+\ell) + \frac{E_0}{2}\begin{bmatrix} 1 \\ +j \end{bmatrix}\exp(j\beta_-\ell) \tag{3}$$

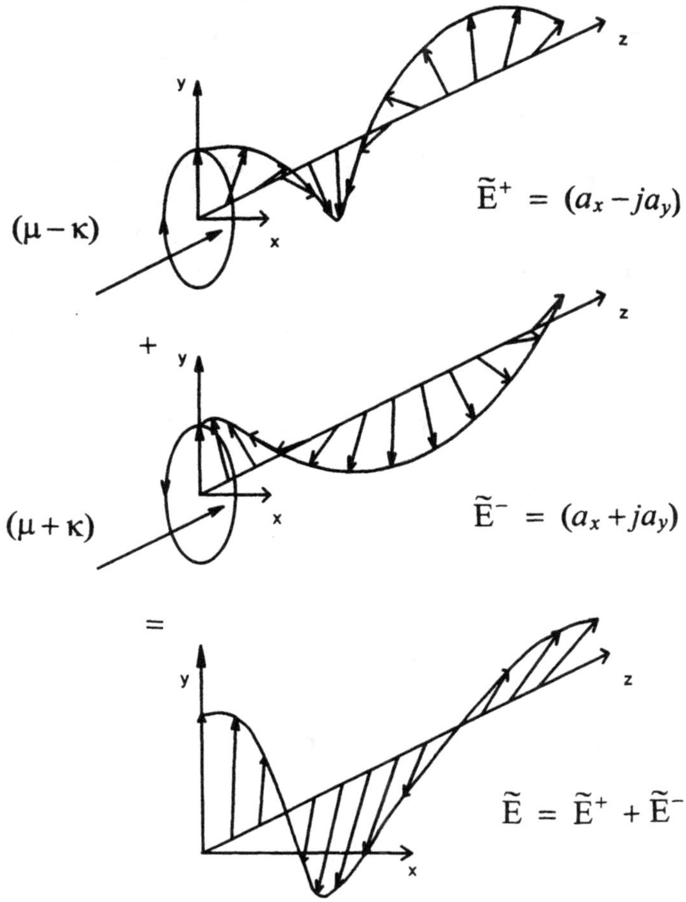

Figure 10.3 Faraday rotation (electric field)

This equation is satisfied at the origin by

$$E_0\begin{bmatrix}1\\0\end{bmatrix}=\frac{E_0}{2}\begin{bmatrix}1\\-j\end{bmatrix}+\frac{E_0}{2}\begin{bmatrix}1\\+j\end{bmatrix} \tag{4}$$

One possible configuration is obtained by introducing a ferrite rod along the axis of the waveguide. Another is to introduce a ferrite or garnet ring insert along the waveguide. Still another is to place ferrite tiles on each flat face of the ridges. While the rotation per unit length of such an arrangement is less than that met with an axial ferrite rod configuration it has the merit of being more amenable in the development of high peak and average power devices. To proceed with the design of this class of component it is necessary to have a description of the degenerate or demagnetised propagation constant of the waveguide and also its split or magnetised propagation constants. The finite element method is again one means of making such calculations.

10.3 Scattering matrix of Faraday rotation section

One way to model the Faraday effect is to treat the rotator section as a 4-port nonreciprocal directional coupler. The entries of its scattering matrix with an input at port 1 may then be expressed in terms of the properties of the counter-rotating reflection ($\rho_\pm$) and transmission variables ($\tau_\pm$) of the system by

$$S_{11} = \frac{\rho_- + \rho_+}{2} \tag{5a}$$

$$S_{21} = \frac{\rho_- - \rho_+}{-j2} \tag{5b}$$

$$S_{31} = \frac{\tau_- + \tau_+}{2} \tag{5c}$$

$$S_{41} = \frac{\tau_- - \tau_+}{j2} \tag{5d}$$

If the magnetised line is matched to the demagnetised one by a stepped impedance transformer, then the reflection coefficients $\rho_\pm$ are given in terms of the normal mode phase constants by

$$\rho_\pm \approx \frac{-\beta_0 + \beta_\pm}{\beta_0 + \beta_\pm} \tag{6}$$

The transmission coefficients $\tau_\pm$ are related in the usual way to the reflection coefficients $\beta_\pm$ by

$$\tau_\pm = (1 - \rho_\pm \rho_\pm^*)^{\frac{1}{2}} \exp(-j\beta_\pm \ell) \tag{7}$$

It is readily verified that equations (5a)–(5d) satisfy the unitary condition

$$S_{11}S_{11}^* + S_{21}S_{21}^* + S_{31}S_{31}^* + S_{41}S_{41}^* = 1 \tag{8}$$

For symmetric splitting the scattering parameters in equations (5a)–(5d) become

$$S_{11} \approx 0 \tag{9a}$$

$$S_{21} \approx j\frac{\beta_- - \beta_+}{2\beta_0} \tag{9b}$$

$$S_{31} \approx \left[1 - \left(\frac{\beta_- - \beta_+}{2\beta_0}\right)^2\right]^{\frac{1}{2}} \cos\left(\frac{\beta_- - \beta_+}{2}\ell\right) \exp(-j\beta_0\ell) \tag{9c}$$

$$S_{41} \approx \left[1 - \left(\frac{\beta_- - \beta_+}{2\beta_0}\right)^2\right]^{\frac{1}{2}} \sin\left(\frac{\beta_- - \beta_+}{2}\ell\right) \exp(-j\beta_0\ell) \tag{9d}$$

which also satisfies the unitary condition.

One consequence of this result is that matching port 1 in a 4-port nonreciprocal network is not sufficient to decouple port 2. To do so by at least 20 dB it is necessary to have

$$\frac{\beta_- - \beta_+}{2\beta_0} \leqslant 0.10 \tag{10}$$

The above condition imposes an upper bound on the normalised splitting and a lower bound on the overall length of the device.

A scrutiny of the preceding scattering parameters indicates that a wave propagating a certain distance in one direction is rotated through an angle θ with respect to its original polarisation. When it is reflected to its starting point it is again rotated by θ. The nature of this phenomenon may be understood by recognising that the signs of β_+ and β_- are interchanged for propagation in the $-z$ direction and, furthermore, that the phase constants of $\beta_\pm$ are separately exchanged. The total rotation of the reflected wave is therefore 2θ with respect to the outgoing wave, i.e. it is not rotated back to its original orientation. Figure 10.4 illustrates the situation for a 90° section. Thus Faraday rotation is nonreciprocal and leads to a number of nonreciprocal devices.

10.4 Gyrator network

The simplest component that illustrates the nonreciprocal property of a Faraday rotator is the gyrator circuit. This element is a 4-terminal 2-port device that has zero relative phase shift in one direction of propagation and 180 degree in the other. It is characterised by the following scattering matrix:

$$\mathbf{S} = \begin{bmatrix} 0 & -1 \\ 1 & 0 \end{bmatrix} \tag{11}$$

One realisation of this network consists of a rectangular waveguide with a 90 degree twist followed by a 90 degree Faraday rotation section. The

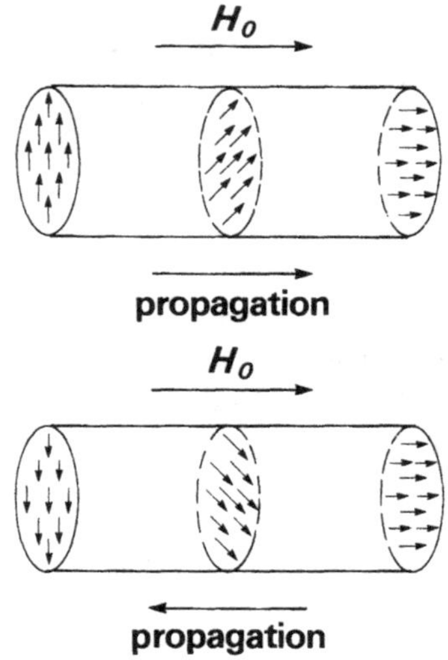

Figure 10.4 Faraday rotation in positive and negative directions of propagation

output port is orientated in the same plane as the input one in the manner indicated in Figure 10.5. A vertically polarised wave propagating from left to right has its polarisation rotated 90 degrees by the twisted rectangular waveguide and a further 90 degrees by the Faraday rotator; it therefore emerges at the output port having been rotated 180 degrees with respect

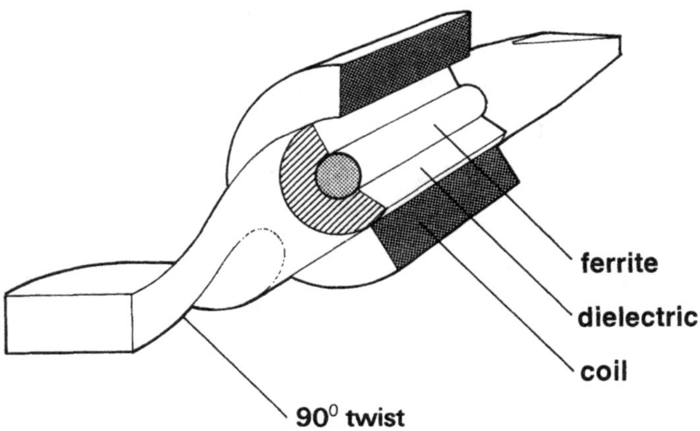

Figure 10.5 Nonreciprocal gyrator circuit

Reprinted with permission, Hogan (1952)

to the input port. A vertically polarised wave propagating in the opposite direction is rotated by the 90 degree Faraday rotator in the same sense as before. In this case, however, the waveguide twist will cancel the Faraday rotation instead of adding to it; the wave therefore displays no rotation in this direction of propagation. The effective length of this gyrator is an odd integral number of half-wavelengths for transmission in one direction of propagation and an even number in the other.

10.5 Gyromagnetic waveguide functional

It is convenient in the variational solution of the gyromagnetic waveguide to use a three-component magnetic field vector formulation of the functional. The use of such a functional avoids spurious solutions that sometimes exist with other formulations. The wave equation is in this instance given easily by

$$\nabla \times \left(\frac{1}{\varepsilon_{\mathrm{f}}} \nabla \times \mathbf{H} \right) - k_0^2[\mu]\mathbf{H} = 0 \tag{12}$$

One solution to this sort of equation is obtained by constructing the functional of the problem region and recognising that the field which extremises the functional is also a solution of the wave equation. One mathematical means of constructing a functional is to premultiply the wave equation by the complex conjugate of the field variable prior to integrating the ensuing quantity over the surface of the problem region. This gives

$$\mathrm{F} = \iint_s \{\mathbf{H}^* \cdot \nabla \times (\varepsilon_{\mathrm{f}}^{-1} \nabla \times \mathbf{H}) - \mathbf{H}^* k_0^2[\mu]\mathbf{H}\}\, \mathrm{d}s \tag{13}$$

The functional obtained in this way embodies a natural boundary condition at any electric or magnetic wall of the problem region. To reveal this property it is rewritten in a slightly different form by using some standard matrix identities. The required result is

$$\mathrm{F} = \iint_s \{(\nabla \times \mathbf{H}^*) \cdot (\varepsilon_{\mathrm{f}}^{-1} \nabla \times \mathbf{H}) - k_0^2 \mathbf{H}^*[\mu]\mathbf{H}\}\, \mathrm{d}s$$

$$- \oint_c \{\mathbf{H}^* \times (\varepsilon_{\mathrm{f}}^{-1} \nabla \times \mathbf{H})\} \cdot \mathbf{n}\, \mathrm{d}c \tag{14}$$

c is the contour of the problem region s and $\mathbf{n}$ is the outward unit vector normal to c. The necessary matrix relationships employed in establishing this result are

$$\nabla \cdot (\mathbf{A} \times \mathbf{B}) = \mathbf{B} \cdot (\nabla \times \mathbf{A}) - \mathbf{A} \cdot (\nabla \times \mathbf{B})$$

$$\mathbf{A} \cdot (\mathbf{B} \times \mathbf{C}) = \mathbf{C} \cdot (\mathbf{A} \times \mathbf{B}) = \mathbf{B} \cdot (\mathbf{C} \times \mathbf{A})$$

$$\iint_s \nabla \cdot \mathbf{A} \, \mathrm{d}s = \oint_c \mathbf{A} \cdot \mathbf{n} \, \mathrm{d}c$$

It may be demonstrated that the contour integral term appearing in the functional is zero at both electric and magnetic walls. It is hereafter not given any further consideration. The functional to be extremised is therefore specified by

$$\mathrm{F} = \iint_s \{(\nabla \times \mathbf{H}^*) \cdot (\varepsilon_{\mathrm{f}}^{-1} \nabla \times \mathbf{H}) - k_0 \mathbf{H}^* [\mu] \mathbf{H}\} \, \mathrm{d}s \tag{15}$$

This equation may now be reduced to a matrix equation by using the Rayleigh-Ritz procedure. The first step in this development amounts to replacing the unknown components of the vector field in the functional by a suitable polynomial approximation with arbitrary coefficients. The second step adjusts the unknown coefficients by extremising the functional. The detailed procedure is dealt with in Chapter 19.

The finite element method produces in practice spurious solutions for which $(\nabla \cdot \mathbf{H}) \neq 0$. One means of alleviating this problem, to a large extent, is to introduce a penalty term into the functional. One possibility is to add the quantity $(\nabla \cdot \mathbf{H}^*)(\nabla \cdot \mathbf{H})$ to the functional. Introducing this penalty term into the preceding functional gives the modified version for calculation purposes,

$$\mathrm{F} = \iint_s \{(\nabla \times \mathbf{H}^*) \cdot (\varepsilon_{\mathrm{f}}^{-1} \nabla \times \mathbf{H}) - k_0^2 \mathbf{H}^* [\mu] \mathbf{H} + (\nabla \cdot \mathbf{H}^*)(\nabla \cdot \mathbf{H})\} \, \mathrm{d}s \tag{16}$$

It is usual in this type of problem to split the field and vector operations into longitudinal and transverse components:

$$\mathbf{H} = \mathbf{H}_{\mathrm{t}} + a_z H_z \tag{17}$$

$$\nabla = \nabla_{\mathrm{t}} - j a_z \beta \tag{18}$$

$$[\mu] = [\mu_{\mathrm{t}}] + [\mu_z] \tag{19}$$

where

$$[\mu_t] = \begin{bmatrix} \mu_{xx} & \mu_{xy} & 0 \\ \mu_{yx} & \mu_{yy} & 0 \\ 0 & 0 & 0 \end{bmatrix} \tag{20}$$

$$[\mu_{zz}] = \begin{bmatrix} 0 & 0 & 0 \\ 0 & 0 & 0 \\ 0 & 0 & \mu_{zz} \end{bmatrix} \tag{21}$$

Introducing these relationships into the functional in equation (14) gives

$$\mathrm{F} = \int\int_s \{\varepsilon_f^{-1}|\nabla_t \times \mathbf{H}_t|^2 - k_0^2[\mathbf{H}_t^*[\mu_t]\mathbf{H}_t + H_z^*\mu_{zz}H_z]$$

$$+ \varepsilon_f^{-1}|\nabla_t H_z + j\beta\mathbf{H}_t|^2\}\,\mathrm{d}s \tag{22}$$

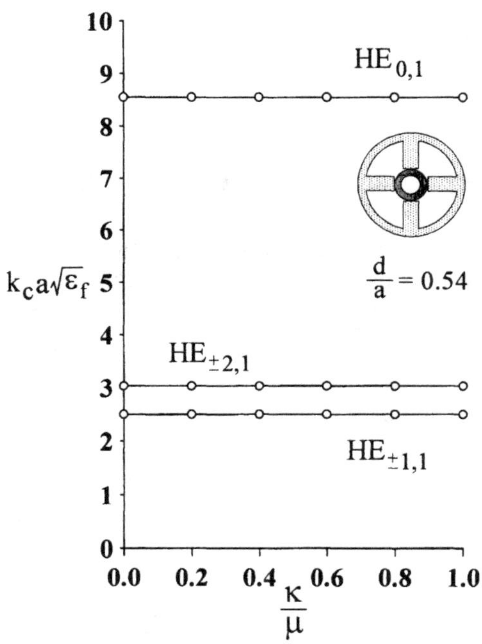

Figure 10.6 Cut-off space of quadruple ridge waveguide using ferrite ring insert

Reprinted with permission, Dillion & Gibson (1993, IEEE MTT-S Digest)

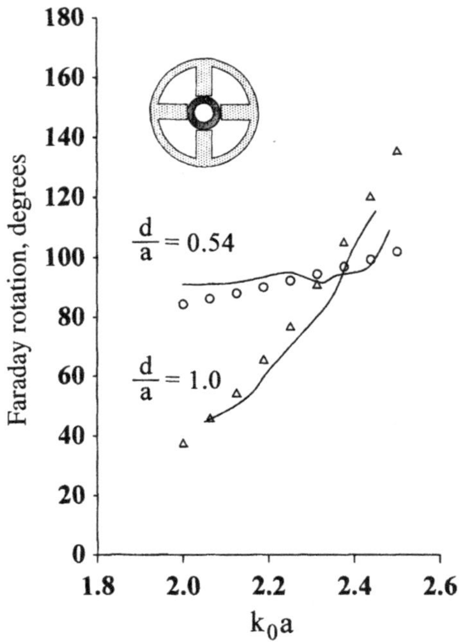

Figure 10.7 Faraday rotation of quadruple ridge waveguide using ferrite ring insert

Reprinted with permission, Dillion & Gibson (1993, IEEE MTT-S Digest)

10.6 Ridge waveguide using gyromagnetic ring

The purpose of this section is to summarise some calculations on the cut-off space and propagation of a ridge structure containing a gyromagnetic ring in contact with the four ridges. The calculations are based on a magnetic vector FEM formulation of the problem region. Figure 10.6 depicts, first of all, the cut-off space for its first three $HE_{m,n}$ modes. It indicates, in keeping with theory, that the cut-off space displays no splitting for this family of modes. Figure 10.7 separately indicates the rotation per unit length of two arrangements for one typical value of gyrotropy.

10.7 Quadruple ridge waveguide using gyromagnetic tiles

A more practical ridge waveguide geometry is a quadruple one with ferrite tiles cemented on each ridge. It is suitable for the construction of devices

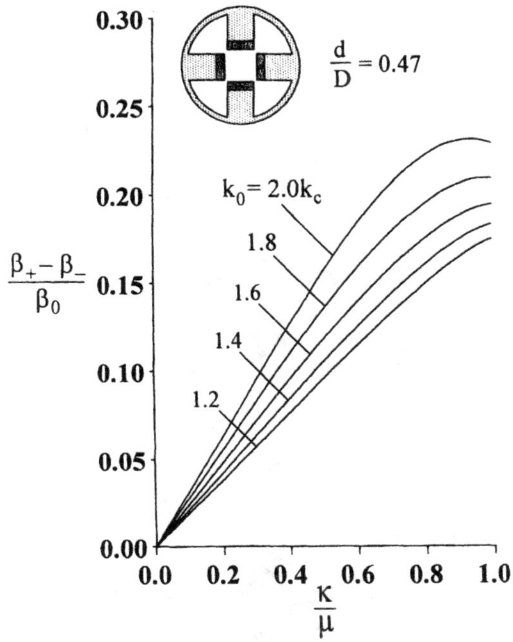

Figure 10.8 Split phase constants in quadruple ridge waveguide using gyromagnetic tiles

Reprinted with permission, McKay & Helszajn (2000)

with large CW power ratings. The relationship between the split phase constants and the gyrotropy is indicated in Figure 10.8 for one geometry. The standing wave patterns for the first pair of split modes and the first symmetric mode are indicated in Figure 10.9 for two typical values of gyrotropy.

10.8 Faraday rotation isolator

The operation of this type of isolator can be understood with the help of Figure 10.10. The rotator prototype is matched to a rectangular waveguide at each end by a taper or a quarter-wave transformer in a round waveguide. Resistance vanes are inserted in the round waveguide sections in a plane perpendicular to the electric fields of the input and output rectangular waveguides. A signal incident at the input port will be perpendicular to the first resistance card, and after a clockwise rotation through 45 degrees will also be perpendicularly polarised with respect to the card at the output port.

a

Figure 10.9a Standing wave solutions in quadruple ridge waveguide using gyromagnetic tiles

(*a*) $\kappa/\mu = 0.50$, $k_0 a\sqrt{\varepsilon_f} = 8.0$; (*b*) $\kappa/\mu = 1.0$, $k_0 a\sqrt{\varepsilon_f} = 8.0$

Reprinted with permission, McKay & Helszajn (2000)

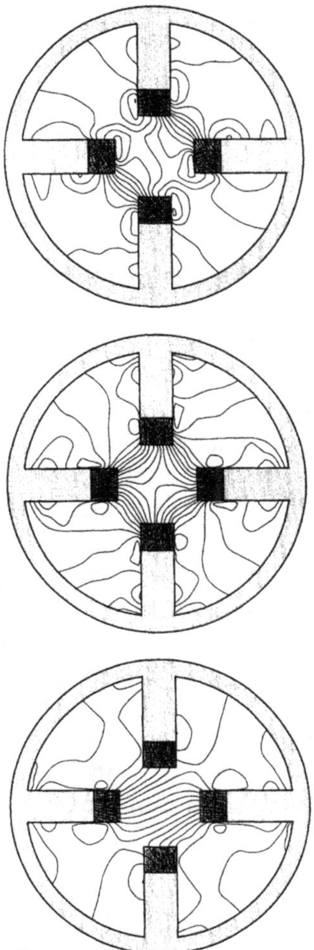

b

Figure 10.9b Standing wave solutions in quadruple ridge waveguide using gyromagnetic tiles

(*a*) $\kappa/\mu = 0.50$, $k_0 a\sqrt{\varepsilon_f} = 8.0$; (*b*) $\kappa/\mu = 1.0$, $k_0 a\sqrt{\varepsilon_f} = 8.0$

Reprinted with permission, McKay & Helszajn (2000)

It will, therefore, be transmitted without attenuation through the isolator. A reflected signal at the output port is likewise perpendicular to the card there, but after rotation through 45 degrees in a clockwise sense will now be in the plane of the input vane where it is attenuated. The wavelength of the transformer section is approximately the geometric mean of the wavelength of the rectangular waveguide and that of the isotropic round waveguide containing the ferrite rod. This device may also be used as an amplitude modulator by suitably varying the direct magnetic field in an appropriate manner.

10.9 Four-port Faraday rotation circulator

Another important application of the Faraday section is in the realisation of the 4-port Faraday rotation circulator. A schematic diagram of this component is indicated in Figure 10.11. In this device, power entering port 1 emerges at port 2, and so on in a cyclic manner. The physical arrangement is similar to the Faraday rotation isolator except that the sections containing the resistance vanes are replaced by two-mode transducers. Such transducers allow orthogonal linearly polarised waves to be applied to the round waveguide section. The Faraday rotator is again a 45-degree section. A wave entering port 1 with its electric field vertically polarised is rotated clockwise by 45 degrees by the ferrite rotator and emerges at port 2. A wave entering port 2 is also rotated clockwise, so that its electric field is now horizontally polarised at the input of the first two-mode transducer; it therefore emerges at port 3. Similarly, transmission occurs from port 3 to port 4, port 4 to port 1, and so on. This arrangement may also be used as a switch by reversing the direction of the direct magnetic field.

10.10 Nonreciprocal Faraday rotation-type phase shifter

A feature of a gyromagnetic medium is that if the excitation to a typical rotation section corresponds to one of its normal modes, no Faraday rotation will occur. Instead, the wave travels in the same normal mode through the rotator section, but phase shifted through either $\beta_+ z$ or $\beta_- z$. This principle can be utilised to design a nonreciprocal phase shifter. One arrangement is illustrated in Figure 10.12. It uses two reciprocal quarter-wave plates at either end of the Faraday rotation section. The first quarter plate converts a linearly polarised input wave into a positive circularly polarised wave at

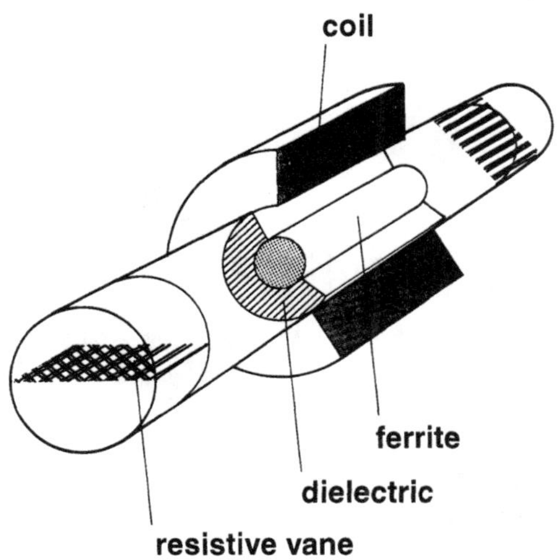

Figure 10.10 Schematic diagram of Faraday rotation isolator

the input of the rotator section. This wave is then phase shifted through $\beta_+ z$ radians in the rotator section. The phase-shifted circularly polarised wave is reconverted to a linearly polarised wave at the output by the second quarter-wave plate. In the reverse direction of propagation the circularly polarised wave is in the opposite sense and the wave is therefore phase shifted through $\beta_- z$. The overall assembly therefore behaves as a nonreciprocal phase shifter. A forward wave can, of course, also be switched from $\beta_+ z$ to $\beta_- z$ by reversing the direct magnetic field on the rotator section.

10.11 Faraday rotation in dual-mode triple ridge waveguide

While the triple ridge gyromagnetic waveguide can support planes of pure Faraday rotation on its axis it cannot decouple one pair of ridges from another. The only arrangement for which a wave between one pair of ridges will produce a corresponding wave at periodic planes between either of the other two pairs corresponds to the notion of an ideal circulator. One prerequisite for this situation is that the waveguide supports propagation of a pair of counter-rotating degenerate modes and one symmetric mode along its axis. The other is that the degeneracy of the counter-rotating modes should be removed by the gyrotropy of the waveguide. The concept of Faraday rotation in a triple ridge waveguide is therefore only of value in this restricted class of circuit and is otherwise inappropriate.

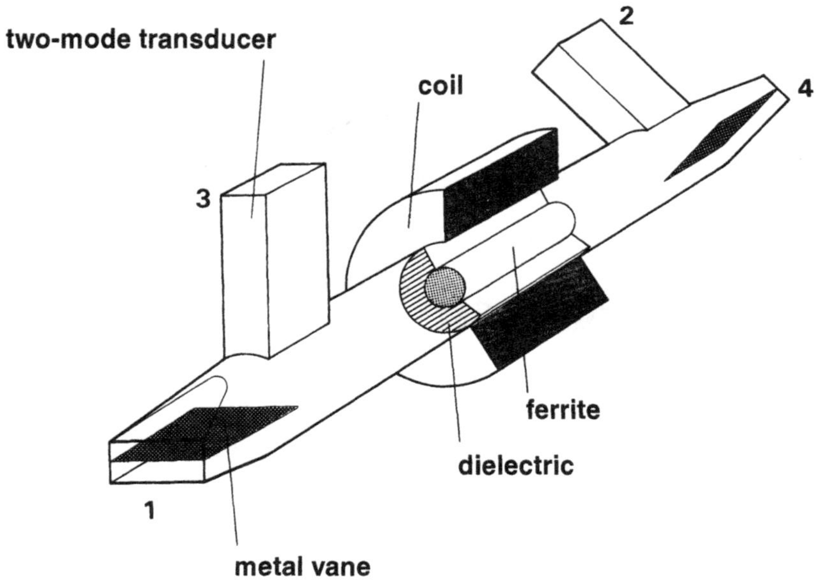

Figure 10.11 Schematic diagram of Faraday rotation circulator

A physical understanding of propagation on this waveguide starts by decomposing a single generator setting or incident wave between one pair of ridges into a linear combination of three normal modes comprising triplets of generator settings between each possible pair of ridges:

$$\begin{bmatrix}1\\0\\0\end{bmatrix} = \frac{1}{3}\begin{bmatrix}1\\1\\1\end{bmatrix} + \frac{1}{3}\begin{bmatrix}1\\\alpha\\\alpha^2\end{bmatrix} + \frac{1}{3}\begin{bmatrix}1\\\alpha^2\\\alpha\end{bmatrix}$$

where

$$\alpha = \exp(j120)$$

$$\alpha^2 = \exp(j240)$$

It proceeds by splitting the degenerate pair of counter-rotating modes by the gyrotropy of the waveguide. This gives

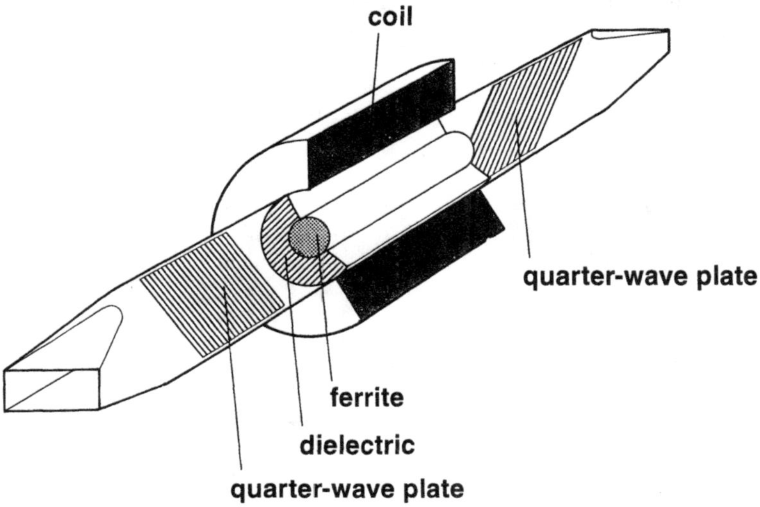

Figure 10.12 Schematic diagram of Faraday rotation phase shifter

$$a_1 = \frac{1\exp[-j\theta_0] + 1\exp[-j(\theta_1+\theta_+)] + 1\exp[-j(\theta_1+\theta_-)]}{3}$$

$$a_2 = \frac{1\exp[-j\theta_0] + \alpha\exp[-j(\theta_1+\theta_+)] + \alpha^2\exp[-j(\theta_1+\theta_-)]}{3}$$

$$a_3 = \frac{1\exp[-j\theta_0] + \alpha^2\exp[-j(\theta_1+\theta_+)] + \alpha\exp[-j(\theta_1+\theta_-)]}{3}$$

To couple one pair of ridges to a second one it is necessary to have

$$a_1 = 0$$

$$a_2 = 1$$

$$a_3 = 0$$

One solution to this problem region is

$$\theta_1 = \theta_0 + \pi$$

$$\theta_+ = -\theta_- = \frac{\pi}{3}$$

In the absence of the third normal mode the wave will be in general elliptically polarised along the waveguide.

If the gyrotropy is removed, then

$$\theta_+ = \theta_- = 0$$

and $a_1 = \frac{1}{3}$, $a_2 = \frac{2}{3}$ and $a_3 = \frac{2}{3}$.

Chapter 11

Characterisation of discontinuity effects in single ridge waveguide

11.1 Introduction

The transition between any two different waveguides is of some interest in the design of filters and other waveguide components. One canonical representation of such a discontinuity is a lumped element susceptance in cascade with an ideal transformer. The purpose of this chapter is to summarise some experimental data on a transition between an ordinary waveguide and a single ridge one. This sort of data may be experimentally extracted by making separate measurements on the external quality factor and resonant frequency of a half-wave long prototype. The network variables of the overall arrangement are separately established by using the ABCD notation. The chapter includes a careful characterisation of this problem region for each possible definition of impedance in the waveguide. The work outlined here omits, in keeping with some prior art, the effect of the lumped element susceptance of the discontinuity on the description of the external quality factor but, in keeping with some previous work, retains it in that of the midband frequency. It differs, however, in that it does not restrict the turns ratio of the ideal transformer to unity. The value of the turns ratio involves the choice of impedance employed in the calculation; the normalised lumped element susceptance is independent of its definition. This sort of problem is of interest in the design of half-wave plates, filters and matching networks. It is often dealt with by resorting to a mode matching technique. While this and other techniques are able to model such discontinuities without difficulty these are usually part of a general computer package and are not dedicated to extracting specific data.

11.2 ABCD parameters of 2-port step discontinuity

The nature of the step discontinuity between any two waveguides is of interest in the design of microwave components. The problem region under consideration is that between a regular and ridge waveguide. It is depicted in Figure 11.1. One canonical 2-port topology which has its origin in a variational approach is indicated in Figure 11.2. It consists of a shunt susceptance (B) across the primary winding of an ideal transformer with a turns ratio n. The values entering into the description of its elements may be established either through measurement or calculation. One arrangement which may be used to extract the discontinuity effects at the junction between a ridge and a standard rectangular waveguide is a half-wave long section of waveguide between standard rectangular waveguides. Figure 11.3 shows the overall equivalent circuit of the half-wave filter section utilised here. In the experimental approach the characteristic admittance and the electrical length of the half-wave prototype are the given variables and the susceptance and turns-ratio are the unknown ones; the experimental quantities are the quality factor and the frequency. The notation employed in this work is ABCD.

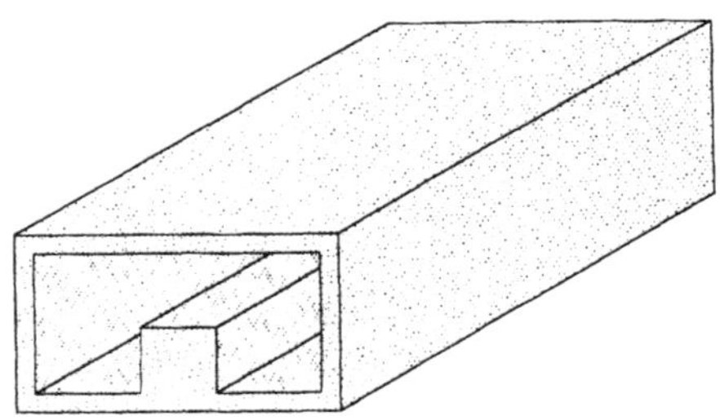

Figure 11.1 Half-wave ridge prototype

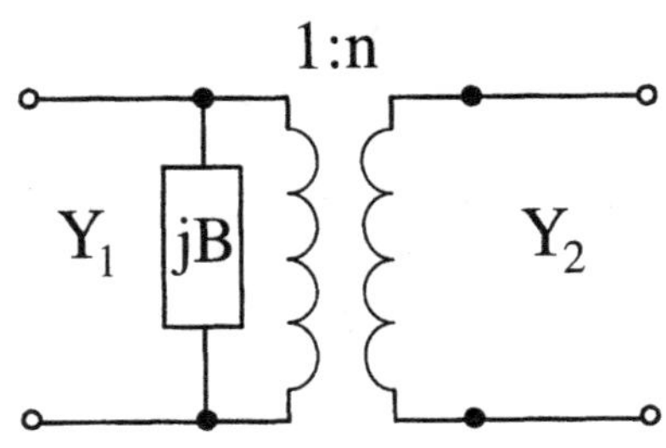

Figure 11.2 Canonical equivalent circuit of step discontinuity

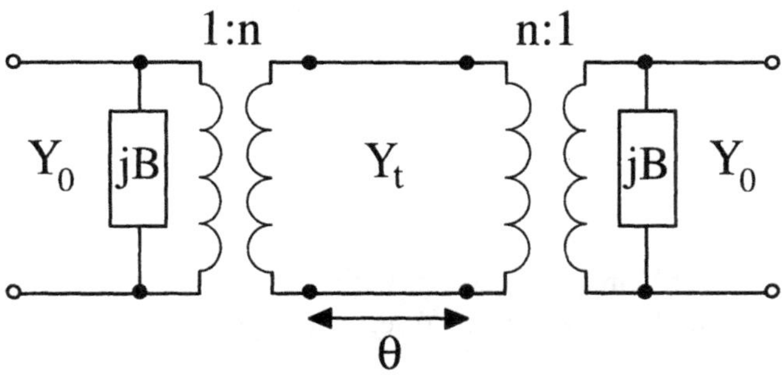

Figure 11.3 Canonical equivalent circuit of half-wave long ridge prototype

The individual ABCD matrices associated with a simple shunt element (B), an ideal transformer (n) and a uniform ridge waveguide with a characteristic admittance (Y_r) are

$$\begin{bmatrix} A & B \\ C & D \end{bmatrix} = \begin{bmatrix} 1 & 0 \\ jB & 1 \end{bmatrix} \tag{1a}$$

$$\begin{bmatrix} A & B \\ C & D \end{bmatrix} = \begin{bmatrix} n & 0 \\ 0 & \dfrac{1}{n} \end{bmatrix} \tag{1b}$$

$$\begin{bmatrix} A & B \\ C & D \end{bmatrix} = \begin{bmatrix} \cos\theta & \dfrac{j\sin\theta}{Y_r} \\ jY_r\sin\theta & \cos\theta \end{bmatrix} \tag{1c}$$

respectively.

θ is defined by

$$\theta \approx \left[1 + \frac{1}{2}\left(\frac{\beta}{\beta_r} - \frac{\beta_r}{\beta}\right)\right](\beta_r \ell) \tag{2}$$

The derivation of this quantity starts with the definition of the radian angle θ,

$$\theta = \beta\ell \tag{3}$$

It is continued by writing the phase constant β at finite frequency in terms of that at the midband

$$\beta = \beta_r\left(\frac{\beta}{\beta_r}\right) \tag{4}$$

Expanding this quantity about the midband phase constant β_r gives

$$\beta = \beta_r\left[1 + \left(\frac{\beta - \beta_r}{\beta_r}\right)\right] \tag{5}$$

This relationship is also sometimes written as

$$\beta \approx \beta_r\left[1 + \frac{1}{2}\left(\frac{\beta}{\beta_r} - \frac{\beta_r}{\beta}\right)\right] \tag{6}$$

It is instructive, before proceeding with the construction of the overall ABCD matrix of the topology in question, to construct that of the three inner sections:

$$\begin{bmatrix} A & B \\ C & D \end{bmatrix} = \begin{bmatrix} \frac{1}{n} & 0 \\ 0 & n \end{bmatrix} \begin{bmatrix} \cos\theta & \frac{j\sin\theta}{Y_r} \\ jY_r\sin\theta & \cos\theta \end{bmatrix} \begin{bmatrix} n & 0 \\ 0 & \frac{1}{n} \end{bmatrix} \tag{7}$$

or

$$A = \cos\theta \tag{8a}$$

$$B = \frac{j\sin\theta}{(n^2 Y_r)} \tag{8b}$$

$$C = j(n^2 Y_r)\sin\theta \tag{8c}$$

$$D = \cos\theta \tag{8d}$$

A scrutiny of this result suggests that the effect of the ideal transformers amounts to scaling the characteristic admittance of the transmission line by a factor n^2. The ABCD matrix of the overall arrangement is now specified by

$$A = \cos\theta - \frac{B\sin\theta}{(n^2 Y_r)} \tag{9a}$$

$$B = \frac{j\sin\theta}{(n^2 Y_r)} \tag{9b}$$

$$C = j\left[2B\cos\theta + (n^2 Y_r)\sin\theta - \frac{B^2\sin\theta}{(n^2 Y_r)}\right] \tag{9c}$$

$$D = \cos\theta - \frac{B\sin\theta}{(n^2 Y_r)} \tag{9d}$$

The above entries satisfy the reciprocity condition below, as is readily verified:

$$AD - BC = 1 \tag{10}$$

In the vicinity of $\beta_r \ell = p\pi$

$$\sin\theta \approx -\left(\frac{\beta}{\beta_r} - \frac{\beta_r}{\beta}\right)\left(\frac{p\pi}{2}\right) \tag{11a}$$

$$\cos\theta \approx -1 \tag{11b}$$

11.3 Frequency response

The validity of the entries of the equivalent circuit of the half-wave filter section may be verified by constructing its frequency response. The amplitude squared transmission coefficient ($\gamma\gamma^*$) of a symmetrical network for which $A = D$ is given in terms of its ABCD parameters by

$$\gamma\gamma^* = \frac{1}{1 - \frac{1}{4}(B - C)^2} \tag{12}$$

The amplitude squared reflection coefficient ($\rho\rho^*$) is deduced in terms of the amplitude squared transmission coefficient by using the unitary condition

$$\gamma\gamma^* + \rho\rho^* = 1 \tag{13}$$

Scrutiny of equation (12) indicates that the condition for perfect transmission coincides with

$$BY_0^2 = C \tag{14}$$

The normalisation factor Y_0^2 ensures that the units of equation (14) are consistent.

11.4 Characterisation of half-wave long ridge waveguide test-set

One test-set which may be employed to extract some remarks about the discontinuity effect of the ridge waveguide is the half-wave long ridge structure. The problem region in question is sometimes solved by setting the turns ratio of the ideal transformer in its equivalent circuit to unity and by neglecting the shunt susceptance in constructing the external quality factor. In the approximation adopted here the effect of the ideal transformer is retained but that of the shunt susceptance on the quality factor is again disregarded. The external quality factor (Q_{ex}) or normalised susceptance slope parameter (b') of such a degree-1 section is described in the absence of fringing effects at the input and output planes of the section by

$$Q_{ex} \approx \left(\frac{\pi}{2}\right)\left(\frac{n^2 Y_r(\omega)}{Y_0(\omega)} - \frac{Y_0(\omega)}{n^2 Y_r(\omega)}\right)\left(\frac{\lambda_{gr}}{\lambda_0}\right)^2 \tag{15}$$

$Y_r(\omega)$ is the admittance of the ridge waveguide at finite frequency, $Y_0(\omega)$ is that of the input and output rectangular waveguides,

$$Y_r(\omega) = Y_r(\infty)\left(\frac{\lambda_0}{\lambda_{gr}}\right) \tag{16}$$

$$Y_0(\omega) = Y_0(\infty)\left(\frac{\lambda_0}{\lambda_{g0}}\right) \tag{17}$$

$Y_r(\infty)$ and $Y_0(\infty)$ take on the values $Y_{VI}(\infty)$, $Y_{PV}(\infty)$ and $Y_{PI}(\infty)$, respectively, and λ_{gr} and λ_{g0} are the wavelengths in the ridge and rectangular waveguides.

The external Q-factor of the circuit is defined in the usual way by

$$Q_{ex} = \frac{\omega}{2Y_0}\left.\frac{\partial B_{in}}{\partial \omega}\right|_{\omega=\omega_0} \tag{18}$$

One convenient way to evaluate the input immittance of a cascade arrangement of a number of sections is to use the ABCD notation. The admittance of such a network terminated in a load Y_0 is a standard result in the literature:

$$Y_{in}(\omega) = \frac{C + DY_0}{A + BY_0} \tag{19}$$

The real and imaginary part of $Y_{in}(\omega)$ may be deduced by using the symmetry and reciprocity properties of the ABCD parameters and by noting that A and D are pure real numbers and that C and B are pure imaginary ones. This gives

$$Y_{in}(\omega) = \frac{(2A^2 - 1)Y_0 + A(C - BY_0^2)}{A^2 + B^2 Y_0^2} \tag{20}$$

The first term in the numerator polynomial of $Y_{in}(\omega)$ represents its real part and the second one its imaginary part.

The input admittance in the vicinity of $\beta_r \ell = p\pi$ in the absence of the fringing susceptances at either the input or output terminals of the network is therefore

$$Y_{in}(\omega) \approx Y_0(\omega)\left[1 + j\left(\frac{n^2 Y_r(\omega)}{Y_0(\omega)} - \frac{Y_0(\omega)}{n^2 Y_r(\omega)}\right)\left(\frac{\beta}{\beta_r} - \frac{\beta_r}{\beta}\right)\left(\frac{p\pi}{2}\right)\right] \tag{21}$$

where

$$\beta_r^2 = k_0^2 - k_{cr}^2$$

$$\beta^2 = k^2 - k_{cr}^2$$

k_{cr} is the cut-off number of the ridge waveguide.

The real part of the complex admittance at the input terminals in the vicinity of $\beta_r \ell = p\pi$, in keeping with simple transmission line theory, is merely that at the output terminals.

Since the cut-off numbers entering into the descriptions of $Y_0(\omega)$ and $Y_r(\omega)$ in standard and ridge waveguides are finite both quantities are dispersive. It is therefore strictly speaking necessary, in order to evaluate Q_{ex}, to write B_{in} in the form

$$B_{in} = uv$$

where

$$u = Y_0(\omega)\left(\frac{n^2 Y_r(\omega)}{Y_0(\omega)} - \frac{Y_0(\omega)}{n^2 Y_r(\omega)}\right)$$

$$v = \left(\frac{\beta}{\beta_r} - \frac{\beta_r}{\beta}\right)\left(\frac{p\pi}{2}\right)$$

and to form

$$\frac{\partial B_{in}}{\partial \omega} = u\frac{\partial v}{\partial \omega} + v\frac{\partial u}{\partial \omega}$$

A scrutiny of $\partial u/\partial \omega$ for both d/b and β_r/β in the vicinity of unity suggests that the product of these two quantities is small. It is neglected here. This readily gives the required result in equation (15).

Solving the relationship in equation (15) for $n^2 Y_r(\omega)/Y_0(\omega)$ gives

$$\frac{n^2 Y_r(\omega)}{Y_0(\omega)} = \left(\frac{Q_{ex}}{\pi}\right)\left(\frac{\lambda_0}{\lambda_{gr}}\right)^2 + \sqrt{\left(\frac{Q_{ex}}{\pi}\right)^2\left(\frac{\lambda_0}{\lambda_{gr}}\right)^4 + 1} \tag{22}$$

The turns ratio n for each definition of impedance may therefore be readily evaluated once Q_{ex} and the other quantities are at hand. The parameters of the ridge waveguide may be calculated by referring to Chapter 3.

The midband frequency condition may be established exactly and straightforwardly by making use of the fact that it coincides with the condition in equation (14). The required result is

$$\tan\theta = \frac{2b}{\dfrac{b^2}{(n^2 y_r)} + \dfrac{1}{(n^2 y_r)} - (n^2 y_r)} \tag{23}$$

It reduces to the classic condition associated with two susceptances separated by a transmission line when $y_r = 1$ and $n = 1$:

$$\tan\theta = \frac{2}{b} \tag{24}$$

The characteristic equation in equation (23) is recognised, for calculation purposes, as a quadratic in the required unknown b,

$$\frac{b^2 \tan\theta}{(n^2 y_r)} - 2b + \left[\frac{1}{(n^2 y_r)} - (n^2 y_r)\right]\tan\theta = 0 \tag{25}$$

The two unknowns of the problem region may now be evaluated once β/β_r and Q_{ex} are at hand. The meaning of the normalised variable utilised here is understood:

$$b = \frac{B}{Y_0(\omega)} \tag{26}$$

$$y_r = \frac{Y_r(\omega)}{Y_0(\omega)} \tag{27}$$

The angle θ is defined in equation (11).

11.5 Experimental characterisation

The experimental characterisation of a typical step using a half-wave long test piece involves a measurement of its frequency and its quality factor. Figure 11.4 indicates the relationship between the frequency of this sort of arrangement for four different inserts and the gap factor with $s/a = 0.50$. Some numerical calculations based on the mode matching method (MMM) are separately superimposed on these data. These calculations represent an upper bound on the measured data. Figure 11.5 gives the relationship between the wavelength and the gap factor.

The aspect ratios (b/a) of both the mating and ridge waveguides used in this work coincide with that of the WR 62 waveguide. The results obtained here apply, therefore, to a single ridge structure with an aspect ratio of 0.50 rather than 0.45 usually associated with this kind of waveguide.

The quality factor is obtained from a knowledge of the 20dB return loss points by treating the network as a degree-1 one-port LCR circuit:

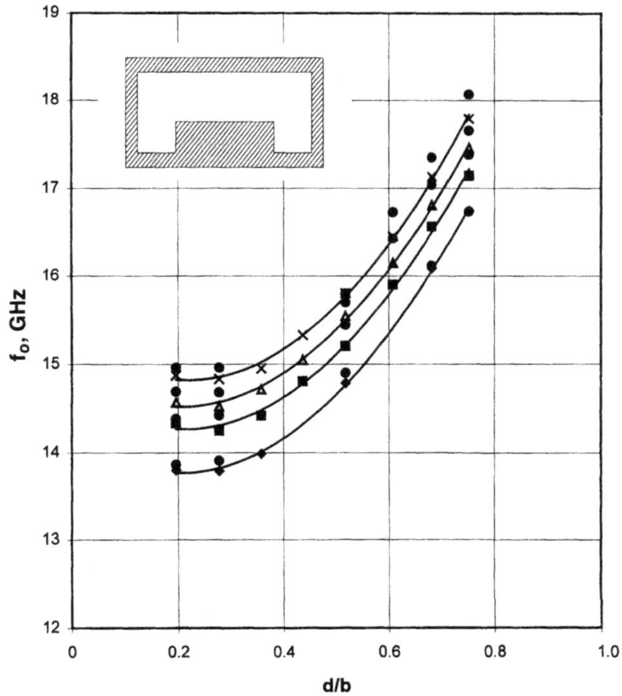

Figure 11.4 Relationship between frequency and aspect ratio of half-wave long ridge waveguide

Reprinted with permission, Helszajn (2000)

◆ $L = 8.5\,\mathrm{mm}$, ■ $L = 8.0\,\mathrm{mm}$, △ $L = 7.75\,\mathrm{mm}$, × $L = 7.5\,\mathrm{mm}$, ● MMM

$$Q_{\mathrm{L}} = \frac{(\mathrm{VSWR} - 1)}{2\delta_0\sqrt{\mathrm{VSWR}}} \tag{28}$$

where

$$2\delta_0 = \frac{\omega_2 - \omega_1}{\omega_0}$$

Figure 11.6 depicts the experimental relationship between the quality factor of this type of element and its aspect ratio. This database was obtained in a WR 62 waveguide. When the gap factor d/b equals unity the frequency is equal to that of a standard WR 62 waveguide. One feature of these data is that the external quality factor is essentially independent of the overall length of the test piece. Since each test piece resonates at a different frequency the elements entering into the description of a typical discontinuity are a flat function of frequency over the variables investigated here. In obtaining these data no attention has been given to the quality of the

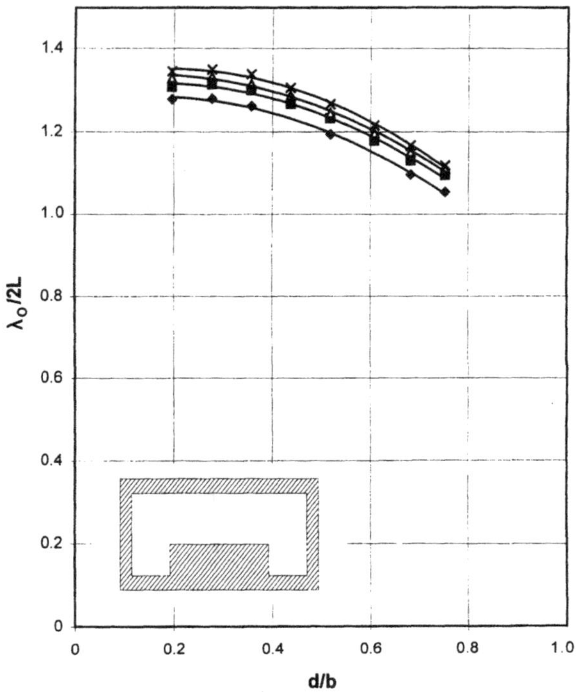

Figure 11.5 Relationship between wavelength and aspect ratio of half-wave long ridge waveguide

Reprinted with permission, Helszajn (2000)

◆ $L = 8.5\,\text{mm}$, ■ $L = 8.0\,\text{mm}$, △ $L = 7.75\,\text{mm}$, × $L = 7.5\,\text{mm}$

coaxial to waveguide transitions employed in the test set. The possibility of some interaction between a typical transition and a typical discontinuity cannot be ruled out.

The turns ratio of the ideal transformer may be calculated once the frequency and the quality factor of the experimental prototype are available. Figures 11.7–11.9 show the relationships between the gap factor and the turns ratio of the ideal transformer for each definition of waveguide impedance. Each result is deduced by curve fitting the theoretical and experimental results established here. Since the turns ratio can be absorbed into the definition of the impedance of the ridge section its size may be used to determine the definition of impedance that is most appropriate between any two particular waveguides.

The normalised susceptance of such a discontinuity is indicated in Figure 11.10. It is essentially independent of the choice of the waveguide impedance.

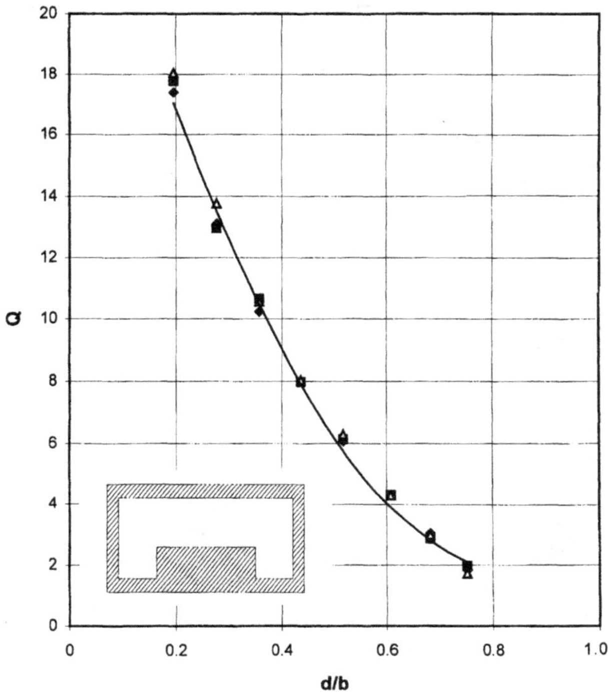

Figure 11.6 Relationship between loaded Q-factor and aspect ratio of half-wave long ridge waveguide

Reprinted with permission, Helszajn (2000)

◆ $L = 8.5\,\text{mm}$, ■ $L = 8.0\,\text{mm}$, △ $L = 7.5\,\text{mm}$

11.6 Symmetrical short section

One equivalent circuit of a short symmetrical obstacle between regular waveguides which neglects any interaction between the discontinuities at each side is a tee-circuit. The equivalent circuit in question is indicated in Figure 11.11. It is of some importance in the design of immittance inverters which enter into the description of directly coupled bandpass filters. The element values of its series and shunt reactances may be readily expressed in terms of the canonical representation of a typical single step. One means of doing so may be achieved by comparing the odd and even impedances of the two arrangements. The derivation of the required equivalence starts by constructing the even and odd impedances of the tee-circuit:

$$Z_{\text{even}} = j(X_s + 2X_p) \tag{29a}$$

$$Z_{\text{odd}} = jX_s \tag{29b}$$

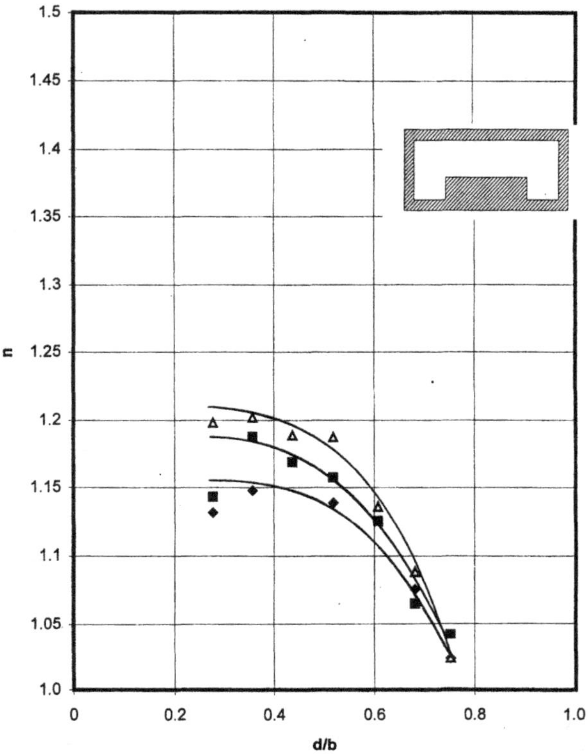

Figure 11.7 Experimental relationship between gap factor (d/b) and turns ratio (n) of ideal transformer for Z_{PV}

Reprinted with permission, Helszajn (2000)

$L =$ ◆ 8.5 mm, ■ 8.0 mm, △ 7.5 mm

or

$$jX_s = Z_{odd} \tag{30a}$$

$$jX_p = \frac{Z_{even} - Z_{odd}}{2} \tag{30b}$$

The required result is now established once the even and odd impedances of the original 2-port circuit are deduced in terms of the original variables:

$$\frac{1}{Z_{even}} = j\left[\omega C + n^2 Y_r(\omega) \cot \frac{\theta}{2}\right] \tag{31a}$$

$$\frac{1}{Z_{odd}} = j\left[\omega C - n^2 Y_r(\omega) \tan \frac{\theta}{2}\right] \tag{31b}$$

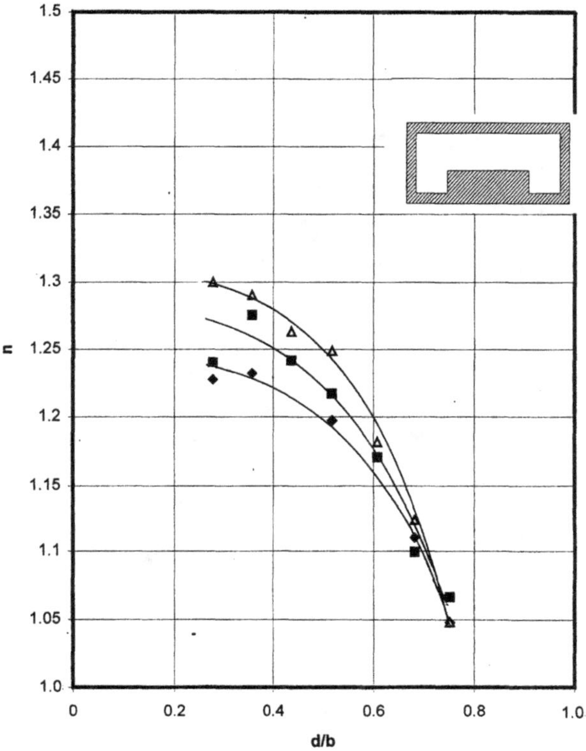

Figure 11.8 Experimental relationship between gap factor (d/b) and turns ratio (n) of ideal transformer for Z_{PI}

Reprinted with permission, Helszajn (2000)

$L =$ ◆ 8.5 mm, ■ 8.0 mm, △ 7.5 mm

Another equivalent circuit that is often encountered in this sort of problem is the π-circuit in Figure 11.12. The required relationships are in this instance given simply, in terms of the even and odd admittances of the circuit, by

$$jB_p = Y_{even} \tag{32a}$$

$$jB_s = 2(Y_{odd} - Y_{even}) \tag{32b}$$

The even and odd immittance parameters are connected by

$$Y_{even} = \frac{1}{Z_{even}} \tag{33a}$$

$$Y_{odd} = \frac{1}{Z_{odd}} \tag{33b}$$

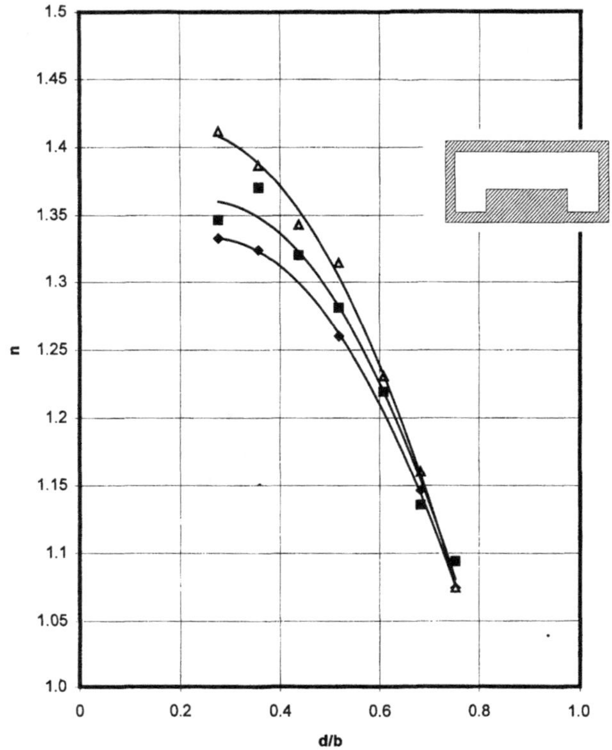

Figure 11.9 Experimental relationship between gap factor (d/b) and turns ratio (n) of ideal transformer for Z_{VI}

Reprinted with permission, Helszajn (2000)

$L =$ ◆ 8.5 mm, ■ 8.0 mm, △ 7.5 mm

The even mode parameters of any 2-port symmetric circuit are 1-port variables which may be obtained by placing equal amplitude in-phase voltage generators at each port. The odd mode variables are obtained by placing equal amplitude out-of-phase generators there. The even and odd mode immittances may be separately shown to coincide with the eigenvalues of the related immittance matrices.

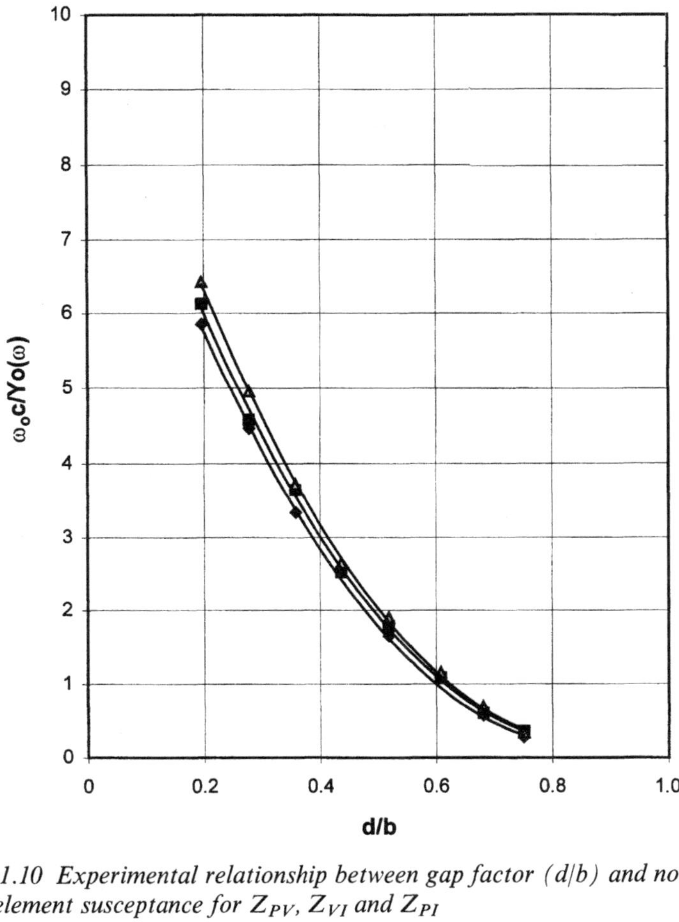

Figure 11.10 Experimental relationship between gap factor (d/b) and normalised lumped element susceptance for Z_{PV}, Z_{VI} *and* Z_{PI}

Reprinted with permission, Helszajn (2000)

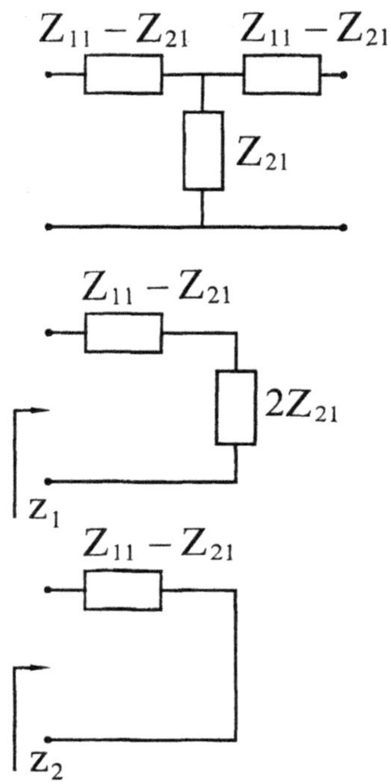

Figure 11.11 Symmetrical tee-circuit

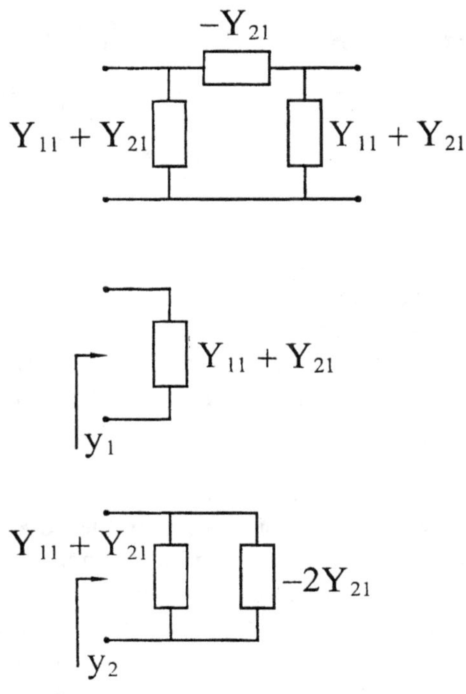

Figure 11.12 Symmetrical pi-circuit

Chapter 12

Ridge cross-guide directional coupler

M. McKay and J. Helszajn

12.1 Introduction

An important microwave component is the 4-port directional coupler. The two classic arrangements consist of a cross-guide structure and a distributed version. The relationships between the scattering variables in this sort of network are fixed by the unitary condition. Such a network has the properties that it is a matched device with one adjacent port decoupled from any incident port. The object of this chapter is to present some calculations on the coupling and directivity of a number of cross-guide couplers in a ridge waveguide. The four possibilities consist of one arrangement with both the primary and secondary waveguides in a double ridge, a similar arrangement with primary and secondary waveguides in a single ridge, and configurations with the primary and secondary waveguides in either single and double ridge or double and single ridge, respectively. Figure 12.1 illustrates the four topologies. The coupling geometry usually consists of either a circular or cross-slot aperture. The calculations are based on Bethe's small-hole aperture theory in conjunction with the approximate closed-form expressions for the fields in the waveguides introduced in Chapter 4. Apart from discrepancies in the vicinity of the ridge discontinuities these agree well with some Finite Element Method (FEM) calculations in the same chapter. The closed-form formulations may therefore be utilised for engineering purposes.

12.2 Operation of cross-guide directional coupler

The 4-port cross-guide directional coupler is a classic component in microwave engineering. It is defined as a matched 4-port network with one adjacent port decoupled from any input port. The various topologies in a

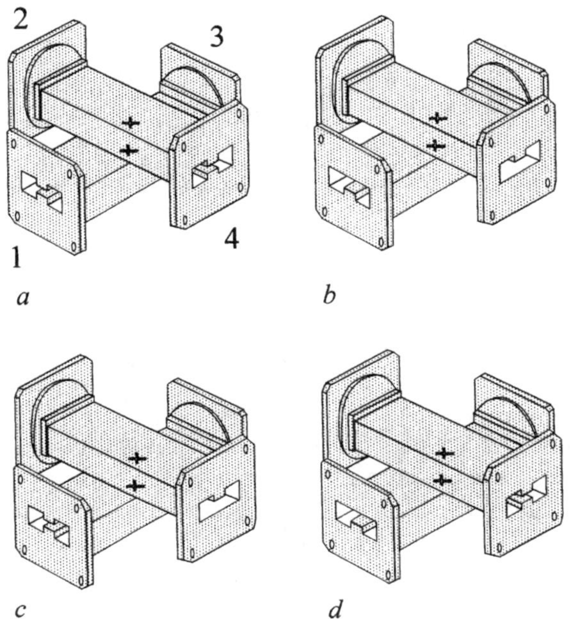

Figure 12.1 Schematic diagrams of four possible topologies of cross-guide ridge directional coupler

ridge waveguide are illustrated in Figure 12.1. The waveguide section containing ports 1 and 3 is sometimes referred to as the primary waveguide, while that containing ports 2 and 4 is denoted the secondary waveguide. A scrutiny of the symmetry of this sort of network suggests the possibility of reducing the entries of the scattering matrix to linear combinations of odd and even modes. This may be done by taking ports 1 and 2 as a typical pair of input ports and ports 3 and 4 as typical output ports. The scattering matrix of the ideal, symmetric and lossless network with port 2 decoupled from port 1 is given in the usual way by

$$[S] = \begin{bmatrix} 0 & 0 & S_{31} & S_{41} \\ 0 & 0 & S_{41} & S_{31} \\ S_{31} & S_{41} & 0 & 0 \\ S_{41} & S_{31} & 0 & 0 \end{bmatrix} \tag{1}$$

The relationship between the transmitted (S_{31}) and coupled (S_{41}) coefficients is given by the unitary condition,

$$S_{31}S_{31}^* + S_{41}S_{41}^* = 1 \tag{2a}$$

$$S_{31}S_{41}^* + S_{31}^*S_{41} = 0 \tag{2b}$$

One solution is

$$S_{31} = \alpha \tag{3a}$$

$$S_{41} = j\beta \tag{3b}$$

The performance of any nonideal directional coupler is defined in terms of coupling and directivity factors:

$$\text{coupling factor} = 20\log_{10}(S_{41})\ \text{dB} \tag{4a}$$

$$\text{directivity} = 20\log_{10}\left(\frac{S_{41}}{S_{21}}\right)\ \text{dB} \tag{4b}$$

The operation of the cross-guide directional coupler using a pair of circular apertures may be understood by examining the path lengths connecting the two waveguides in Figure 12.2. Since the path lengths connecting $a_1 \rightarrow a_2 \rightarrow b_2$ and $b_1 \rightarrow c_1 \rightarrow c_2$ are equal, then a wave incident on port 1 will be emergent at port 4. In a similar fashion the path lengths $a_1 \rightarrow a_2$ and $b_1 \rightarrow c_1 \rightarrow c_2 \rightarrow d_2$ differ by $\lambda_g/2$ so that a wave incident at port 1 is decoupled from port 2. The operation of this cross-guide coupler is frequency dependent, with perfect directivity being obtained only when the spacing between the apertures is exactly $\lambda_g/4$. The coupling between

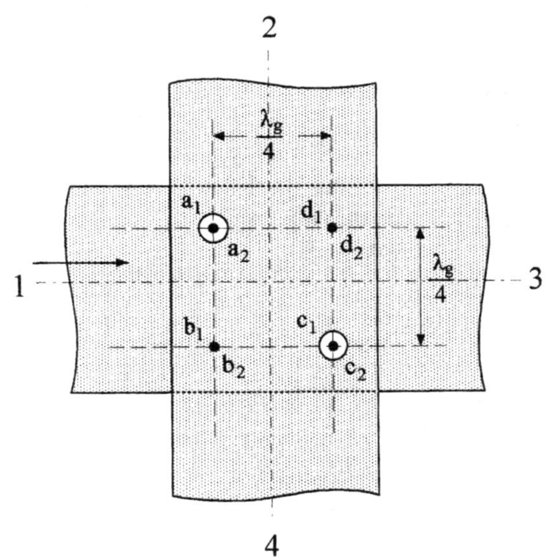

Figure 12.2 Operation of cross-guide coupler using circular apertures

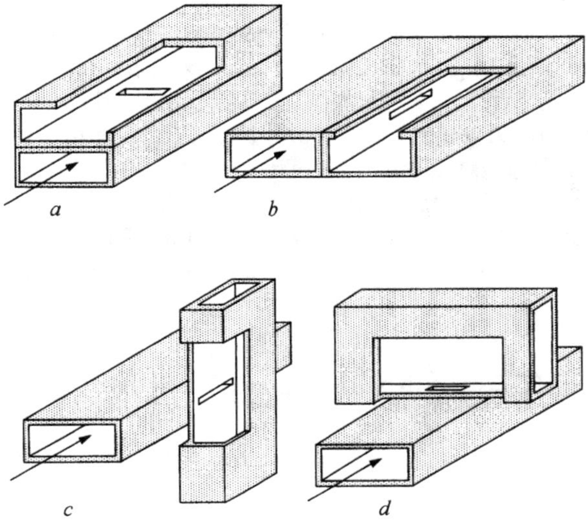

Figure 12.3 Coupling apertures in rectangular waveguide

a Series-series; *b* shunt-shunt; *c* shunt-series; *d* series-shunt

any two waveguides can be associated with one of four standard topologies. The name associated with each kind is summarised in Figure 12.3. The series-series topology is that of interest here.

12.3 Bethe's small-hole coupling theory

Coupling between two waveguides by means of an aperture is a classic problem in the literature. The coupling in the case of an infinitesimal common wall is determined by the fields in the primary and secondary waveguides and the equivalent electric and magnetic dipoles of the aperture. In the arrangement under consideration, the electric ($\mathbf{p}_e$) and magnetic ($\mathbf{m}_t$, $\mathbf{m}_\ell$) dipoles are defined in terms of electric and magnetic polarisabilities and the unperturbed fields in the primary waveguide:

$$\mathbf{p}_e = \mathbf{a}_n \varepsilon_0 P_e E_{0n} \tag{5a}$$

$$\mathbf{m}_t = \mathbf{a}_t M_t H_{0t} \tag{5b}$$

$$\mathbf{m}_\ell = \mathbf{a}_\ell M_\ell H_{0\ell} \tag{5c}$$

$\mathbf{a}_t$ and $\mathbf{a}_\ell$ represent unit vectors along the principal axes (t, ℓ) of the aperture and $\mathbf{a}_n$ denotes a unit normal to the aperture. P_e, M_t and M_ℓ are scalar quantities corresponding to the electric and axial magnetic polarisabilities of the aperture, respectively. Table 12.1 summarises these for three typical geometries. H_{0t} and $H_{0\ell}$ denote the components of the unperturbed tangential magnetic field (H_0) along the principal axes of the aperture in the primary waveguide. E_{0n} represents the unperturbed electric field normal to the centre of the aperture in the same waveguide. The forward and backward transmission coefficients of any mode in the secondary waveguide are given by

$$A_m = \frac{-j\omega}{4P}\{\mathbf{p}_e E_{mn}^- + \mu_0 \mathbf{m}_t H_{mt}^- + \mu_0 \mathbf{m}_\ell H_{m\ell}^-\} \tag{6a}$$

and

$$B_m = \frac{-j\omega}{4P}\{\mathbf{p}_e E_{mn}^+ + \mu_0 \mathbf{m}_t H_{mt}^+ + \mu_0 \mathbf{m}_\ell H_{m\ell}^+\} \tag{6b}$$

H_{mt}, $H_{m\ell}$ and E_{mn} correspond to the tangential magnetic fields and normal electric field of the propagating mode in the secondary waveguide. The plus and minus signs indicate propagation along the positive and negative z directions, respectively. P denotes the power flow and the calculation proceeds by normalising the fields at every frequency in the primary and secondary waveguides such that

Table 12.1 Electric and magnetic polarisabilities of circular and slot apertures

Aperture	Section	M_ℓ	M_t	P_e
Circular hole		$\frac{d^3}{6}$	$\frac{d^3}{6}$	$\frac{d^3}{12}$
Longitudinal slot	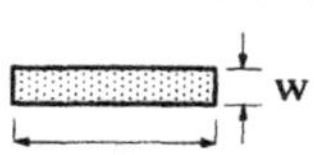	$\frac{\pi\ell^3/24}{\ln(4\ell/w-1)}$	$\frac{\pi\ell w^2}{24}$	$\frac{\pi\ell w^2}{24}$
Transverse slot	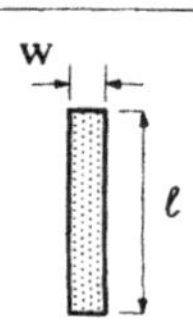	$\frac{\pi\ell w^2}{24}$	$\frac{\pi\ell^3/24}{\ln(4\ell/w-1)}$	$\frac{\pi\ell w^2}{24}$

$$P = \frac{1}{2} \text{Re} \int (\mathbf{E} \times \mathbf{H}^*) \cdot \mathbf{n} \, \mathrm{d}s = 1 \tag{7}$$

The effect of finite wall thickness is treated elsewhere in this chapter. It is determined by visualising the aperture as a cut-off section of waveguide with a length equivalent to the thickness of the wall.

12.4 The 0-degree crossed-slot aperture

Commercial cross-guide couplers often employ single or pairs of orthogonal slots which are located on the main diagonal and aligned with the direction of propagation (0 degree) or at 45 degrees to it. The first configuration is treated here and the second in the next section. Figure 12.4 depicts the four possible arrangements in a rectangular waveguide. In keeping with convention it is assumed that port 2 is the decoupled port so that the topologies

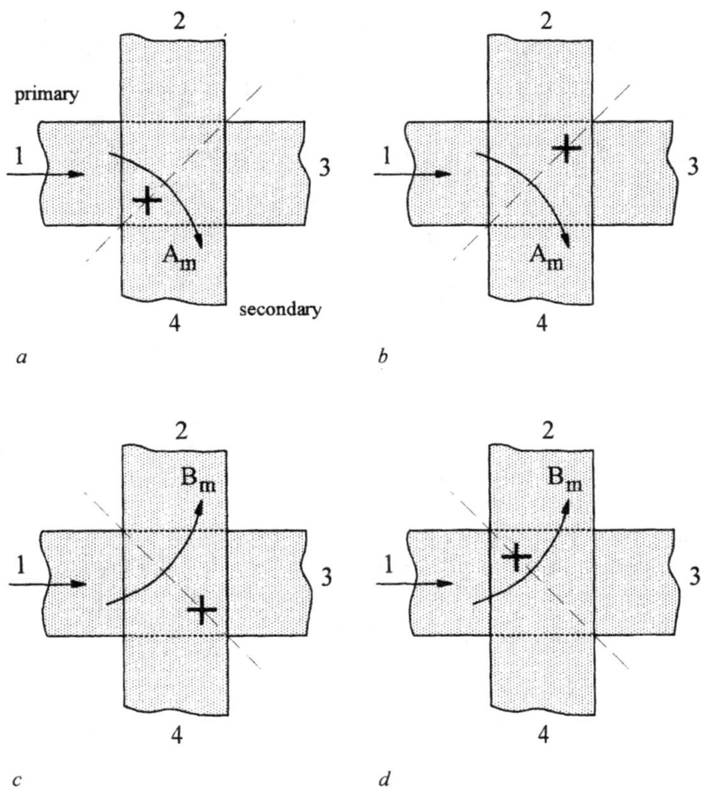

Figure 12.4 Cross-guide coupler using a single 0-degree crossed-slot aperture

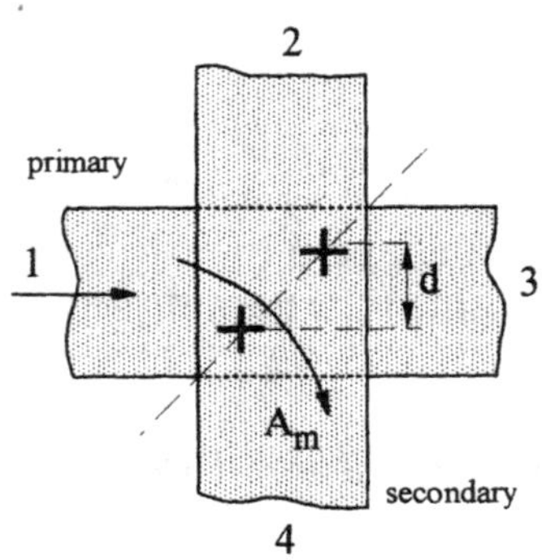

Figure 12.5 Cross-guide coupler using a pair of 0-degree crossed-slot apertures

in Figures 12.4*a* and *b* are the geometries of interest here. In practice, to increase the coupling between the primary and secondary waveguides the two arrangements are often combined in the manner shown in Figure 12.5. The case in which the primary and secondary waveguides are at an arbitrary orientation is understood but outside the remit of this work.

The calculation of A_m and B_m in the secondary waveguide starts by evaluating the electric and magnetic dipoles along the axes of the primary waveguide. For a slot orientated along the direction of propagation of the primary waveguide, the magnetic dipole moments along the axes of the secondary waveguide are given by

$$\mathbf{m}_{x\mathrm{s}} = M_\ell H_{z\mathrm{p}} \tag{8a}$$

$$\mathbf{m}_{z\mathrm{s}} = -M_t H_{x\mathrm{p}} \tag{8b}$$

The corresponding results in the case of a perpendicular slot are

$$\mathbf{m}_{x\mathrm{s}} = M_t H_{z\mathrm{p}} \tag{9a}$$

$$\mathbf{m}_{z\mathrm{s}} = -M_\ell H_{x\mathrm{p}} \tag{9b}$$

Introducing these quantities into the governing equations for A_m and B_m and assuming narrow slots ($M_t \approx 0$) gives

$$A_m = \frac{-jk_0 P_e E_{y\mathrm{p}} E_{y\mathrm{s}}}{4\eta_0 P} + \frac{jk_0 \eta_0 M_\ell}{4P}\{H_{x\mathrm{p}} H_{z\mathrm{s}} + H_{z\mathrm{p}} H_{x\mathrm{s}}\} \tag{10a}$$

and

$$B_m = \frac{-jk_0 P_e E_{y\mathrm{p}} E_{y\mathrm{s}}}{4\eta_0 P} + \frac{jk_0 \eta_0 M_\ell}{4P}\{H_{x\mathrm{p}} H_{z\mathrm{s}} - H_{z\mathrm{p}} H_{x\mathrm{s}}\} \tag{10b}$$

12.5 The 0-degree crossed-slot aperture in rectangular waveguide

To familiarise the reader with the calculations met in connection with this sort of device, the standard problem of two rectangular waveguides propagating the dominant TE_{10} mode will be tackled first prior to dealing with the ridge structure. The field variables are given in this instance by

$$H_z = A_{10}\left(\frac{k_c}{\eta_0 k_0}\right)\cos\left(\frac{\pi x}{a}\right) \tag{11a}$$

$$H_x = jA_{10}\left(\frac{\beta}{\eta_0 k_0}\right)\sin\left(\frac{\pi x}{a}\right) \tag{11b}$$

$$E_y = -\left(\frac{k_0\eta_0}{\beta}\right)H_x \tag{11c}$$

The time and spatial variation $\exp j(\omega t - \beta z)$ is understood. The above field description assumes that the amplitude of E_y rather than H_z is independent of frequency. This permits the power flow in the waveguide to be written in the form below

$$P(\omega) = P(\infty)\left(\frac{\beta}{k_0}\right) \tag{12a}$$

where

$$P(\infty) = A_{10}^2\left(\frac{ab}{4\eta_0}\right) \tag{12b}$$

Introducing these quantities into A_m and B_m for the arrangement in Figure 12.4*a* gives

$$A_1 = \frac{-M_\ell k_c}{ab}\sin\left(\frac{2\pi x}{a}\right) + \frac{jP_e k_0^2}{ab\beta}\sin^2\left(\frac{\pi x}{a}\right) \tag{13a}$$

$$B_1 = \frac{jP_e k_0^2}{ab\beta}\sin^2\left(\frac{\pi x}{a}\right) \tag{13b}$$

When the aperture is located on the other diagonal (Figure 12.4*c*),

$$A_1 = \frac{jP_e k_0^2}{ab\beta}\sin^2\left(\frac{\pi x}{a}\right) \tag{14a}$$

$$B_1 = \frac{M_\ell k_c}{ab}\sin\left(\frac{2\pi x}{a}\right) + \frac{jP_e k_0^2}{ab\beta}\sin^2\left(\frac{\pi x}{a}\right) \tag{14b}$$

The transmitted and coupled waves are dependent on the position of the aperture due to the fact that H_z reverses across the symmetry plane of the waveguide.

To increase the degree of coupling between the primary and secondary waveguides, two rather than one apertures are often used. The transmission coefficients are in this instance dependent on the phase difference, $\exp(-j2\beta d)$, between the apertures. By noting that the magnetic dipoles of the two apertures in the direction of propagation of the primary waveguide are in antiphase, the transmitted and coupled coefficients in the case of narrow slots ($P_e \approx 0$) are given by

$$|S_{41}| = \frac{2M_\ell k_c}{ab} \sin\left(\frac{\pi d}{a}\right) \sin(\beta d) \tag{15a}$$

$$|S_{21}| = 0 \tag{15b}$$

Optimum coupling with minimum variation over the waveband occurs when $d = a/2$ and $\beta = k_c$:

$$|S_{41}|_{\max} = \frac{2M_\ell k_c}{ab} \tag{16}$$

The maximum slot length is $L = a/2$.

12.6 The 0-degree crossed-slot aperture in single ridge waveguide

The derivation of A_m and B_m in a ridge waveguide proceeds in a like manner to that utilised in the case of the rectangular waveguide. A knowledge of the magnetic field in the plane of the aperture and the electric field perpendicular to it is sufficient for the purpose of calculation. Approximate closed form field expressions for the TE modes in the trough region have been given in Chapter 4 by

$$H_x = \sum_{n=0,1,2\ldots}^{N} -jA_n\left(\frac{\beta\alpha_n}{\eta_0 k_0}\right) \sin(\alpha_n x) \cos\left(\frac{n\pi y}{b}\right) \tag{17a}$$

$$H_z = \sum_{n=0,1,2\ldots}^{N} -A_n\left(\frac{k_c^2}{\eta_0 k_0}\right) \cos(\alpha_n x) \cos\left(\frac{n\pi y}{b}\right) \tag{17b}$$

$$E_y = -\left(\frac{k_0 \eta_0}{\beta}\right) H_x \tag{17c}$$

where

$$A_n = \frac{-\Gamma_n \cos\left(\frac{k_c s}{2}\right)}{\alpha_n(n\pi)\sin\left[\frac{\alpha_n(a-s)}{2}\right]}\left\{\sin\left[\left(\frac{n\pi}{2b}\right)(b+d)\right] - \sin\left[\left(\frac{n\pi}{2b}\right)(b-d)\right]\right\} \tag{17d}$$

and

$$\alpha_n^2 = k_c^2 - \left(\frac{n\pi}{b}\right)^2 \tag{17e}$$

$\Gamma_n = 1(n=0)$, $\Gamma_n = 2(n>0)$. An amplitude term A_{10} premultiplying each field variable is understood.

An approximate expression for the power flow of the dominant mode is given by equation (12a). The Poynting vector at infinite frequency is

$$\mathbf{P}(\infty) = \frac{A_{10}^2}{\eta_0}\left(\frac{ab}{2\pi}\right)\left\{2m\left(\frac{b}{a}\right)\left(\frac{d}{b}\right)\left(\frac{a}{\lambda_c}\right)\cos^2\theta_2 \ln \operatorname{cosec}\left(\frac{\pi d}{2b}\right) + \frac{\theta_2}{2}\right.$$
$$\left. + \frac{\sin 2\theta_2}{4} + \left(\frac{d}{b}\right)\left(\frac{\cos\theta_2}{\sin\theta_1}\right)^2\left[\frac{\theta_1}{2} - \frac{\sin 2\theta_1}{4}\right]\right\}\left(\frac{d}{b}\right)\left(\frac{\lambda_c}{a}\right) \tag{18}$$

where

$$\theta_1 = \pi\left(1 - \frac{s}{a}\right)\left(\frac{a}{\lambda_c}\right)$$

$$\theta_2 = \pi\left(\frac{s}{a}\right)\left(\frac{a}{\lambda_c}\right)$$

$m = 1$ for double ridge waveguide and $m = 2$ for single ridge waveguide.

Figure 12.6 indicates the variation of the coupling and directivity between two single ridge waveguides using a single 0-degree crossed-slot aperture.

12.7 The 45-degree crossed-slot aperture

One practical method for increasing the coupling between any two waveguides is to orientate the apertures at 45 degrees with respect to the primary waveguide. This configuration is indicated in Figure 12.7. The advantage of this topology is that longer slot lengths and higher coupling values may be obtained. The magnetic field in the primary waveguide is elliptically polarised with different senses at apertures 1 and 2. Resolving the magnetic

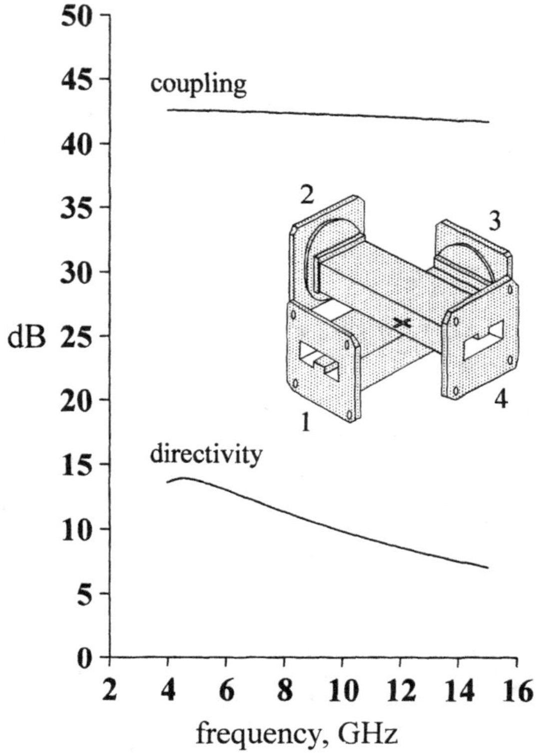

Figure 12.6 Coupling and directivity between two single ridge waveguides using a single 0-degree crossed slot aperture

dipoles of apertures 1 and 2 along the axes of the secondary waveguide gives

$$\mathbf{m}_{xs} = \sqrt{2}M_\ell H_{\ell\,\max} \cos\alpha \tag{19a}$$

$$\mathbf{m}_{zs} = -j\sqrt{2}M_\ell H_{\ell\,\max} \sin\alpha \tag{19b}$$

and

$$\mathbf{m}_{xs} = -\sqrt{2}M_\ell H_{\ell\,\max} \cos\alpha \tag{19c}$$

$$\mathbf{m}_{zs} = -j\sqrt{2}M_\ell H_{\ell\,\max} \sin\alpha \tag{19d}$$

where

$$H_{\ell\,\max} = H_{t\,\max} = \frac{1}{\sqrt{2}}\sqrt{H_{zp}^2 - H_{xp}^2} \tag{20a}$$

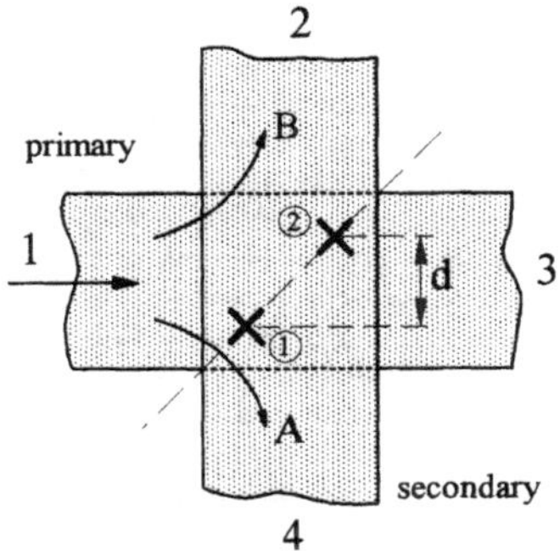

Figure 12.7 Cross-guide coupler using a pair of 45-degree crossed-slot apertures

and

$$\alpha = \tan^{-1}\left(\frac{-jH_{xp}}{H_{zp}}\right) \tag{20b}$$

In the case of aperture 1, substituting the above into the governing equations gives

$$A_1 = \frac{jk_0\eta_0 M_\ell H_{\ell\,\max}}{2\sqrt{2}P}\{H_{xs}\cos\alpha + jH_{zs}\sin\alpha\} - \frac{jk_0 P_e E_{yp} E_{ys}}{4\eta_0 P} \tag{21a}$$

$$B_1 = \frac{jk_0\eta_0 M_\ell H_{\ell\,\max}}{2\sqrt{2}P}\{-H_{xs}\cos\alpha + jH_{zs}\sin\alpha\} - \frac{jk_0 P_e E_{yp} E_{ys}}{4\eta_0 P} \tag{21b}$$

The above hold for aperture 2, but with a 180 degree phase change in the coupling due to the magnetic field. This may be understood by recognising that H_z reverses across the symmetry plane of the waveguide and the x-directed magnetic dipoles of the apertures are in antiphase.

12.8 Circular polarisation in rectangular and ridge waveguides

A scrutiny of the forward and backward waves in the secondary waveguide of the directional coupler indicates that these are closely related to the polarisation of the magnetic field in both waveguides and to the intensities of the fields. One simple solution is to place the aperture where the alternating magnetic field is circularly polarised. This solution is realisable in a regular rectangular waveguide and in a single ridge waveguide. It is not, however, possible in a double ridge waveguide. In the former cases

$$H_{\ell\,\max} = H_{t\,\max} = |H_{xp}| = |H_{zp}| \tag{22a}$$

and

$$\alpha = \frac{\pi}{4} \tag{22b}$$

In the case of narrow slots it is usual to ignore the contribution due to the electric dipoles. This then gives, for apertures 1 and 2,

$$A_1 = \frac{jk_0\eta_0 M_\ell |H_{xp}| H_{xs}}{2P} \tag{23a}$$

$$B_1 = 0 \tag{23b}$$

and

$$A_1 = \frac{-jk_0\eta_0 M_\ell |H_{xp}| H_{xs}}{2P} \tag{24a}$$

$$B_1 = 0 \tag{24b}$$

respectively.

In a rectangular waveguide, the magnitudes of the transmitted waves when two apertures are located at the positions of circular polarisation on either side of the symmetry plane are given by

$$|S_{41}| = \frac{4M_\ell\beta}{ab}\cos^2\left(\frac{\pi d}{2a}\right)\sin(\beta d) \tag{25a}$$

$$|S_{21}| = 0 \tag{25b}$$

In general, the polarisation will only be unity at a single frequency. Elsewhere, it is elliptically polarised and the general formulation in equation (21) must be employed.

12.9 Rectangular and ridge waveguide cross-guide couplers using 45-degree crossed-slot apertures

This section presents some calculations on a number of cross-guide directional couplers using 45-degree crossed-slot apertures in both rectangular and ridge waveguides. To proceed with a calculation it is necessary to fix the location of the apertures. The position of the apertures and the geometry have not, however, been optimised at this time. These are arbitrarily fixed at $x/a = 0.25$ and $w/a = 0.02$.

Figures 12.8–12.10 indicate the relationships between the coupling/directivity and slot length for various combinations of primary and secondary waveguides at three typical frequencies. The internal dimensions of

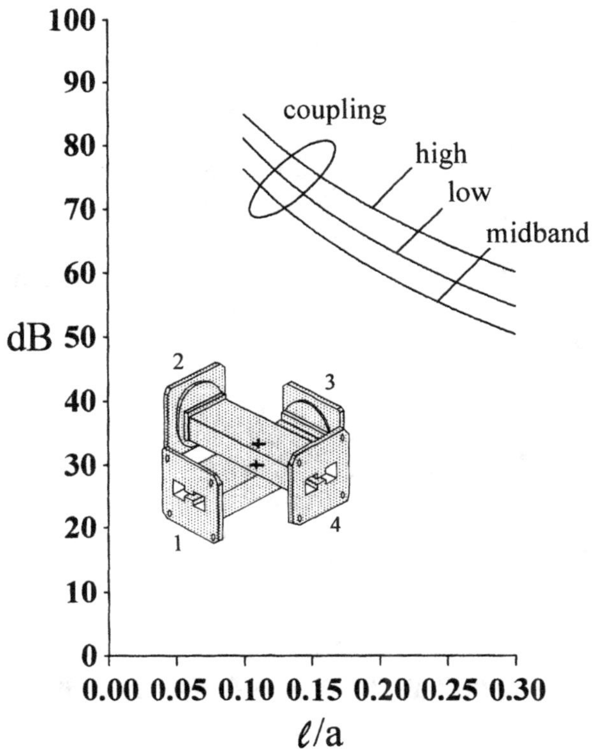

Figure 12.8 Coupling of cross-guide coupler in double ridge waveguide using a pair of 45-degree crossed-slot apertures ($x/a = 0.25$, $w/a = 0.02$)

Reprinted with permission, McKay *et al.* (1999)

the double ridge waveguide (WRD 580) are defined by $a = 19.82$mm, $b/a = 0.474$, $s/a = 0.256$ and $d/b = 0.324$. The single ridge waveguide (WRS 580) is defined by $b/a = 0.469$, $s/a = 0.256$ and $d/b = 0.396$. A comparison between the data in Figures 12.8 and 12.9 indicates that for a given slot length the degree of coupling is largest between two single ridge waveguides. This may be understood by recognising that this arrangement produces the largest values of magnetic fields at the apertures. Figure 12.10 indicates the result in the case of the geometry consisting of either a single and double, or double and single ridge arrangement. Figure 12.11 indicates one typical result in the regular rectangular geometry. Figure 12.12 indicates the directivities of the various arrangements for $\ell/a = 0.25$. The nature of the fields involved in the calculations is indicated in Table 12.2.

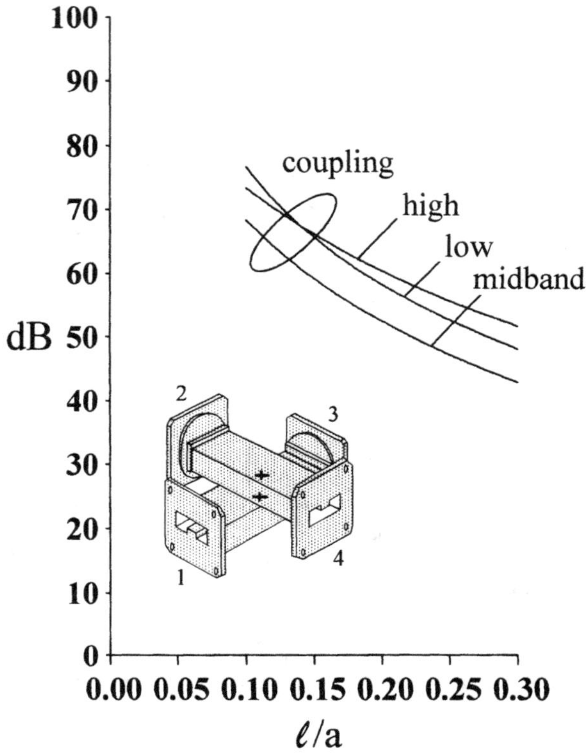

Figure 12.9 Coupling of cross-guide coupler in single ridge waveguide using a pair of 45-degree crossed-slot apertures ($x/a = 0.25$, $w/a = 0.02$)

Reprinted with permission, McKay *et al.* (1999)

12.10 Coupling via waveguide walls of finite thickness

In practice the thickness of the common wall cannot be neglected. One historic approximation to this problem is to represent the aperture by a short section of evanescent waveguide with the appropriate cross-section (Cohn, 1952). The total attenuation of the aperture is then given by

$$\alpha_t(\text{dB}) = L_t + L_a \tag{26}$$

The nature of L_t has its origin in the relationship between E_{in} and E_{out} on a section of transmission line,

$$E_{out} = E_{in} \exp(-\alpha t) \tag{27a}$$

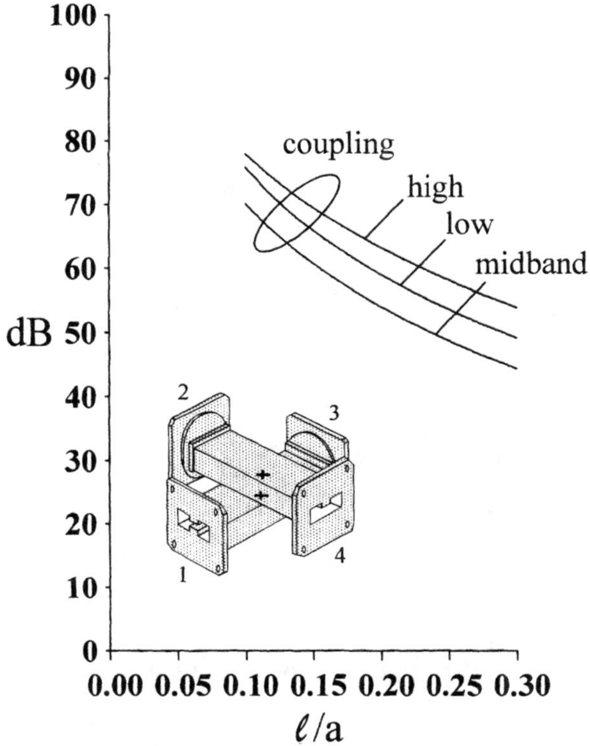

Figure 12.10 Coupling of cross-guide coupler between single ridge/double ridge primary waveguide and double ridge/single ridge secondary waveguide using a pair of 45-degree crossed-slot apertures ($x/a = 0.25$, $w/a = 0.02$)

Reprinted with permission, McKay *et al.* (1999)

where

$$\alpha = \frac{2\pi}{\lambda'_g} \tag{27b}$$

and λ'_g is the waveguide wavelength of the aperture,

$$\left(\frac{2\pi}{\lambda'_g}\right)^2 = \left(\frac{2\pi}{\lambda_c}\right)^2 - \left(\frac{2\pi}{\lambda_0}\right)^2 \tag{27c}$$

The required result is

$$L_t = 54.6A\left(\frac{1}{\lambda_0\lambda_c}\right)(\lambda_0^2 - \lambda_c^2)^{\frac{1}{2}}t \text{ dB} \tag{28}$$

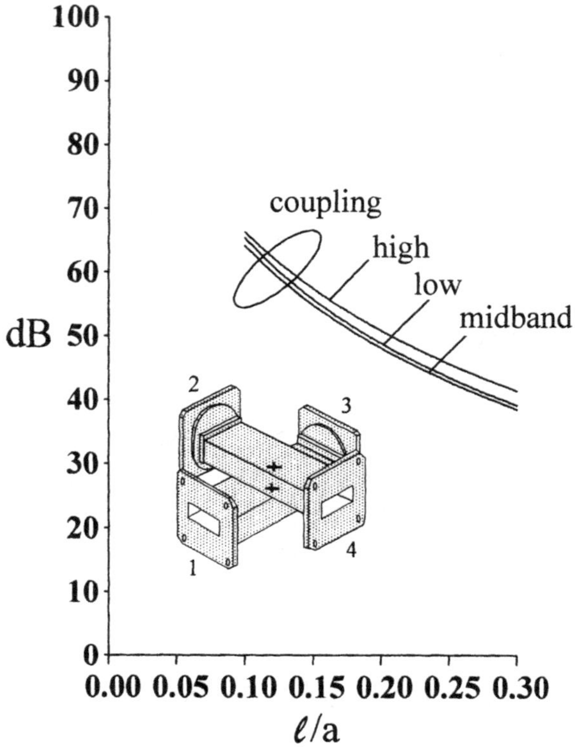

Figure 12.11 Coupling in rectangular waveguide (WR 137 waveguide) cross-guide coupler using a pair of 45-degree crossed-slot apertures ($x/a = 0.25$, $w/a = 0.02$)

Reprinted with permission, McKay *et al.* (1999)

The factor A is an arbitrary constant and accounts for the interaction of the local fields on either side of the wall. L_a accounts for the change in slot attenuation when the resonant length of the slot becomes an appreciable portion of the operating wavelength

$$L_a = -20\log_{10}\left[1 - \left(\frac{\lambda_c}{\lambda_0}\right)^2\right] \text{dB} \tag{29}$$

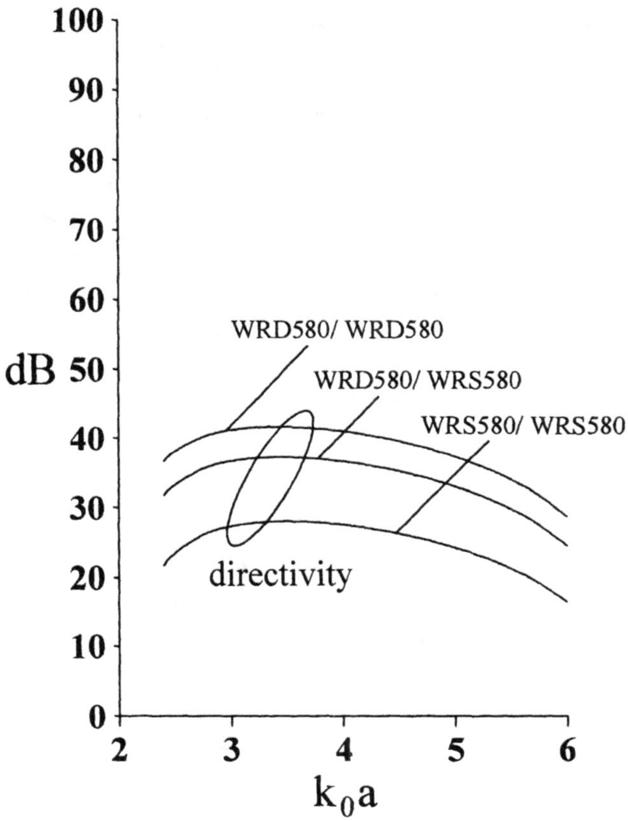

Figure 12.12 Directivities in various ridge couplers ($x/a = 0.25$, $w/a = 0.02$, $\ell/a = 0.25$)

Reprinted with permission, McKay *et al.* (1999)

Table 12.2 Ellipticity of magnetic field in single and double ridge waveguide ($x/a = 0.25$)

	Single ridge		Double ridge	
k_0a	$\frac{H_x}{H_z}$	α	$\frac{H_x}{H_z}$	α
2.407	0.836	0.696	0.192	0.189
4.212	2.577	1.201	0.556	0.507
6.017	3.974	1.324	0.852	0.706

Chapter 13

Directly coupled filter circuits using immittance inverters

13.1 Introduction

The design of 2-port microwave filters relies either on an exact synthesis procedure in the t-plane involving one kind of element separated by UEs or on an approximate s-plane technique involving one kind of element separated by immitance inverters. This sort of topology enters in the realisation of directly coupled bandpass filters using half-wave long cavities connected by metal or inductive posts. The circuit is not canonical since the impedance inverters do not contribute to the overall amplitude response of the filter but only provide a practical layout of the circuit elements. The design is completed by physically realising the impedance inverters. The original immittance inverter took the form of a simple quarter-wave impedance transformer. Such filters are known as quarter-wave coupled. The modern version, in which the inverter is realised by a step discontinuity, is known as a directly coupled arrangement. Since the notion of the immittance inverter is central to the design it is given special attention notwithstanding that it is a classic topic in the literature.

13.2 Immittance inverters

The lowpass all-pole filter prototype, which forms the basis for the highpass, bandpass and bandstop filters, is a Cauer type ladder network, whose topology is not always the most practical circuit layout at very high frequencies. A more desirable filter architecture would be one involving only shunt or only series elements spaced by UEs of commensurate length. Immittance inverters provide one means of replacing a Cauer type lowpass ladder network utilising lumped Ls and Cs by one using only Ls and impedance inverters or one using only Cs and admittance inverters.

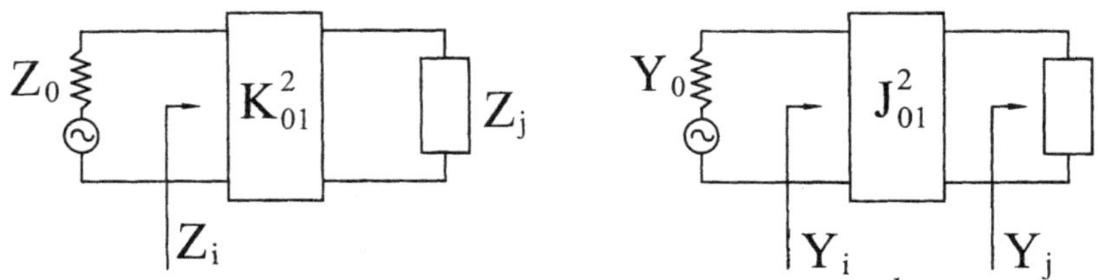

Figure 13.1 Impedance (a) and admittance (b) inverters

The impedance inverter K_{ij} is a frequency invariant 2-port network which transforms an impedance Z_j at one plane into an impedance Z_i at another one:

$$Z_i Z_j = K_{ij}^2 \tag{1}$$

The admittance inverter J_{ij} similarly maps an admittance with one value (Y_j) into another with a value (Y_i) according to

$$Y_i Y_j = J_{ij}^2 \tag{2}$$

The transformations defined by equations (1) and (2) are indicated in Figure 13.1*a* and *b*.

The realisation of some practical lumped element immittance inverters is separately discussed in this chapter.

13.3 Lowpass filters using immittance inverters

Repeated introduction of either the first or second immittance inverter to the lowpass prototype in Figure 13.2*a* maps it to the topology in either Figure 13.2*b* or *c*. The standard topology assumes that the first element of the lowpass prototype is a shunt one. The mapping between the lowpass prototype in Figure 13.2*a* and that in Figure 13.2*b* will now be demonstrated for a degree $n = 3$ network. The derivation starts by expanding the impedance of the network in a first Cauer form,

$$Z_1(s) = \cfrac{1}{sC_1 + \cfrac{1}{sL_2 + \cfrac{1}{sC_3 + 1/1}}} \tag{3}$$

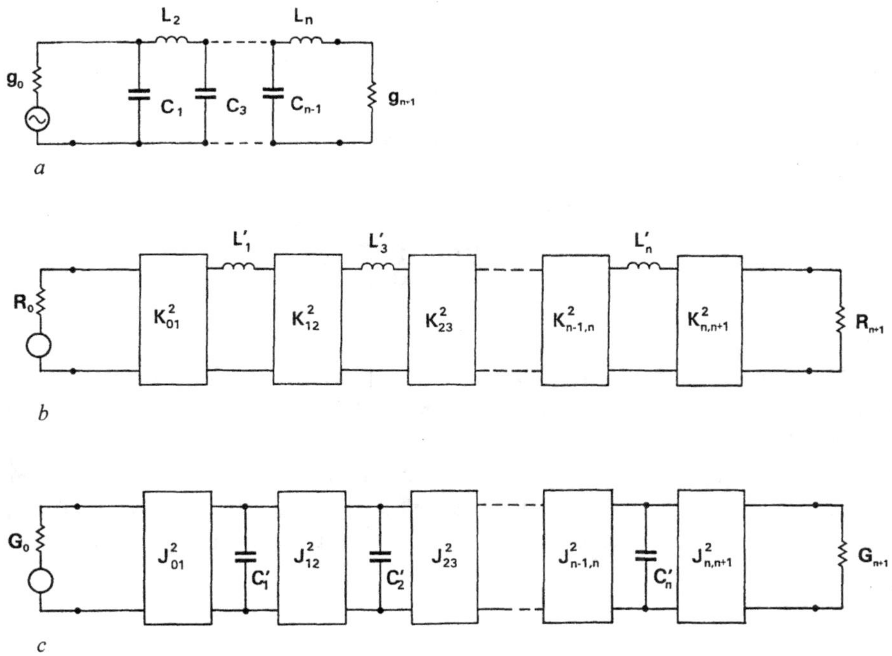

Figure 13.2 Cauer lowpass ladder network using (a) shunt and series elements, (b) series elements and impedance inverters and (c) shunt elements and admittance inverters

The realisation of this impedance function starts by extracting an impedance inverter K_{01} from $Z_1(s)$. This step produces a remainder impedance $Z_1'(s)$ given by

$$Z_1'(s) = \frac{K_{01}^2}{Z_1(s)} \tag{4}$$

Scrutiny of the two preceding equations indicates that the required impedance $Z_1'(s)$ at the output terminals of the immittance inverter is

$$Z_1'(s) = K_{01}^2 \left[sC_1 + \cfrac{1}{sL_2 + \cfrac{1}{sC_3 + \cfrac{1}{1}}} \right] \tag{5}$$

This impedance has the nature of an inductance L' in series with an impedance $Z_2(s)$

$$Z_1'(s) = sL_1' + Z_2(s) \tag{6}$$

where

$$L_1' = K_{01}^2 C_1 \tag{7}$$

and

$$Z_2(s) = \cfrac{K_{01}^2}{sL_2 + \cfrac{1}{sC_3 + \cfrac{1}{1}}} \tag{8}$$

This operation completes the first cycle of the synthesis. It is of note that either L_1' or K_{01} in this arrangement is completely arbitrary. The topology in Figure 13.2*b* is now obtained by repeating this cycle until the degree of the problem is reduced to zero.

The network obtained in this way is scaled to a generator impedance g_0 and a cut-off frequency of 1 rad/s. Making use of the fact that C_1 corresponds to the element g_1 in the lowpass prototype, and impedance scaling,

$$g_1\left(\frac{g_0}{Z_0}\right)$$

indicates that C_1 may be replaced by

$$C_1 \rightarrow g_1\left(\frac{g_0}{Z_0}\right)$$

and that the first inverter K_{01} is fixed according to

$$K_{01} = \sqrt{\frac{L_1' Z_0}{g_0 g_1}} \tag{9}$$

Likewise, making use of the fact that L_2 and C_3 correspond to the elements g_2 and g_3 in the lowpass prototype and impedance scaling these quantities indicate that the former quantities may be replaced by

$$L_2 \rightarrow g_2\left(\frac{Z_0}{g_0}\right)$$

$$C_3 \rightarrow g_3\left(\frac{g_0}{Z_0}\right)$$

and that the impedance inverters K_{12} and K_{23} are set by

$$K_{12} = \sqrt{\frac{L_1' L_2'}{g_1 g_2}} \tag{10}$$

$$K_{23} = \sqrt{\frac{L'_2 L'_3}{g_2 g_3}} \tag{11}$$

respectively.

The general nature for the inner inverters is therefore

$$K_{j,j+1} = \sqrt{\frac{L'_j L'_{j+1}}{g_j g_{j+1}}} \tag{12}$$

Similarly, considerations suggest that the outer inverter is described by

$$K_{n,n+1} = \sqrt{\frac{L'_n Z_{n+1}}{g_n g_{n+1}}} \tag{13}$$

It is of note that the inductors $L'_1, L'_2, \ldots, L'_n$ appearing in the description of the impedance inverters are completely arbitrary. It is also of note that, whereas the network has been impedance scaled, its cut-off frequency is still 1 rad/s. Scrutiny of the entries of the impedance inverters indicates that frequency scaling the network to some other frequency leaves the impedance inverters unchanged. The series inductors are, however, modified ($L'_j \to L'_j/\omega_0$). The possibility of making the impedance inverters the independent variables and the series inductors the dependent ones is also understood.

The derivation of the topology using admittance inverters is left as an exercise for the reader. The result is

$$J_{01} = \sqrt{\frac{C'_1 Y_0}{g_0 g_1}} \tag{14}$$

$$J_{j,j+1} = \sqrt{\frac{C'_j C'_{j+1}}{g_j g_{j+1}}} \tag{15}$$

$$J_{n,n+1} = \sqrt{\frac{C'_n Y_{n+1}}{g_n g_{n+1}}} \tag{16}$$

13.4 Bandpass filters using immittance inverters

The realisation of the bandpass prototype begins by introducing the lowpass frequency transformation in the input impedance of the circuit,

$$s \rightarrow \frac{\omega_0}{BW}\left(\frac{s}{\omega_0} + \frac{\omega_0}{s}\right) \tag{17}$$

The result is

$$Z_1(s) = \cfrac{1}{\cfrac{\omega_0}{BW}\left(\cfrac{s}{\omega_0} + \cfrac{\omega_0}{s}\right)g_1 + \cfrac{1}{\cfrac{\omega_0}{BW}\left(\cfrac{s}{\omega_0} + \cfrac{\omega_0}{s}\right)g_2 + \cfrac{1}{\cfrac{\omega_0}{BW}\left(\cfrac{s}{\omega_0} + \cfrac{\omega_0}{s}\right)g_3 + \cfrac{1}{g_4}}}} \tag{18}$$

This impedance is now synthesised in terms of impedance inverters and series lumped element resonators. If the synthesis starts with the extraction of an impedance inverter K_{01} then the first series lumped element resonator is defined by

$$sL_1' + \frac{1}{sC'} = K_{01}^2\left[\frac{\omega_0}{BW}\left(\frac{s}{\omega_0} + \frac{\omega_0}{s}\right)\right]g_1$$

which satisfies

$$L_1' = \frac{K_{01}^2 g_1}{(BW)} \tag{19}$$

$$C_1' = \frac{(BW)}{K_{01}^2 g_1 \omega_0^2} \tag{20}$$

and

$$\omega_0^2 L_1' C_1' = 1 \tag{21}$$

The first impedance inverter is therefore deduced from either equation (19) or (20) as

$$K_{01}^2 = \frac{L_1'(BW)}{g_1}$$

Impedance scaling this quantity by replacing g_1,

$$g_1 \rightarrow g_1\left(\frac{g_0}{Z_0}\right)$$

gives the required result,

$$K_{01} = \sqrt{\frac{x_1 w Z_0}{g_0 g_1}} \tag{22}$$

w is a bandwidth parameter

$$w = \frac{BW}{\omega_0} \tag{23}$$

and x_j is the reactance slope parameter of the series resonator

$$x_j = \frac{\omega_0}{2} \frac{\partial X_j}{\partial \omega}\bigg|_{\omega=\omega_0} = \omega_0 L'_j \tag{24}$$

The definitions of the other inverters follow in a like manner:

$$K_{j,j+1} = w\sqrt{\frac{x_j x_{j+1}}{g_j g_{j+1}}} \tag{25}$$

$$K_{n,n+1} = \sqrt{\frac{x_n w Z_{n+1}}{g_n g_{n+1}}} \tag{26}$$

Z_0 and Z_{n+1} are the input and output impedances, respectively. Once the immittance inverters are set from a knowledge of L'_j, C'_j is calculated from the resonance condition in equation (21).

The dual equations for the arrangement in Figure 13.3*c* employing admittance inverters are

$$J_{01} = \sqrt{\frac{b_1 w Y_0}{g_0 g_1}} \tag{27}$$

$$J_{j,j+1} = w\sqrt{\frac{b_j b_{j+1}}{g_j g_{j+1}}} \tag{28}$$

$$J_{n,n+1} = \sqrt{\frac{b_n w Y_{n+1}}{g_n g_{n+1}}} \tag{29}$$

and b_j is the susceptance slope parameter of the shunt resonator,

$$b_j = \frac{\omega_0}{2} \frac{\partial Y_j}{\partial \omega}\bigg|_{\omega=\omega_0} = \omega_0 C'_j \tag{30}$$

Y_0 and Y_{n+1} are the input and output conductances, respectively.

13.5 Immittance inverters

Four immittance inverters are illustrated in Figure 13.4. Although each circuit requires negative elements for its realisation these can be absorbed

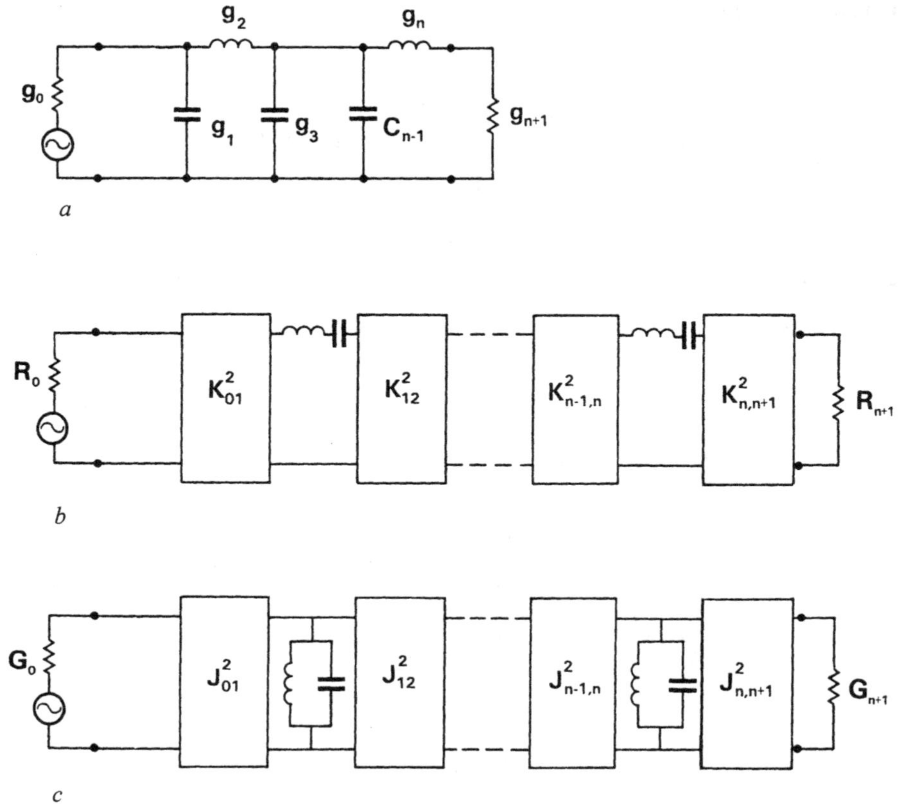

Figure 13.3 Cauer lowpass ladder network using (a) shunt and series elements and (b) series elements and impedance inverters, and (c) shunt elements and admittance inverters

in adjacent positive elements. The equivalence between any of these circuits and an ideal immittance inverter consisting of a UE of characteristic impedance K will now be demonstrated. This may be done by establishing a one-to-one equivalence between the ABCD parameters of the two topologies. The derivation begins with the definition of the ABCD matrix of a uniform transmission line of characteristic impedance Z_0 and electric length θ:

$$\begin{bmatrix} A & B \\ C & D \end{bmatrix} = \begin{bmatrix} \cos\theta & jZ_0 \sin\theta \\ \dfrac{j\sin\theta}{Z_0} & \cos\theta \end{bmatrix} \tag{31}$$

Evaluating this relationship at

$$\theta = 90^\circ$$

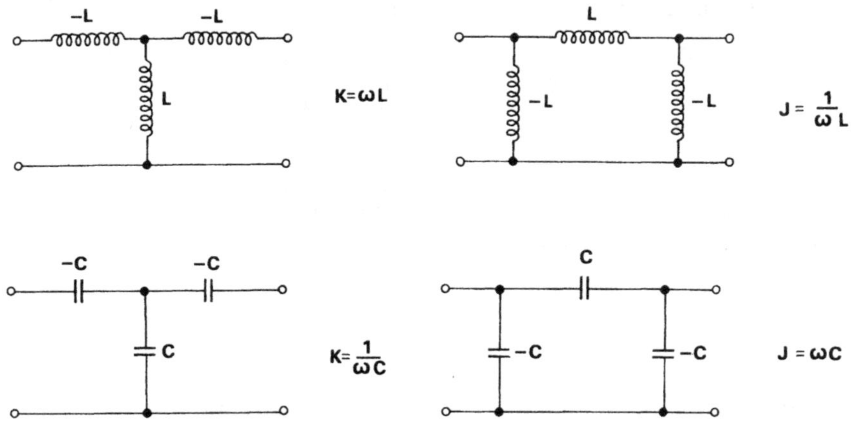

Figure 13.4 Schematic diagrams of immittance inverters

gives

$$\begin{bmatrix} A & B \\ C & D \end{bmatrix} = \begin{bmatrix} 0 & jK \\ \dfrac{j}{K} & 0 \end{bmatrix} \tag{32}$$

where

$$K = Z_0 \tag{33}$$

The derivation continues by forming a one-to-one equivalence between such a unit element and any of the possible circuits illustrated in Figure 13.4. Taking the T network consisting of series impedances (Z) and a shunt admittance (Y) by way of an example and recalling the nature of the overall ABCD matrix of such a network gives:

$$A = 1 + ZY \tag{34a}$$

$$B = Z(2 + ZY) \tag{34b}$$

$$C = Y \tag{34c}$$

$$D = 1 + ZY \tag{34d}$$

In the situation considered here,

$$Z = \frac{1}{j\omega C} \tag{35a}$$

$$Y = j\omega C \tag{35b}$$

Evaluating the ABCD parameters under these conditions indicates that

$$\begin{bmatrix} A & B \\ C & D \end{bmatrix} = \begin{bmatrix} 0 & \dfrac{j}{\omega C} \\ j\omega C & 0 \end{bmatrix} \tag{36}$$

A comparison between the entries of this matrix and that of the ideal impedance inverter suggests that the two are equivalent, provided

$$K = \frac{1}{\omega C} \tag{37}$$

The equivalences between the topologies of the other possible inverters and a quarter-wave long UE are understood easily.

13.6 Practical inverter

A much used practical immittance inverter is a T or π circuit with positive elements embedded between negative lengths of uniform transmission lines. The negative lengths of line are then absorbed in any connecting line. Such T and π circuits are often met in the descriptions of metal posts, irises or similar discontinuities in waveguides or other transmission lines. In practice, the series elements in the T circuit and the shunt ones in the π circuit are often neglected. The required conditions for the T circuits in Figure 13.5 will now be derived under this assumption. Forming the ABCD matrix of this circuit disregarding the series elements gives

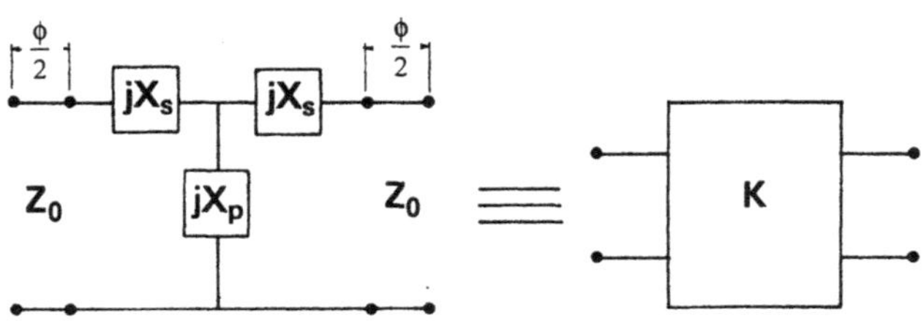

Figure 13.5 Schematic diagram of practical impedance inverter

$$K = Z_0 \left| \tan \frac{\phi}{2} + \tan^{-1}\left(\frac{X_s}{Z_0}\right) \right|$$

$$\phi = -\tan^{-1}\left(\frac{2X_p}{Z_0} + \frac{X_s}{Z_0}\right) - \tan^{-1}\left(\frac{X_s}{Z_0}\right)$$

$$\begin{bmatrix} A & B \\ C & D \end{bmatrix} = \begin{bmatrix} \cos\phi - \dfrac{\sin\phi}{2X_p/Z_0} & -jZ_0\left(\sin\phi + \dfrac{\cos\phi - 1}{2X_p/Z_0}\right) \\ \dfrac{j}{Z_0}\left(\sin\phi + \dfrac{\cos\phi + 1}{2X_p/Z_0}\right) & \cos\phi - \dfrac{\sin\phi}{2X_p/Z_0} \end{bmatrix} \tag{38}$$

Comparing this matrix with that of the ideal impedance inverter indicates that

$$\frac{j}{Z_0}\left(\sin\phi + \frac{\cos\phi + 1}{2X_p/Z_0}\right) = \frac{j}{K} \tag{39a}$$

$$\cos\phi - \frac{\sin\phi}{2X_p/Z_0} = 0 \tag{39b}$$

or

$$\tan\phi = \frac{2X_p}{Z_0} \tag{40}$$

$$\frac{K}{Z_0} = \left|\tan\frac{\phi}{2}\right| \tag{41}$$

Eliminating ϕ between these two relationships also gives

$$\frac{Z_0}{X_p} = \frac{Z_0}{K} - \frac{K}{Z_0} \tag{42}$$

The dual problem is illustrated in Figure 13.6.

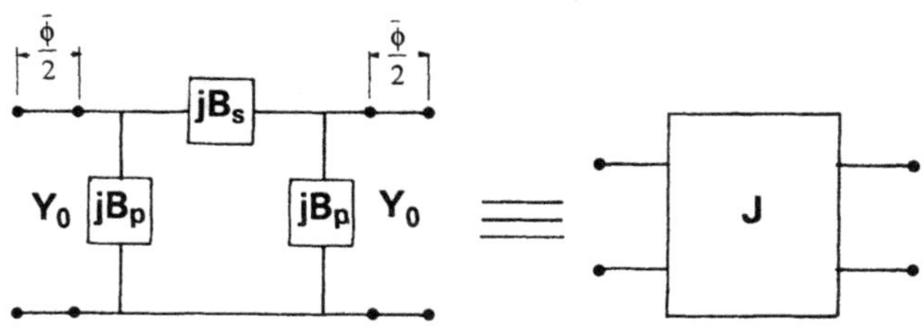

Figure 13.6 Schematic diagram of practical admittance inverter

$$J = Y_0\left|\tan\frac{\phi}{2} + \tan^{-1}\left(\frac{B_p}{Y_0}\right)\right|$$

$$\phi = \tan^{-1}\left(\frac{2B_s}{Y_0} + \frac{B_p}{Y_0}\right) - \tan^{-1}\left(\frac{B_p}{Y_0}\right)$$

13.7 Immittance inverters using evanescent mode waveguide

One means of realising an immittance inverter is to embody an evanescent mode waveguide between its fictitious transmission lines. The purpose of this section is to deduce the equivalent circuit of this sort of waveguide. One possibility is the π-arrangement comprising three inductors depicted in Figure 13.7*a*; the other is the dual T-arrangement in Figure 13.7*b*. The derivation of the π-topology starts by using the standard identities for the branch elements of the structure:

$$Y_1 = Y_2 = Y_0 \coth\left(\frac{\gamma\ell}{2}\right) \tag{43}$$

$$Y_3 = Y_0 \sinh(\gamma\ell) \tag{44}$$

Y_0 is the wave admittance of the waveguide, γ is its propagation constant and ℓ is the length of a typical section.

It proceeds by noting that the propagation constant (γ) of a waveguide below cut-off is real for all frequencies from the origin to the cut-off frequency:

$$\gamma^2 = \left(\frac{2\pi}{\lambda_c}\right)^2 - \left(\frac{2\pi}{\lambda_0}\right)^2 \tag{45}$$

It is also recognised that the waveguide admittance (Y_0) is a pure imaginary quantity under the same circumstances:

$$Y_0 = \frac{\gamma}{-j\omega\mu_0} \tag{46}$$

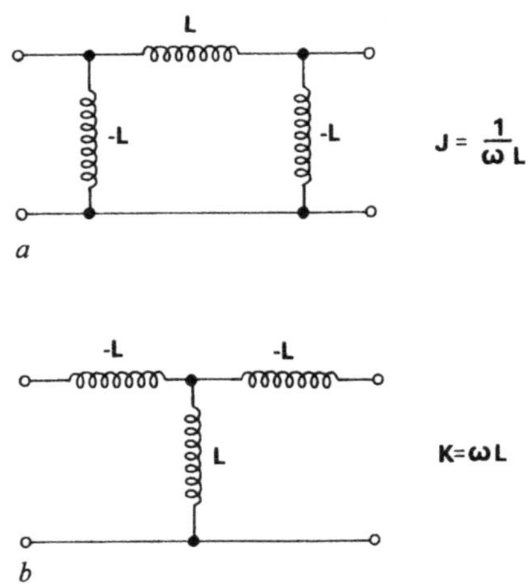

Figure 13.7 π-equivalent circuit (a) and T-equivalent circuit (b) of cut-off waveguide

Introducing these relationships into the branch immittances in equations (43) and (44) of the π prototype indicates that each element is a nearly frequency-independent inductance as asserted.

This result is obtained by first expanding the hyperbolic function in series form:

$$\sinh A = A + \frac{A^3}{3!} + \frac{A^5}{5!}$$

$$\coth A = \frac{1}{A - \frac{A^3}{3} + \frac{2A^5}{15}}$$

The derivation of the element values of the T-equivalent circuit proceeds in a dual manner.

13.8 E-plane filter

A classic directly coupled bandpass filter is a ridge or finline arrangement based on a lowpass lumped element prototype consisting of a series of half-wave long cavities connected by metal or inductive posts. Figure 13.8 indicates one practical construction. One possible equivalent circuit for this type of structure is illustrated in Figure 13.9. It is obtained by representing each discontinuity by an equivalent inductive T-circuit and each cavity by a uniform transmission line. The design of this type of filter proceeds by absorbing the reactive T-circuits into ideal impedance inverters prior to forming a one-to-one equivalence between it and a suitable bandpass prototype consisting of series resonators and ideal impedance inverters. The series

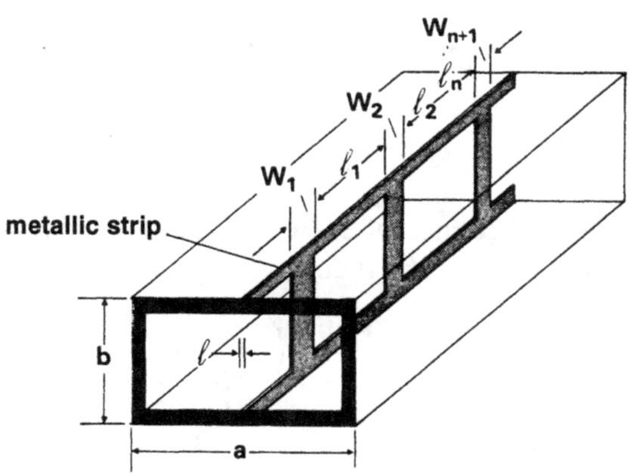

Figure 13.8 Schematic diagram of waveguide bandpass filters using metal septa

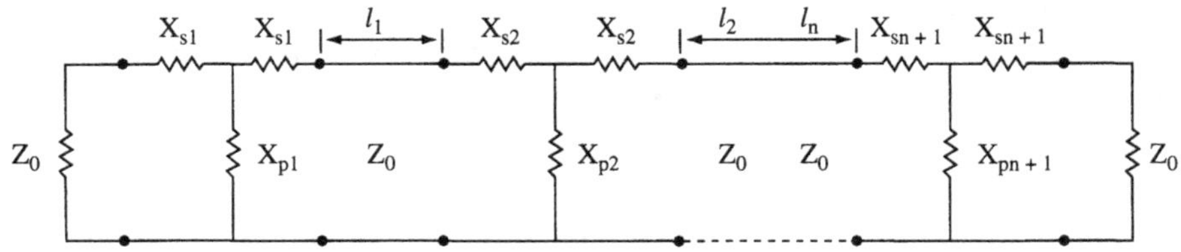

Figure 13.9 Equivalent circuit of bandpass filter using metal septa

resonators of the bandpass prototype are in this instance implemented by half-wave long distributed cavity resonators. The impedance inverters of the network are often realised by neglecting the series elements of the metal posts and embedding the shunt ones between suitable negative lengths of uniform transmission lines in the manner discussed in the preceding section. These additional lengths of transmission lines are then absorbed in the half-wave long cavities. The design is complete once the *K*s are specified from the filter specification.

Combining equations (22), (25) and (26) with equation (42) gives the required design equations

$$\frac{Z_0}{X_{01}} = Z_0\left(\frac{g_0 g_1}{x_1 w Z_0}\right)^{\frac{1}{2}} - \frac{1}{Z_0}\left(\frac{x_1 w Z_0}{g_0 g_1}\right)^{\frac{1}{2}} \tag{47a}$$

$$\frac{Z_0}{X_{j,j+1}} = Z_0\left(\frac{g_j g_{j+1}}{x_j x_{j+1} w^2}\right)^{\frac{1}{2}} - \frac{1}{Z_0}\left(\frac{x_j x_{j+1} w^2}{g_j g_{j+1}}\right)^{\frac{1}{2}} \tag{47b}$$

$$\frac{Z_0}{X_{n,n+1}} = Z_0\left(\frac{g_n g_{n+1}}{x_n w Z_{n+1}}\right)^{\frac{1}{2}} - \frac{1}{Z_0}\left(\frac{x_n w Z_0}{g_n g_{n+1}}\right)^{\frac{1}{2}} \tag{47c}$$

The reactance slope parameter of a half-wave section of characteristic impedance Z_j is

$$X_j = \frac{\pi Z_j}{2} \tag{48}$$

If the characteristic impedance of each half-wave section is made equal to the characteristic impedance Z_0, then

$$\frac{K_{01}^2}{Z_0^2} = \left(\frac{\pi}{2}\right)\left(\frac{w}{g_0 g_1}\right) \tag{49a}$$

$$\frac{K_{j,j+1}^2}{Z_0^2} = \left(\frac{\pi}{2}\right)^2\left(\frac{w^2}{g_j g_{j+1}}\right) \tag{49b}$$

$$\frac{K_{n,n+1}^2}{Z_0^2} = \left(\frac{\pi}{2}\right)\left(\frac{w}{g_n g_{n+1}}\right) \tag{49c}$$

and

$$\frac{Z_0}{X_{01}} = \left(\frac{2g_0 g_1}{\pi w}\right)^{\frac{1}{2}} - \left(\frac{\pi w}{2g_0 g_1}\right)^{\frac{1}{2}} \tag{50a}$$

$$\frac{Z_0}{X_{j,j+1}} = \frac{2}{\pi}\left(\frac{g_j g_{j+1}}{w^2}\right)^{\frac{1}{2}} - \frac{\pi}{2}\left(\frac{w^2}{g_j g_{j+1}}\right)^{\frac{1}{2}} \tag{50b}$$

$$\frac{Z_0}{X_{n,n+1}} = \left(\frac{Z_0}{Z_{n+1}}\right)^{\frac{1}{2}}\left(\frac{2g_n g_{n+1}}{\pi w}\right)^{\frac{1}{2}} - \left(\frac{Z_{n+1}}{Z_0}\right)^{\frac{1}{2}}\left(\frac{\pi w}{2g_n g_{n+1}}\right)^{\frac{1}{2}} \tag{50c}$$

The reactance of each step may therefore be evaluated once the immittance inverters have been fixed by the specification of the filter. This fixes the dimensions of the steps.

Once the details of the steps have been evaluated it only remains to calculate the spacing between the discontinuities. This condition is met provided

$$\left(\frac{2\pi}{\lambda_g}\right)\ell_j = \theta_j + \frac{1}{2}(\phi_j + \phi_{j+1}) \tag{51a}$$

with

$$\theta_j = \pi \tag{51b}$$

In calculating θ_j and θ_{j+1} the exact equivalent circuit of the discontinuity is usually employed:

$$\phi_j = -\tan^{-1}\left(\frac{2X_{pj}}{Z_0} + \frac{X_{sj}}{Z_0}\right) - \tan^{-1}\left(\frac{X_{sj}}{Z_0}\right) \tag{51c}$$

The overall structure of this filter in terms of immittance inverters and half-wave long waveguide cavities is indicated in Figure 13.10.

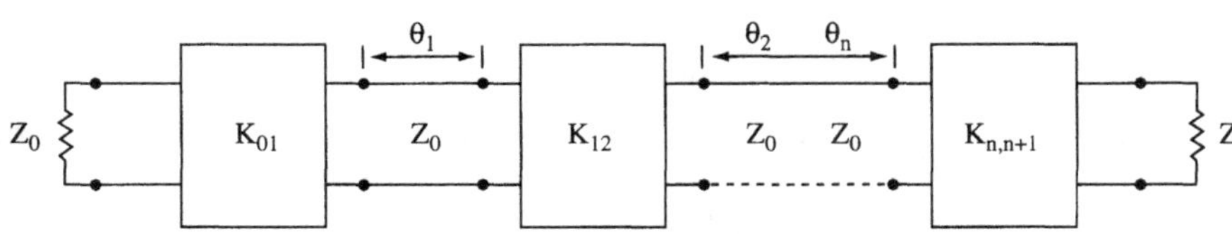

Figure 13.10 Equivalent circuit of bandpass filter using impedance inverters

13.9 Element values of lowpass prototypes

The element values for lowpass filter networks with Butterworth and Chebȳshev amplitude approximations are classic problems in the literature. The order of the lowpass Chebȳshev solution is fixed by the attenuation α_1 at ω_1 in the stopband:

$$n = \frac{\cosh^{-1}(\sqrt{A-1}/\varepsilon)}{\cosh^{-1}\left(\dfrac{\omega}{\omega_1}\right)} \tag{52a}$$

where

$$A = \log_{10}^{-1}\left(\frac{\alpha_1}{10}\right) \tag{52b}$$

ε is the ripple level in the passband.

The order of the bandpass filter is specified by

$$n = \frac{\cosh^{-1}\left(\dfrac{\sqrt{A-1}}{\varepsilon}\right)}{\cosh^{-1}\left[\dfrac{\omega_0}{BW}\left(\dfrac{\omega_{3,4}}{\omega_0} - \dfrac{\omega_0}{\omega_{3,4}}\right)\right]} \tag{53}$$

The quantity is fixed by the definition of its upper bandedge.

The recurrence formula for the element values of the Chebȳshev solution is

$$g_1 = \frac{2\sin\left(\dfrac{\pi}{2n}\right)}{\sinh\left(\dfrac{1}{n}\sinh^{-1}\dfrac{1}{\varepsilon}\right)} \tag{54a}$$

and

$$g_i g_{i+1} = \frac{4\sin\left(\dfrac{2i-1}{2n}\pi\right)\sin\left(\dfrac{2i+1}{2n}\pi\right)}{\sinh^2\left(\dfrac{1}{n}\sinh^{-1}\dfrac{1}{\varepsilon}\right) + \sin^2\left(\dfrac{i\pi}{n}\right)} \qquad i = 1, 2, \ldots, n-1 \tag{54b}$$

The filter is symmetrical for n odd.

$$g_1 = g_n, g_2 = g_{n-1}, \ldots, g_i = g_{n-i+1} \cdots \tag{55}$$

If n is odd, the terminating resistance g_{n+1} is

$$g_{n+1} = 1 \tag{56}$$

If n is even, the terminating resistance g_{n+1} is

$$g_{n+1} = (\varepsilon + \sqrt{1+\varepsilon^2})^2 \tag{57}$$

or

$$g_{n+1} = \frac{1}{(\varepsilon + \sqrt{1+\varepsilon^2})^2} \tag{58}$$

13.10 Frequency response of microwave filters

The ABCD notation is admirably suited for the analysis of microwave filters. The transmission (τ) and reflection (ρ) parameters of a reactance network are once more given by

$$\tau\tau^* = \frac{1}{1 + \frac{1}{4}(A-D)^2 - \frac{1}{4}(B-C)^2} \tag{59}$$

$$\rho\rho^* = \frac{\frac{1}{4}(A-D)^2 - \frac{1}{4}(B-C)^2}{1 + \frac{1}{4}(A-D)^2 - \frac{1}{4}(B-C)^2} \tag{60}$$

Scrutiny of these two relationships indicates that these satisfy the unitary condition,

$$\rho\rho^* + \tau\tau^* = 1 \tag{61}$$

The nature of these parameters is compatible with the definition of the characteristic function met in the description of the transfer function of this class of networks:

$$K(\omega^2) = \tfrac{1}{4}(A-D)^2 - \tfrac{1}{4}(B-C)^2 \tag{62}$$

It is also recalled that, for reactance networks, A and D are real numbers and B and C are pure imaginary ones. For symmetric networks $A = D$.

As a simple illustration of the analysis of this type of circuit using the ABCD notation it is used to construct the insertion loss function of a low-pass $n = 3$ doubly terminated prototype. Here

$$A = 1 - 2\omega^2$$

$$B = j2\omega$$

$$C = j(2\omega - 2\omega^3)$$

$$D = 1 - 2\omega^2$$

Introducing these quantities into equation (59) gives the required result,

$$\tau\tau^* = \frac{1}{1 + (\omega^2)^3}$$

in keeping with the Butterworth lowpass generating function and the definition of the characteristic function.

Chapter 14

Ridge waveguide filter design using mode matching method

M. McKay and J. Helszajn

14.1 Introduction

The specification of a conventional waveguide filter is usually incorporated in a ladder lumped element network which is then realised in terms of immittance inverters and one kind of element. The synthesis of this type of topology is a standard problem in the literature. A typical inverter is realised by introducing a suitable discontinuity into the waveguide and embedding it into negative line lengths. This is usually achieved by foreshortening the lengths of each cavity. To proceed with the design, it is necessary to have some representation of the discontinuities involved. One means of characterising discontinuities in waveguides is the MMM (mode matching method). To overcome the effects of interaction between discrete discontinuities on the overall filter performance a global transmission matrix is constructed and its overall specification is optimised by resorting to a suitable optimisation subroutine. The chapter includes the layout of one lowpass filter which relies on cut-off waveguide sections for its inverters. It separately includes the design of one bandpass filter using cut-off waveguide sections for its inverters and another employing inductive septa for the realisation of the inverters.

14.2 Mode matching method

To proceed with the design, it is necessary to have some representation of the discontinuities involved. One classic numerical procedure is the mode matching method (MMM). It involves expanding the forward and reflected fields in each region in terms of local modes and thereafter satisfying the boundary conditions at the junction between the two. One solution is to retain the same number of modes on either side of the discontinuity. The application of the orthogonality condition of each family of modes in each region produces a matrix equation from which the unknown coefficients may be evaluated. The modes in each waveguide may be established by using the FEM or some other closed form formulation. The scattering matrix of the discontinuity may be evaluated once the fields in both waveguides are at hand.

While the topology of the E-plane metal filter is not strictly speaking a ridge structure, it has some of its features. Since it represents a classic application of the mode matching method, its topology will be used to outline the approach. It amounts to forming the fields in each region on either side of the discontinuity. Figure 14.1 illustrates the problem region in question.

The transverse fields to the left of the discontinuity ($z < 0$) are given by

$$E_y = \sum_{n=1}^{M} [A_n^+ \phi_{an}(x) \exp(-\gamma_{an} z) + A_n^- \phi_{an}(x) \exp(\gamma_{an} z)] \tag{1}$$

$$H_x = \sum_{n=1}^{M} Y_{an}[A_n^+ \phi_{an}(x) \exp(-\gamma_{an} z) - A_n^- \phi_{an}(x) \exp(\gamma_{an} z)] \tag{2}$$

$\phi_{an}(x)$ represents a typical normal mode, $A_n^\pm$ represent the amplitudes of the forward and backward waves and are the unknowns of the problem region, and γ_{an} is the propagation constant. The wave admittance is

$$Y_{an} = \frac{\gamma_{an}}{j\omega\mu_0} \tag{3}$$

The normal modes satisfy the orthogonality condition

$$\int A_m \phi_{am}(x) A_n \phi_{an}(x) \, \mathrm{d}x = \delta_{mn} \tag{4}$$

$\delta_{mn} = 1$ for $m = n$, and 0 for $m \neq n$.

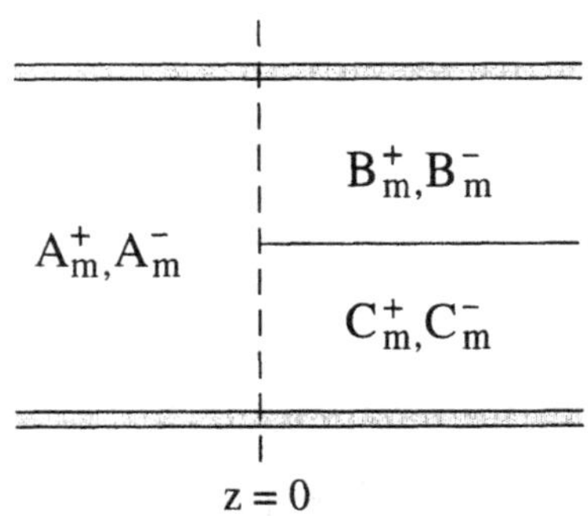

Figure 14.1 Bifurcated waveguide

This condition implies that the amplitude squared of a typical mode integrated over the problem region is unity and that the integral of the product of two different modes over the same surface is zero.

If the geometry in question also supports a discontinuity along the y-direction, then both E_x and H_y exist. This is the situation met in the case of a ridge waveguide.

The fields to the right of the discontinuity ($z > 0$) are given by

$$E_y = \sum_{n=1}^{K} [B_n^+ \phi_{bn}(x) \exp(-\gamma_{bn} z) + B_n^- \phi_{bn}(x) \exp(\gamma_{bn} z)] \tag{5}$$

$$H_x = \sum_{n=1}^{K} Y_{bn}[B_n^+ \phi_{bn}(x) \exp(-\gamma_{bn} z) - B_n^- \phi_{bn}(x) \exp(\gamma_{bn} z)] \tag{6}$$

and

$$E_y = \sum_{n=1}^{L} [C_n^+ \phi_{cn}(x) \exp(-\gamma_{cn} z) + C_n^- \phi_{cn}(x) \exp(\gamma_{cn} z)] \tag{7}$$

$$H_x = \sum_{n=1}^{L} Y_{cn}[C_n^+ \phi_{cn}(x) \exp(-\gamma_{cn} z) - C_n^- \phi_{cn}(x) \exp(\gamma_{cn} z)] \tag{8}$$

for regions $0 < x < b$ and $b < x < a$, respectively,

ϕ_{bn} and ϕ_{cn} are the normal modes in regions B and C, and $B^\pm$ and $C^\pm$ are the amplitudes of the forward and backward modes in the two regions and are the other unknowns of the problem region. The normal modes again satisfy the orthogonality condition. The wave admittances are defined in a like manner to that in region A.

The derivation of the required result now proceeds by satisfying the continuity conditions at the plane of the discontinuity:

$$\sum_{n=1}^{M} (A_n^+ + A_n^-)\phi_{an}(x) = \sum_{n=1}^{K} (B_n^+ + B_n^-)\phi_{bn}(x) \qquad 0 < x < b \tag{9}$$

$$\sum_{n=1}^{M} (A_n^+ - A_n^-)Y_{an}\phi_{an}(x) = \sum_{n=1}^{K} (B_n^+ - B_n^-)Y_{bn}\phi_{bn}(x) \qquad 0 < x < b \tag{10}$$

$$\sum_{n=1}^{M} (A_n^+ + A_n^-)\phi_{an}(x) = \sum_{n=1}^{L} (C_n^+ + C_n^-)\phi_{cn}(x) \qquad b < x < a \tag{11}$$

$$\sum_{n=1}^{M} (A_n^+ - A_n^-)Y_{an}\phi_{an}(x) = \sum_{n=1}^{L} (C_n^+ - C_n^-)Y_{cn}\phi_{cn}(x) \qquad b < x < a \tag{12}$$

Each preceding equation constitutes a doubly infinite set of linear equations for the unknown modal coefficients. The required linear equations for the unknown coefficients are now simplified by making use of the orthogonality condition between the modes in each region one at a time.

Multiplying both sides of each continuity condition in the interval $0 < x < b$ by ϕ_{am} and integrating over the problem region gives

$$(A_m^+ + A_m^-) = \sum_{n=1}^{K} H_{mn}(B_n^+ + B_n^-) \tag{13}$$

$$Y_{an}(A_m^+ - A_m^-) = \sum_{n=1}^{K} Y_{bn} H_{mn}(B_n^+ - B_n^-) \tag{14}$$

where

$$H_{mn} = \int_0^b \phi_{am}(x)\phi_{bn}(x)\,\mathrm{d}x \tag{15}$$

The corresponding result in the interval $b < x < a$ is

$$(A_m^+ + A_m^-) = \sum_{n=1}^{K} Y_{cn} K_{mn}(C_n^+ + C_n^-) \tag{16}$$

$$Y_{an}(A_m^+ - A_m^-) = \sum_{n=1}^{K} Y_{cn} K_{mn}(C_n^+ - C_n^-) \tag{17}$$

where

$$K_{mn} = \int_b^a \phi_{am}(x)\phi_{cn}(x)\,\mathrm{d}x \tag{18}$$

The two preceding pairs of equations may be combined into a single condition over the complete interval $0 < x < a$ of the problem region:

$$A_m^+ + A_m^- = \sum_{n=1}^{K} H_{mn}(B_n^+ + B_n^-) + \sum_{n=1}^{L} K_{mn}(C_n^+ + C_n^-), \quad m = 1, 2, \ldots, M \tag{19}$$

$$Y_{am}(A_m^+ - A_m^-) = \sum_{n=1}^{K} Y_{bn} H_{mn}(B_n^+ - B_n^-) + \sum_{n=1}^{L} Y_{cn} K_{mn}(C_n^+ - C_n^-), \quad m = 1, 2, \ldots, M \tag{20}$$

Applying the orthogonality conditions of the modes in regions B and C likewise gives

$$\sum_{n=1}^{M} H_{mn}(A_n^+ + A_n^-) = B_m^+ + B_m^-, \qquad m = 1, 2, \ldots, K \tag{21}$$

$$\sum_{n=1}^{M} H_{mn} Y_{an}(A_n^+ - A_n^-) = Y_{bn}(B_m^+ - B_m^-), \qquad m = 1, 2, \ldots, K \tag{22}$$

$$\sum_{n=1}^{M} K_{mn}(A_n^+ + A_n^-) = C_m^+ + C_m^-, \qquad m = 1, 2, \ldots, L \tag{23}$$

$$\sum_{n=1}^{M} K_{mn} Y_{an}(A_n^+ - A_n^-) = Y_{cn}(C_m^+ - C_m^-), \qquad m = 1, 2, \ldots, L \tag{24}$$

The preceding six simultaneous equations may now be solved for the family of unknown coefficients $A_n^+, A_n^-, B_n^+, B_n^-, C_n^+$ and C_n^-. Writing these relationships in matrix form gives

$$\mathbf{A}^+ + \mathbf{A}^- = \mathbf{H}(\mathbf{B}^+ + \mathbf{B}^-) + \mathbf{K}(\mathbf{C}^+ + \mathbf{C}^-) \tag{25}$$

$$\mathbf{A}^+ - \mathbf{A}^- = \mathbf{Z}_a\mathbf{H}\mathbf{Y}_b(\mathbf{B}^+ - \mathbf{B}^-) + \mathbf{Z}_a\mathbf{K}\mathbf{Y}_c(\mathbf{C}^+ - \mathbf{C}^-) \tag{26}$$

$$\mathbf{H}^{\mathrm{T}}(\mathbf{A}^+ + \mathbf{A}^-) = \mathbf{B}^+ + \mathbf{B}^- \tag{27}$$

$$\mathbf{Z}_b\mathbf{H}^{\mathrm{T}}\mathbf{Y}_z(\mathbf{A}^+ - \mathbf{A}^-) = \mathbf{B}^+ - \mathbf{B}^- \tag{28}$$

$$\mathbf{K}^{\mathrm{T}}(\mathbf{A}^+ + \mathbf{A}^-) = \mathbf{C}^+ + \mathbf{C}^- \tag{29}$$

$$\mathbf{Z}_c\mathbf{K}^{\mathrm{T}}\mathbf{Y}_z(\mathbf{A}^+ - \mathbf{A}^-) = \mathbf{C}^+ - \mathbf{C}^- \tag{30}$$

The **A**s and **B**s are column vectors given by

$$\mathbf{A}^+ = \begin{bmatrix} A_1^+ \\ A_2^+ \\ \vdots \\ A_M^+ \end{bmatrix}, B^+ = \begin{bmatrix} B_1^+ \\ B_2^+ \\ \vdots \\ B_K^+ \end{bmatrix}, C^+ = \begin{bmatrix} C_1^+ \\ C_2^+ \\ \vdots \\ C_L^+ \end{bmatrix}$$

$$\mathbf{A}^- = \begin{bmatrix} A_1^- \\ A_2^- \\ \vdots \\ A_M^- \end{bmatrix}, B^- = \begin{bmatrix} B_1^- \\ B_2^- \\ \vdots \\ B_K^- \end{bmatrix}, C^- = \begin{bmatrix} C_1^- \\ C_2^- \\ \vdots \\ C_L^- \end{bmatrix}$$

H is a matrix of size $M \times K$ with generic element H_{mn} as defined before, while **K** is a matrix of size $M \times L$ with generic element K_{mn}. The superscript T denotes the transpose operation, and Y_i and $Z_i (i = a, b, c)$ are diagonal matrices with diagonal elements Y_{in} and $Z_{in} (i = a, b, c)$.

A microwave passive network may be represented by an immittance (impedance or admittance) matrix, a scattering matrix, an ABCD formulation or a transmission matrix. There is therefore more than one possible formulation of the problem region under discussion. The reader is referred to the original literature for more details. Methods which avoid repetitive calculations for each sampled frequency in the frequency interval of the specification are of special interest. Immittance matrices are appropriate for the synthesis of equivalent circuits, transmission or ABCD ones for the construction of the overall representation of a number of discontinuities in cascade, and the scattering formulation is suitable for the construction of the overall frequency response.

A feature of the matrices and vectors not considered so far is that these are of different size and that some are not even square. This difficulty may be alleviated by assuming that $K + L = M$ and by partitioning the same. It is, however, of note that there is more than one way to solve the ensuing family of matrix relationships.

14.3 MMM characterisation of 1-port networks

While the MMM permits all the network parameters of a passive circuit to be deduced, the development outlined here is restricted to its parameters at a typical port. This is done for both brevity and clarity. One port variables are in fact all that is needed to characterise symmetrical 2-port networks.

The detailed development of the 1-port problem region begins by putting down the relationships between the incident and reflected waves in regions B and C in the problem region in question:

$$\mathbf{B}_n^- = \rho_{bn} \mathbf{B}_n^+, \qquad n = 1, 2, \ldots, K \tag{31}$$

$$\mathbf{C}_n^- = \rho_{cn} \mathbf{C}_n^+, \qquad n = 1, 2, \ldots, L \tag{32}$$

where

$$\rho_{bn} = \Gamma_{bn} \exp(-2\gamma_{bn}\ell) \tag{33}$$

$$\rho_{cn} = \Gamma_{cn} \exp(-2\gamma_{cn}\ell) \tag{34}$$

Γ_{bn} equals +1 at an electric wall (even mode problem region) and −1 at a magnetic wall (odd mode arrangement).

Introducing the relationships between the forward and reflected waves into equation (25) gives

$$\mathbf{A}^+ + \mathbf{A}^- = \mathbf{H}[\mathbf{I} + \boldsymbol{\rho}_b]\mathbf{B}^+ + \mathbf{K}[\mathbf{I} + \boldsymbol{\rho}_c]\mathbf{C}^+ \tag{35}$$

or

$$\mathbf{A}^+ + \mathbf{A}^- = [\mathbf{H} \mid \mathbf{K}]\left[\begin{array}{c|c} \mathbf{R}_b & 0 \\ \hline 0 & \mathbf{R}_c \end{array}\right]\left[\begin{array}{c} \mathbf{B}^+ \\ \hline \mathbf{C}^+ \end{array}\right] \tag{36}$$

The preceding equation may be put in compact form as

$$\mathbf{A}^+ + \mathbf{A}^- = \mathbf{GRd}^+ \tag{37}$$

$\mathbf{G}, \mathbf{R}$ and $\mathbf{d}^+$ are given easily by inspection, and $\mathbf{R}_b$ and $\mathbf{R}_c$ are diagonal matrices defined by

$$\mathbf{R}_b = \begin{bmatrix} 1+\rho_{b1} & 0 & & & \\ 0 & 1+\rho_{b2} & --- & & \\ & & & 1+\rho_{bK-1} & 0 \\ & & & 0 & 1+\rho_{bK} \end{bmatrix} \tag{38}$$

$$\mathbf{R}_c = \begin{bmatrix} 1+\rho_{c1} & 0 & & & \\ 0 & 1+\rho_{c2} & --- & & \\ & & & 1+\rho_{cL-1} & 0 \\ & & & 0 & 1+\rho_{cL} \end{bmatrix} \tag{39}$$

Writing the reflected waves in regions B and C in equation (26) in terms of the forward ones gives

$$\mathbf{A}^+ - \mathbf{A}^- = \mathbf{Z}_a\mathbf{H}\mathbf{Y}_b[\mathbf{I} - \boldsymbol{\rho}_b]\mathbf{B}^+ + \mathbf{Z}_a\mathbf{K}\mathbf{Y}_c[\mathbf{I} - \boldsymbol{\rho}_c]\mathbf{C}^+ \tag{40}$$

or

$$\mathbf{A}^+ - \mathbf{A}^- = \mathbf{Z}_a[\mathbf{H} \mid \mathbf{K}]\left[\begin{array}{c|c} \mathbf{Y}_b & 0 \\ \hline 0 & \mathbf{Y}_c \end{array}\right]\left[\begin{array}{c|c} \mathbf{R}'_b & 0 \\ \hline 0 & \mathbf{R}'_c \end{array}\right]\left[\begin{array}{c} \mathbf{B}^+ \\ \hline \mathbf{C}^+ \end{array}\right] \tag{41}$$

The preceding equation is given in compact form by

$$\mathbf{A}^{+} - \mathbf{A}^{-} = \mathbf{Z}_a \mathbf{G} \mathbf{Y}_d \mathbf{R}' \mathbf{d}^{+} \tag{42}$$

The meanings of $\mathbf{Y}_d$ and $\mathbf{R}'$ are again obtained by inspection and $\mathbf{R}_b$ and $\mathbf{R}_c$ are diagonal matrices defined by

$$\mathbf{R}_b' = \begin{bmatrix} 1-\rho_{b1} & 0 & & & \\ 0 & 1-\rho_{b2} & --- & & \\ & & & 1-\rho_{bK-1} & 0 \\ & & & 0 & 1-\rho_{bK} \end{bmatrix} \tag{43}$$

$$\mathbf{R}_c' = \begin{bmatrix} 1-\rho_{c1} & 0 & & & \\ 0 & 1-\rho_{c2} & --- & & \\ & & & 1-\rho_{cL-1} & 0 \\ & & & 0 & 1-\rho_{cL} \end{bmatrix} \tag{44}$$

Similarly

$$\mathbf{G}^{\mathrm{T}}(\mathbf{A}^{+} + \mathbf{A}^{-}) = \mathbf{R}\mathbf{d}^{+} \tag{45}$$

$$\mathbf{Z}_d \mathbf{G}^{\mathrm{T}} \mathbf{Y}_a (\mathbf{A}^{+} - \mathbf{A}^{-}) = \mathbf{R}' \mathbf{d}^{+} \tag{46}$$

The ratio of the incoming ($\mathbf{A}^{+}$) and outgoing ($\mathbf{A}^{-}$) waves are now readily obtained by eliminating d^{+} between the preceding relationships. The result is

$$\mathbf{A}^{-} = \mathbf{S}_{11} \mathbf{A}^{+} \tag{47}$$

where

$$\mathbf{S}_{11} = [\mathbf{I} + \mathbf{Z}_a \mathbf{G} \mathbf{Y}_d \mathbf{R} (\mathbf{R}')^{-1} \mathbf{G}^{\mathrm{T}}]^{-1} [\mathbf{I} - \mathbf{Z}_a \mathbf{G} \mathbf{Y}_d \mathbf{R} (\mathbf{R}')^{-1} \mathbf{G}^{\mathrm{T}}] \tag{48}$$

and

$$\mathbf{S}_{11} = \begin{bmatrix} S_{11}^{(1)} & 0 & 0 & 0 & 0 \\ 0 & S_{11}^{(2)} & 0 & 0 & 0 \\ 0 & 0 & S_{11}^{(3)} & 0 & 0 \\ 0 & 0 & 0 & S_{11}^{(M-1)} & 0 \\ 0 & 0 & 0 & 0 & S_{11}^{(M)} \end{bmatrix}$$

14.4 Double septa and thick septum problem regions

The MMM formulation of the infinitely thin septum may be readily extended to that of a finite thickness one met in practical filter structures. This may be done by considering the problem region between a single waveguide and one divided into three regions by two infinitely thin septa. The arrangement in question is indicated in Figure 14.2. The MMM formulation of this geometry proceeds in a like manner to that of the single septum arrangement except that it produces one more set of boundary conditions at the plane of the discontinuity. The topology of the required arrangement is obtained easily by introducing an electric wall in region D at the plane of the discontinuity. The ensuing structure is illustrated in Figure 14.3. It coincides with that of a single thick septum as asserted. The approach outlined here in the case of a single thick septum can readily be generalised to that of a number of septa without any difficulty.

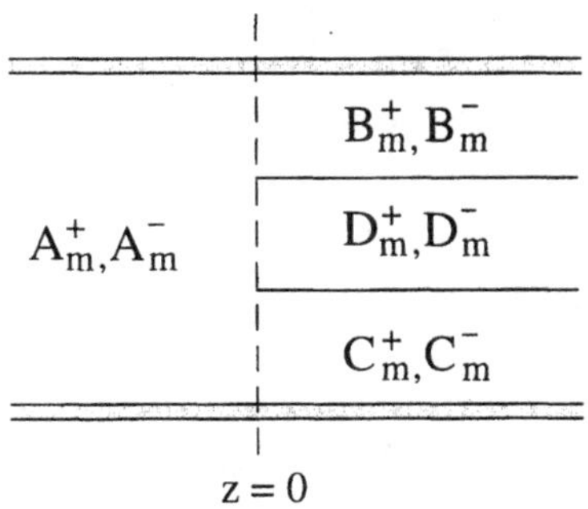

Figure 14.2 Topology of triple septa problem region

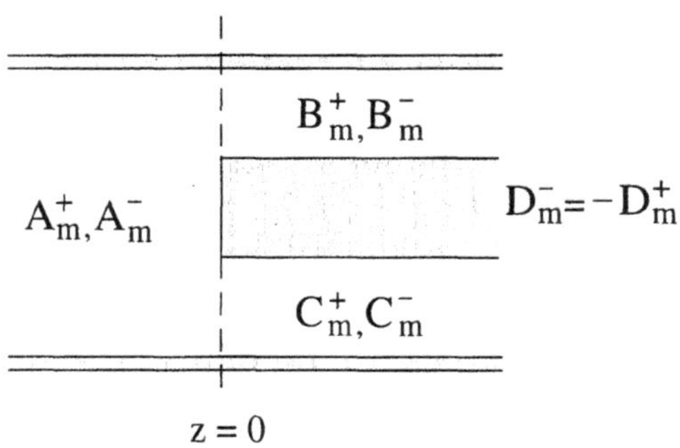

Figure 14.3 Topology of thick septum problem region

14.5 MMM characterisation of symmetrical waveguide discontinuities

A common discontinuity met in the design of many microwave filters is a symmetrical structure such as a septum, bar or evanescent region. The purpose of this section is to specialise the MMM to this situation. This may be best done by using the even and odd mode formulation of any symmetrical network introduced elsewhere in this text. The even and odd mode eigen-networks of such a circuit are revealed by placing electric and magnetic walls at the symmetry plane of the structure. The topology in question is indicated in Figure 14.4 and the two eigen-networks in Figures 14.5*a* and *b*.

The even mode reflection eigenvalues are given with

$$\rho_{b1} = 1 \ldots n = 1, 2, 3, \ldots, K \tag{49a}$$

$$\rho_{b1} = 1 \ldots n = 1, 2, 3, \ldots, L \tag{49b}$$

This gives

$$S_{11} = \rho_{\text{even}} \tag{50}$$

The odd mode reflection eigenvalues are given with

$$\rho_{bn} = -1 \ldots n = 1, 2, 3, \ldots, K \tag{51a}$$

$$\rho_{bn} = -1 \ldots n = 1, 2, 3, \ldots, L \tag{51b}$$

This gives

$$S_{11} = \rho_{\text{odd}} \tag{52}$$

The scattering matrix of the original circuit is then recovered by using the standard relationships below:

$$S_{11} = \frac{\rho_{\text{even}} + \rho_{\text{odd}}}{2} \tag{53a}$$

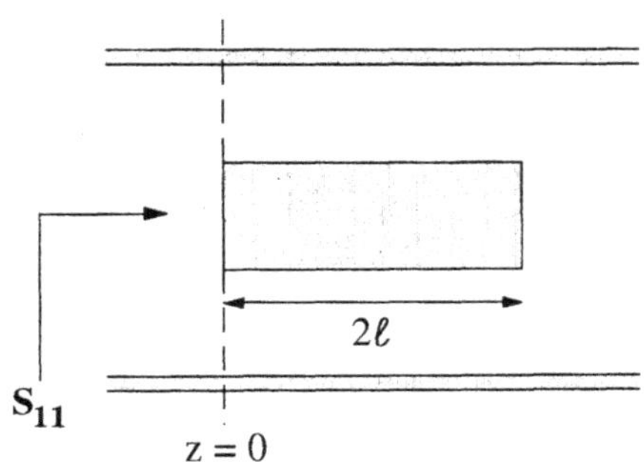

Figure 14.4 Topology of symmetrical septum

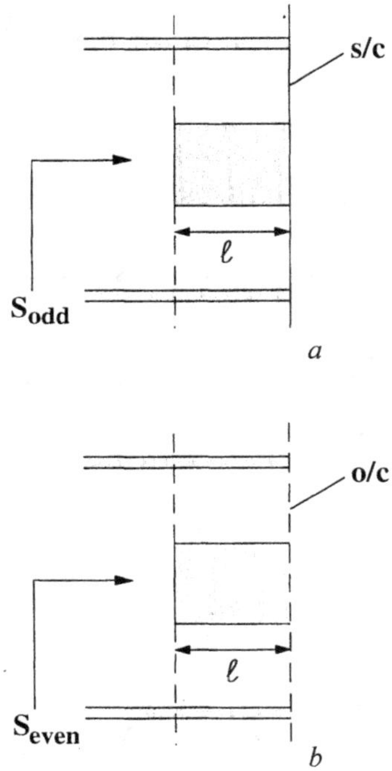

Figure 14.5 Eigen-networks of symmetrical septum

$$S_{21} = \frac{\rho_{even} - \rho_{odd}}{2} \tag{53b}$$

A typical reactance variable is given by using the bilinear relationships between reflection and impedance at a port.

Figure 14.4 indicates the symmetrical septum under discussion and Figure 14.5 shows its even and odd 1-port eigen-networks.

The preceding result may be further simplified by assuming that the incident wave in waveguide A is restricted to the dominant mode. This permits every term in the field vector $\mathbf{A}^+$ except the first to be set to zero. A is therefore reduced to

$$\mathbf{A}^+ = \begin{bmatrix} A_+ \\ 0 \\ \vdots \\ 0 \end{bmatrix} \tag{54}$$

The terms in the vector $\mathbf{A}^-$ in the vicinity of the discontinuity are unaffected by this assumption.

The characterisation of the symmetrical thick septum illustrated in Figure 14.4 proceeds in a like manner to that utilised to tackle the thin geometry.

14.6 Eigensolutions of waveguide sections

To proceed with a calculation it is necessary to evaluate the coupling integrals in equations (15) and (18). This may be done once the eigensolutions of each problem region are established. The problem region consisting of the junction between a regular and a bifurcated waveguide is a classic result in the literature. The transverse fields in waveguide A are given by

$$\phi_{an}(x) = \sin\left(\frac{p\pi x}{a}\right), \qquad 0 < x < a$$

and

$$Y_{an} = \sqrt{\frac{\varepsilon_0}{\mu_0}}\sqrt{1 - \left(\frac{p\lambda_0}{\lambda_c}\right)^2}$$

where

$$\lambda_c = 2a$$

Only symmetrical modes enter in the description of waveguide A. This condition is satisfied by fixing p as

$$p = 2n - 1, \qquad n = 1, 2, 3, \ldots, M$$

The field variable in waveguide B is

$$\phi_{bn}(x) = \sin\left(\frac{2\pi q x}{a - s}\right), \qquad 0 < x < \left(\frac{a - s}{2}\right)$$

and

$$Y_{bn} = \zeta_0\sqrt{1 - \left(\frac{q\lambda_0}{\lambda_c}\right)^2}$$

where

$$\lambda_c = (a - s)$$

and

$$q = n, \ldots, K$$

The eigensolution in waveguide C is

$$\phi_{cn}(x) = \sin\left(\frac{2\pi q x}{a-s}\right), \qquad 0 < x < \left(\frac{a-s}{2}\right)$$

$$Y_{cn} = \zeta_0 \sqrt{1 - \left(\frac{q\lambda_0}{\lambda_c}\right)^2}$$

and

$$\lambda_c = (a-s)$$

where

$$q = n, \ldots, L, \; n = 1, 2, 3, \ldots, L$$

The notation employed here is specified in Figure 14.6.

Once the model fields are formulated, the coupling integrals H_{mn} and K_{mn} may be evaluated. This gives

$$K_{mn} = 0.50 a Y_{ai}$$

$$K_{mn} = 0.50(a-s) Y_{bj}$$

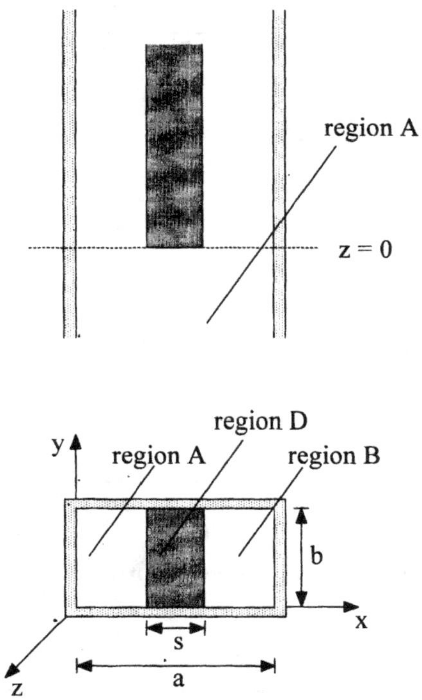

Figure 14.6 Physical variables of thick septum

and

$$K_{mn} = 0.50(a - s)\left(\frac{\sin(f)}{f} - \frac{\sin(g)}{g}\right)Y_{ai}$$

where

$$f = \left[0.50p\left(\frac{a - s}{a}\right) - q\right]\pi$$

and

$$g = \left[0.50p\left(\frac{a - s}{a}\right) + q\right]\pi$$

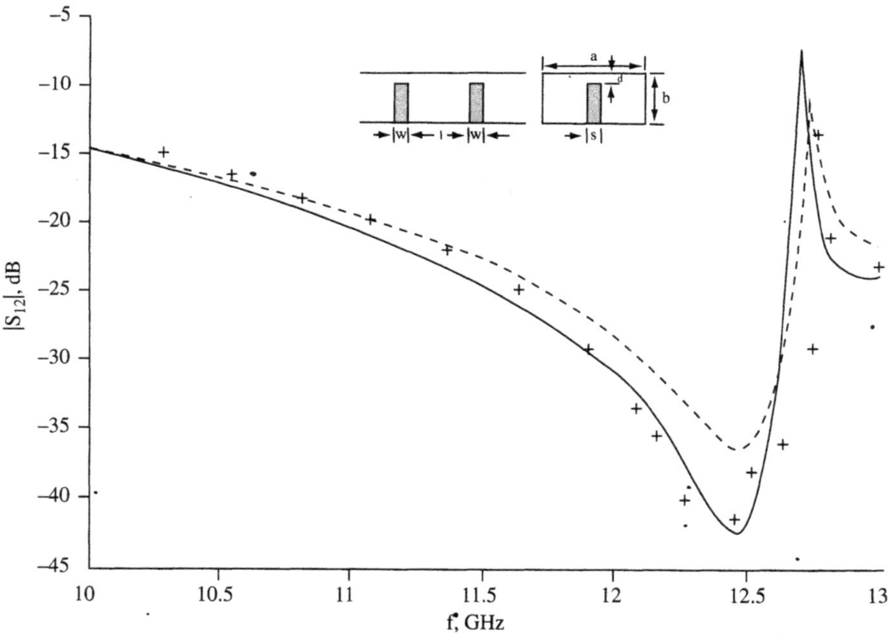

Figure 14.7 Measured and calculated transmission coefficient of a cascaded E-plane ridge guide discontinuity

$a = 22.86$, $b = 10.16$, $d = 4.114$, $w = 1.524$, $l = 12.1192$ mm

— Tao & Baudrand (1991) (- - - theory + + + experiment; Mansour *et al.* (1988))

Reprinted with permission, Tao & Baudrand (1991)

14.7 Immittance inverters

To be able to proceed with the design of filter structures it is necessary both to select a suitable immittance arrangement and separately to characterise it. Two structures that have been found useful for this purpose are the E-plane septum and the evanescent waveguide section. The MMM provides one of a number of means of establishing the equivalent circuits of these sorts of obstacles. Figure 14.7 indicates one typical result.

14.8 E-plane bandpass filters using metal inverters

The ridge waveguide readily lends itself to the realisation of directly coupled bandpass filters based on the use of inverters in cascade with half-wave long ridge conventional waveguide resonators. The inverters are either realised by metal strips across the ridges or by cut-off waveguide sections. The geometry of one arrangement based on the use of septa is indicated in Figure 14.8. The adjustment of this filter is often undertaken by using the MMM in conjunction with a suitable optimisation package. A comparison of the experimental frequency responses of a degree-5 filter in a WRD 750 waveguide before and after optimisation is depicted in Figure 14.9. The topology of an arrangement using cut-off sections of rectangular waveguide of the immittance inverters and its frequency response is separately indicated in Figures 14.10 and 14.11.

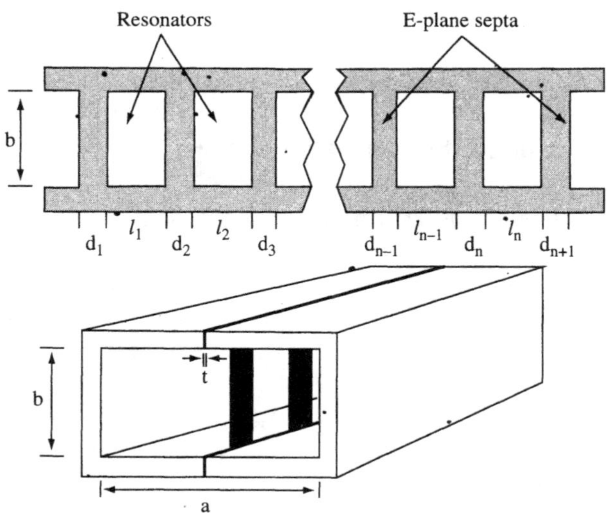

Figure 14.8 Topology of E-plane bandpass filter using septa inverters

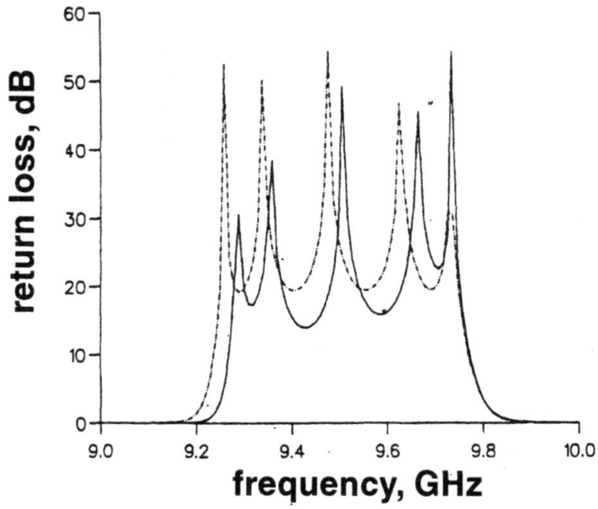

Figure 14.9 Calculated return loss before (solid line) and after (dashed line) optimisation (ridged waveguide filter)

Ridge waveguide gap = 5.00 mm; insert thickness = 0.10 mm;

Septum lengths (mm)	Before optimisation	After optimisation	Resonator lengths (mm)	Before optimisation	After optimisation
$d_1 = d_6$	0.5615	0.4805	$d_{11} = d_{15}$	15.4961	15.3539
$d_2 = d_4$	4.9200	4.6439	$d_{12} = d_{14}$	15.7558	15.6814
$d_3 = d_5$	6.2612	5.9867	d_{13}	15.7507	15.6829

Reprinted with permission, Budimir (1997)

14.9 Lowpass ridge filters using immittance inverters

One practical lowpass filter topology consists of immittance inverters separated by high or low impedance waveguides. One possible immittance inverter met in the design of this sort of filter is a short cut-off rectangular waveguide section. Figure 14.12 depicts a degree-12 layout which has been optimised by using a variational formulation. Its reflection and transmission parameters are reproduced in Figure 14.13.

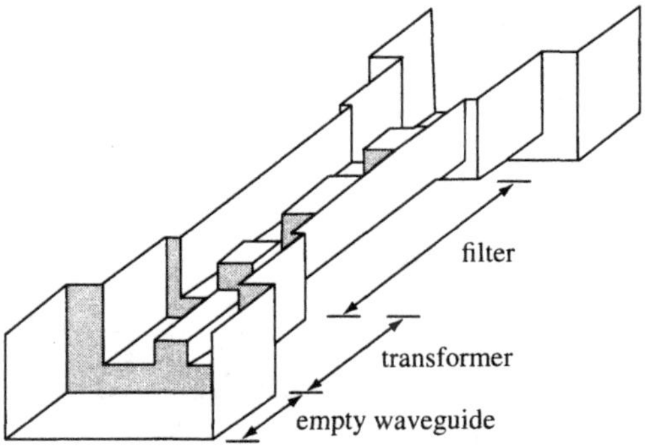

Figure 14.10 Topology of E-plane bandpass filter using evanescent waveguide inverters

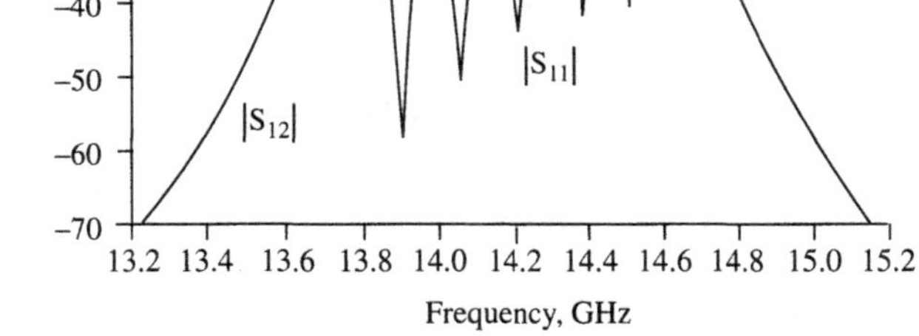

Figure 14.11 Frequency response of E-plane bandpass filter using evanescent waveguide inverters

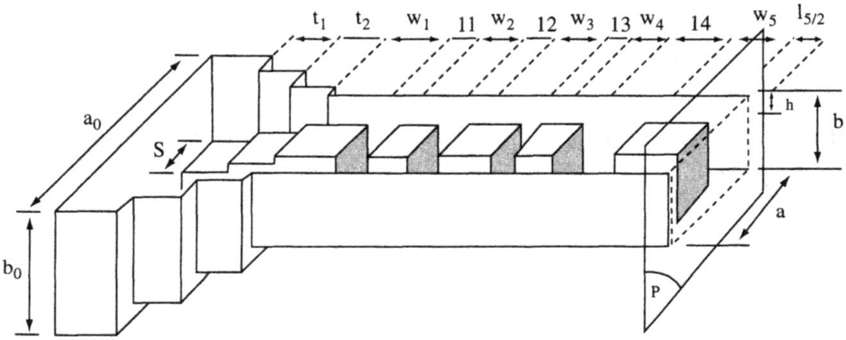

Figure 14.12 Evanescent-mode ridged waveguide lowpass filter

Input waveguide: $a_0 = 19.05, b_0 = 9.52$; first transformer: $a_1 = 14.3, b_1 = 7,$
$h_1 = 4, s_1 = 3.6$; second transformer: $a_2 = 10.7, b_2 = 5.2, h_2 = 1.7, s_2 = 3.6$;
filter section: $a = 8, b = 3.8, h = 0.6, s = 3.6, t_1 = t_4 = 6.15, t_2 = t_3 = 3.97,$
$w_1 = w_0 = 5.21, w_2 = w_9 = 5.24, w_3 = w_8 = 2.47, w_4 = w_7 = 3.29, w_5 = w_6 = 4.44,$
$w_{11} = w_{19} = 0.71, w_{12} = w_{18} = 2.64, w_{13} = w_{17} = 2.96, w_{14} = w_{16} = 2.12,$
$w_{15} = 1.79$

Reprinted with permission, Tao & Baudrand (1991)

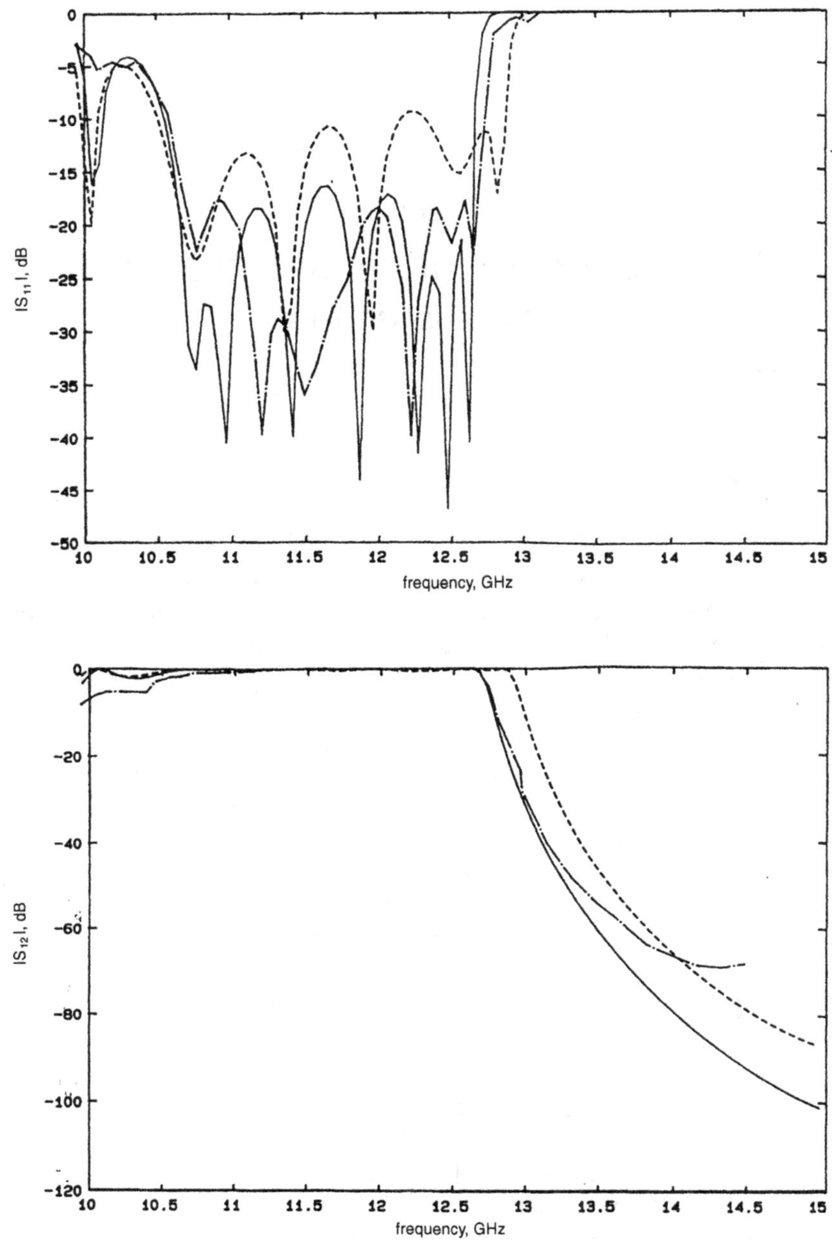

Figure 14.13 Frequency response of evanescent-mode ridged waveguide lowpass filter

Reprinted with permission, Tao & Baudrand (1991)

— 5 accessible modes - - - 1 accessible mode - · - measured data

Chapter 15

Nonreciprocal ridge isolators and phase-shifters

15.1 Introduction

One important class of ridge waveguide hardware is the nonreciprocal 2-port one. Two classic devices are the nonreciprocal phase shifter and the resonance isolator. Each device relies on the two different interactions between a spinning electron with a clockwise or anticlockwise circularly polarised alternating radio frequency magnetic field in a suitably magnetised ferrite medium. If the alternating magnetic field rotates in the same direction as that of the electron spin then the medium exhibits a scalar permeability with one distinct value; if, however, it rotates in the oppposite direction it again behaves as a scalar medium but with a different value. This property of a magnetised ferrite medium is the basis of a number of important non-reciprocal devices. One means of realising such components in a ridge wave-guide is to recognise that such polarisation always exists at the interface between any two different semi-infinite dielectric regions and everywhere outside and that its hand is determined by the direction of propagation. Another significant property of this sort of arrangement is that the senses of rotations are in opposite directions on either side of a dielectric rib. Such dielectric inserts are therefore suitable for the construction of non-reciprocal ferrite devices such as phase shifters and isolators. It is assumed, for simplicity, throughout, that the introduction of thin H- or E-plane ferrite or garnet tiles in the vicinity of the dielectric wall does not in the first instance disturb the polarisation in its neighbourhood. While the double ridge waveguide does not have natural planes of circular polarisation in its trough regions it does have such planes at its electric symmetry wall. An H-plane bifurcation produces one possible carrier for the design of nonreciprocal phase shifters and isolators.

15.2 Nonreciprocal ferrite devices in rectangular waveguide

One of the most useful of the nonreciprocal ferrite devices in the rectangular or ridge waveguide is the ferrite phase shifter. Its operation relies on the existence of natural or enforced circularly polarised counter-rotating alternating magnetic fields on either side of the planes of symmetry of these sorts of waveguides. E- or H-plane ferrite sheets or plates, magnetised by direct magnetic fields perpendicular to the planes of the alternating magnetic fields, exhibit different values of scalar permeabilities if placed at one or the other of these sites. Figures 15.1*a* to *d* indicate typical arrangements in dielectric loaded double ridge waveguides. If ferrite tiles are located on both sides of the symmetry plane of the arrangement then the direct fields must be oppositely orientated at each site for the waveguide to exhibit one value of scalar permeability; the other value is then obtained by reversing the direct fields. Moreover, the hands of the polarisation of the alternate magnetic fields are interchanged if the direction of propagation is reversed; the device is therefore nonreciprocal. The spin motion in a magnetic insulator is indicated in Figure 15.2.

The scalar permeabilities exhibited by the interactions between a spinning electron in a direct magnetic field and the alternating radio frequency signals on either side of the symmetry plane of a single ridge waveguide are summarised in Figures 15.3 and 15.4. The required result may be established easily by using the appropriate tensor permeability,

$$[\mu] = \begin{bmatrix} \mu & 0 & -j\kappa \\ 0 & 1 & 0 \\ j\kappa & 0 & \mu \end{bmatrix}$$

The nature of the tensor permeability is obtained by solving the equation of motion with the direct magnetic field along the y-coordinate and the alternating magnetic field in the x–z plane. μ is an even function of the direct magnetic field and κ is an odd one so that the sign of κ must be reversed if the direction of the direct magnetic field is reversed. μ and κ are in general functions of the direct magnetisation and magnetic field and the frequency of the alternating radio magnetic field.

Another important ferrite device in a rectangular waveguide is the resonance isolator. The geometry of this device is identical to that of the phase shifter configuration except that the direct magnetic field is in this instance equal to that required to establish the ferromagnetic resonance at the operating frequency of the device.

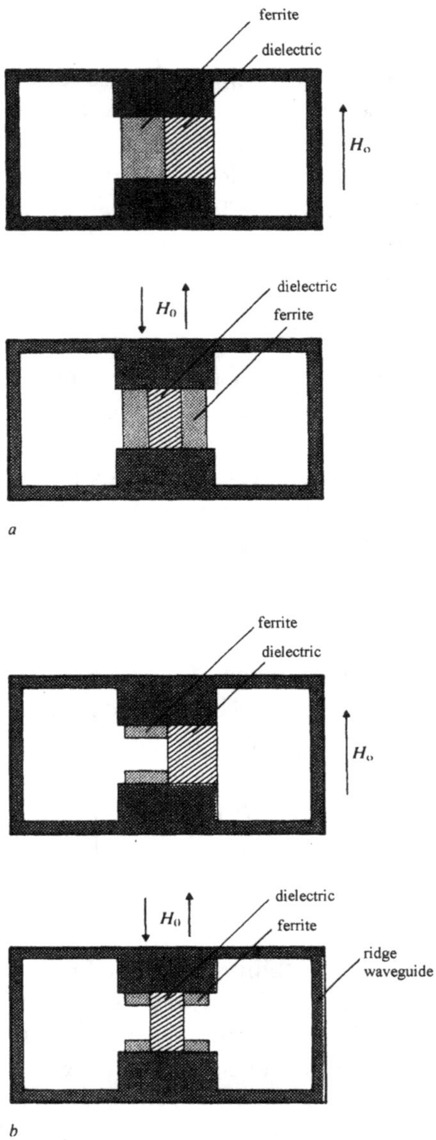

Figure 15.1 Schematic diagrams of ridge waveguide phase shifter using E-plane and H-plane ferrite tiles

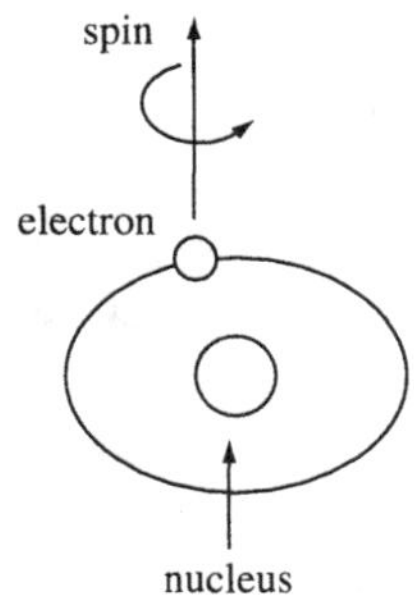

Figure 15.2 Spin motion in magnetic insulator

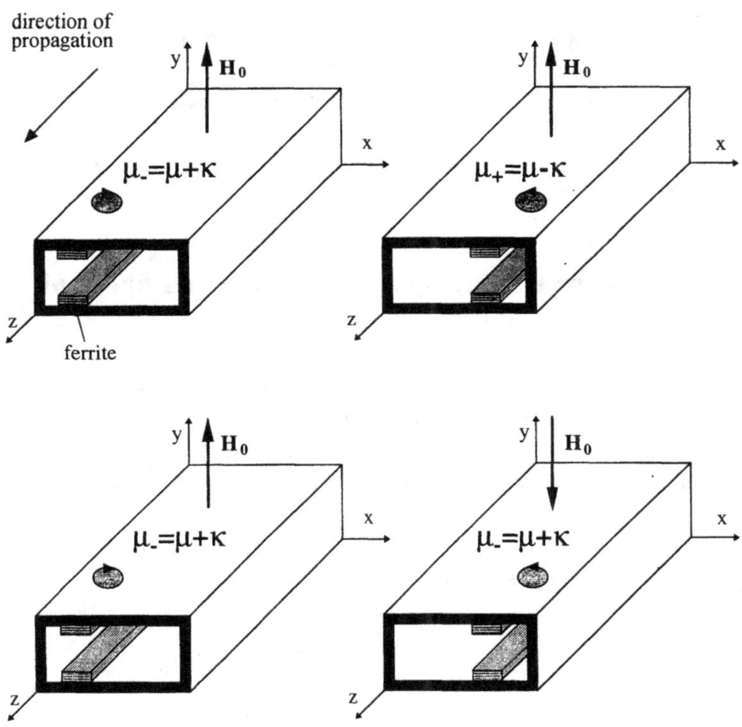

Figure 15.3 Scalar permeabilities in forward direction of propagation in single ridge waveguide loaded with magnetised ferrite tiles

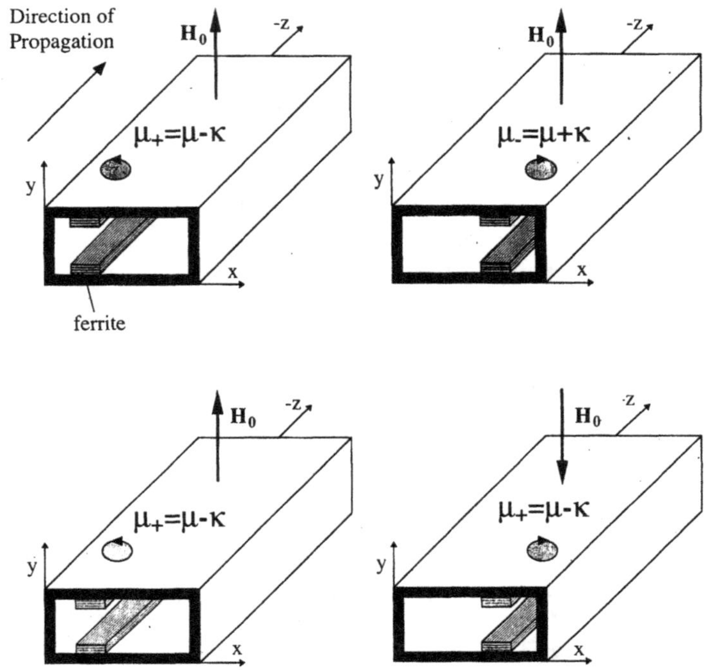

Figure 15.4 Scalar permeabilities in reverse direction of propagation in single ridge waveguide loaded with magnetised ferrite tiles

15.3 Differential phase shift, phase deviation and figure of merit of ferrite phase shifter

The practical specification of a ferrite phase shifter usually involves some compromise between conflicting parameters such as phase deviation over some frequency interval, peak and average power rating, insertion loss, etc. Some typical parameters that are of interest are summarised below prior to investigating some practical devices. Two quantities that are of obvious interest are the phase deviation ($\Delta\theta$) with respect to a 90° bit and the differential phase shift ($\Delta\phi$) per unit length,

$$\frac{\Delta\theta}{\pi/2}$$

and

$$\frac{\Delta\phi}{L}\ \text{rad/m}$$

respectively.

The figure of merit (F) of the device is separately defined in terms of the insertion loss (α) per unit length and the differential phase shift per unit length by

$$F = \frac{\Delta\phi}{\alpha} \text{ rad/dB}$$

For completeness, it is also necessary to spell out the normalised bandwidth (BW) of the device. This is specified in terms of the bandwidth (Δf) and mid-band frequency (f_0) by

$$\frac{\Delta f}{f_0}$$

The insertion phase at deviation may also be of some interest. It is defined below:

$$\frac{\Delta\phi}{\phi_0}$$

Hysteresis effects, temperature, switching speed and voltage standing wave ratio (VSWR) or return loss (RL) are often other parameters of concern.

15.4 90-degree phase shifter in dielectric loaded WRD 200 ridge waveguide

The experimental frequency response of a 90° bit over the 2–4 GHz band in a WRD 200 waveguide is indicated in Figure 15.5. The geometry described here consisted of a WRD 200 ridge waveguide with a dielectric filler extending over half the ridge width with ferrite tiles on the faces of the open half-space. The geometry in question is also indicated in Figure 15.5. The experimental data are obtained by normalising the phase for one orientation of the applied magnetic field and recording the result for the reverse field. The performance of the 90° bit may be summarised by

$$\frac{\Delta\theta}{\pi/2} \approx \pm\frac{3}{90}$$

$$\frac{\Delta\phi}{L} = \frac{\pi/2}{0.120} \text{ rad/m}$$

$$F \approx \frac{\pi/2}{0.50} \text{ rad/dB}$$

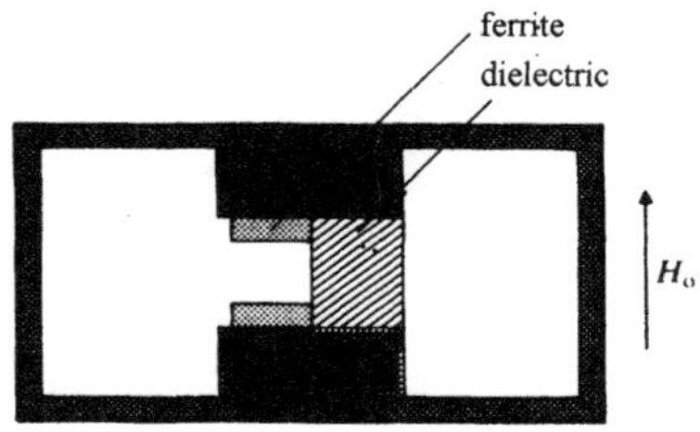

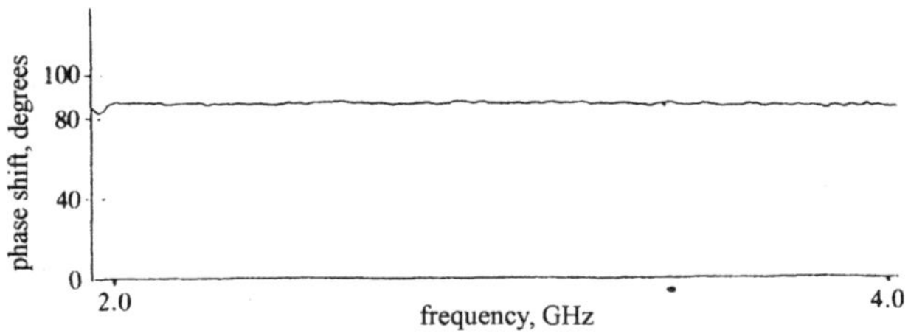

Figure 15.5 Differential phase shift in WRD 200 dielectric loaded ridge waveguide

Reprinted with permission, Hastings & Helszajn (1987)

$$\frac{\Delta f}{f_0} = \frac{2000}{3000}$$

$$\frac{\Delta \phi}{\phi_0} = \frac{90}{860}$$

The figure of merit ϕ of this device reflects, in part, the use of a heavily doped garnet material with a spinwave linewidth (ΔH_k) equal to 0.68 kA/m. The choice of linewidth, in the work, is dictated by the need to suppress spinwave instability (or nonlinear loss) that can occur at high peak power levels in ferrite devices. The other material details are described by $\mu_0 M_0 = 0.0800\text{T}$, $\varepsilon_r = 14.7$, $\Delta H = 5.6\,\text{kA/m}$ and $\tan\delta \approx 0.002$. The direct magnetic field employed at the 90° phase state is ~ 52 kA/m.

One feature of note already remarked on that may have some bearing on this excellent result is that the counter-rotating alternating magnetic fields are nearly circularly polarised over the same frequency interval. Forming β/k_0 in the 2–4 GHz band for a dielectric slab with $\varepsilon_r = 9$, using the data in Chapter 8, gives

$$1.8 \leqslant \frac{\beta}{k_0} \leqslant 2.2$$

The corresponding ellipticity over the same frequency interval is

$$1.20 \geqslant \left|\frac{H_x}{H_z}\right| \geqslant 1.12$$

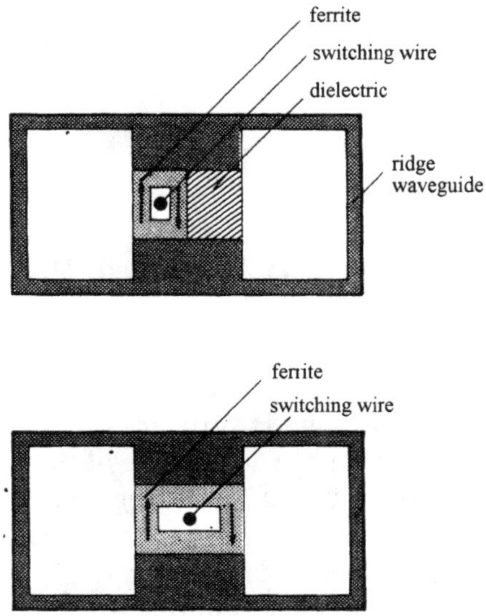

Figure 15.6 Schematic diagrams of nonreciprocal ridge waveguide using closed ferrite magnetic circuits

Another property of this geometry is that its insertion phase (860° at the midband) is quite large compared with the phase deviation of the magnetised bit (±45°) so that perturbation conditions may be assumed to prevail.

The possibility of latching ferrite phase shifters is, of course, always of some interest. Suitable ridge structures are illustrated in Figure 15.6.

15.5 Isolation, insertion loss and figure of merit of resonance isolator

The geometry of the resonance isolator is the same as that of the ferrite phase shifter except for the magnitude of the direct magnetic field. Its operation relies on the two different interactions between the electron spin in a gyromagnetic insulator and counter-rotating alternating magnetic fields at the same frequency. If the electron spin and the alternating magnetic field rotate in the same direction, then there will be strong interaction between the two and the alternating wave will suffer a strong absorption line. If the two rotate in opposite directions then the interaction will be weak or nonexistent and there will be no absorption. This sort of arrangement provides, therefore, one important means of realising a 2-port nonreciprocal network.

A resonance isolator is characterised by its isolation (I), its insertion loss (L) and its return loss (RL). Its figure of merit is defined by

$$F = \frac{I(\text{db})}{L(\text{dB})}$$

15.6 Resonance isolator in dielectric loaded WRD 750 ridge waveguide

The existence of planes of circular polarisation at the open faces of the dielectric insert in a ridge waveguide provides one means of constructing ferrite isolators and phase shifters. The purpose of this section is to describe one resonance isolator in the WRD 750 waveguide. It consists of a magnetised E-plane ferrite plate on one of the two open faces of a dielectric insert extending over the full width of the ridges.

The Kittel or resonance frequency is in this sort of device fixed by

$$\omega_r = \sqrt{\omega_x \omega_y}$$

where

$$\omega_x = (\omega_0 - N_z\omega_m + N_x\omega_m)$$

$$\omega_y = (\omega_0 - N_z\omega_m + N_y\omega_m)$$

The demagnetising factors associated with the ferrite topology utilised here are given approximately by

$$N_x \approx 1$$

$$N_y \approx 0$$

$$N_z \approx 0$$

Hence

$$\omega_r^2 \approx (\omega_0 + \omega_m)\omega_0$$

where

$$\omega_0 = \gamma H_0$$

and

$$\omega_m = \gamma M_0$$

γ is the gyromagnetic ratio (2.21×10^5 rad/s per A/m), ω_0 is the radian frequency (rad/s), H_0 is the direct magnetic field intensity (A/m), M_0 is the saturation magnetisation (A/m) and μ_0 is the free space permeability ($4\pi \times 10^{-7}$ H/m).

Figure 15.7 indicates the frequency responses in the forward and backward directions of propagation of an arrangement in a WRD 750 waveguide. The geometry consists of a double ridge waveguide with a dielectric filler between the ridges with a ferrite tile plastered on one of the dielectric open faces. Since the main purpose of this assembly is to investigate the nature of the polarisation of the radio frequency magnetic field rather than the development of a wideband isolator, a narrow linewidth CVG material has been chosen in this work. Its magnetisation ($\mu_0 M_0$) is equal to 0.1600 T, its relative dielectric constant (ε_r) is equal to 15.3 and its 3 dB and 15 dB linewidths are 0.96 kA/m and 5.18 kA/m, respectively. The relative dielectric constant of the dielectric insert (ε_r) is equal to 9.0. The thickness of the ferrite tile is 0.25 mm and its length is 31.75 mm. The direct magnetic flux density employed in obtaining this result was equal to 0.2150 T. This compares with a calculated value of 0.2050 T. The internal dimensions of the ridge waveguide utilised in this work correspond to those of WRD 750:

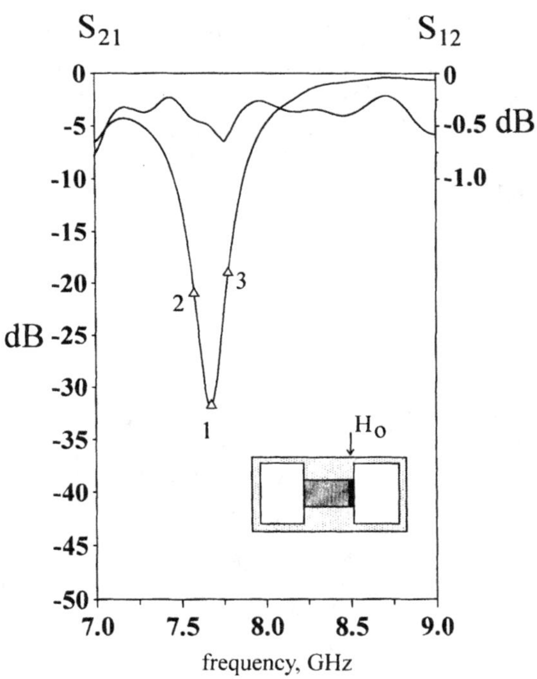

Figure 15.7 Frequency response of resonance isolator in WRD 750 dielectric loaded waveguide

Reprinted with permission, Helszajn & McKay (2000)

$$\frac{b}{a} = 0.45$$

$$\frac{s}{a} = 0.25$$

$$\frac{d}{b} = 0.42$$

The inside dimensions of the isolator section were reduced to $\sim 80\%$ of those of a standard WRD 750 waveguide. The device was separately matched at each end by a dielectric taper. The figure of merit (F) of the overall assembly is 70.

15.7 Resonance isolator in bifurcated ridge waveguide

While the double ridge waveguide does not have natural planes of circular polarisation in its trough regions it does have such planes at its electric symmetry wall. One possible nonreciprocal 2-port structure may be realised with this sort of waveguide by introducing a metal septum there and placing suitably magnetised ferrite tiles on one or both of its sides. Figure 15.8 indicates one arrangement. The direct magnetic field intensity is again given by Kittel's resonance equation except that the demagnetising factors are, in this instance, specified by

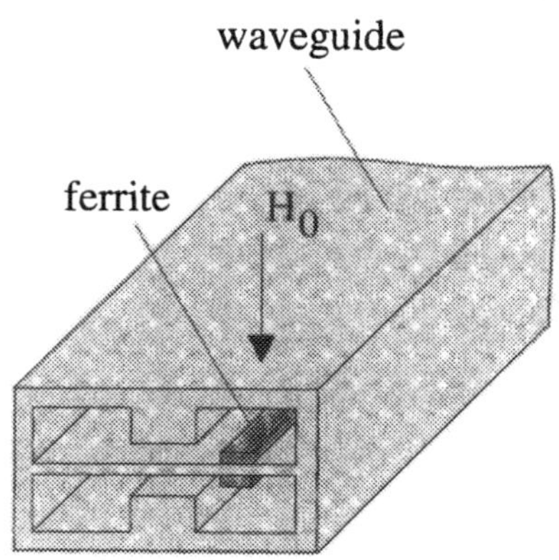

Figure 15.8 Schematic diagram of ferrite isolator using bifurcated WRD 750 ridge waveguide

Reprinted with permission, Helszajn & McKay (1998)

$$N_x \approx 0$$

$$N_y \approx 0$$

$$N_z \approx 1$$

The Kittel frequency is

$$\omega_r \approx (\omega_0 - \omega_m)$$

The frequency response in the forward and backward directions of propagation of an arrangement in a WRD 750 waveguide is shown in Figure 15.9. It consists of ferrite tiles mounted on either side of a metal septum of thickness 0.762 mm at the electric wall symmetry plane. The length of each ferrite tile is 25 mm and its cross-section is 0.815 × 3.510 mm. The ferrite employed was a CVG material with a magnetisation ($\mu_0 M_0$) of 0.1600 T and a relative dielectric constant (ε_r) of 15.0.

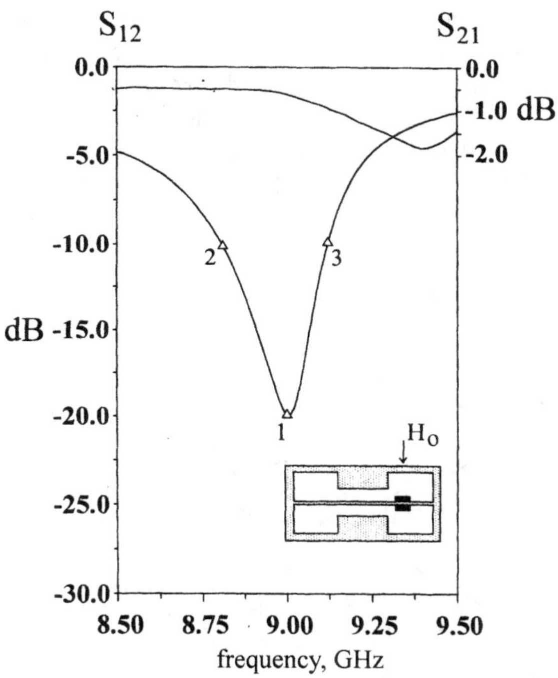

Figure 15.9 Frequency response of resonance isolator in bifurcated WRD 750 ridge waveguide

Reprinted with permission, Helszajn & McKay (1998)

The inside dimensions of the isolator section were reduced to ~80% of those of a standard WRD 750 waveguide. The figure of merit (F) of the overall assembly is 11. The frequency range of the waveguide is 7.50–18.00 GHz.

15.8 Differential phase shift circulator

One important application of the differential ferrite phase shifter is in the construction of the differential phase-shift circulator. It consists of a folded magic-tee and a 3 dB side-wall hybrid between which is placed a dual section of waveguide containing nonreciprocal 45° ferrite phase shifters. The phase shifter in each waveguide is oppositely magnetised. Hence, the total differential phase shift is 90°. The operation of this type of circulator can be understood with the help of Figure 15.10. A wave incident on port 1, at the H arm of the magic-tee, divides equally in phase between the two waveguides. One of the waves is then phase shifted through $+\pi/4$ rad. This wave is then incident on the side-wall hybrid and produces two component waves at ports 2 and 4. The wave at port 2 has now been shifted through $\pi/4$ and the one at port 4 by $3\pi/4$. Similarly, the other wave is phase shifted through $-\pi/4$ through the ferrite phase shifter, and after passing through the side-wall hybrid, produces two more component waves at ports 2 and 4. The one at port 2 is phase shifted through $+\pi/4$ and the one at port 4 by $-\pi/4$. The two waves at port 4 are now π rad out of phase and therefore cancel, while those at port 2 are in phase and

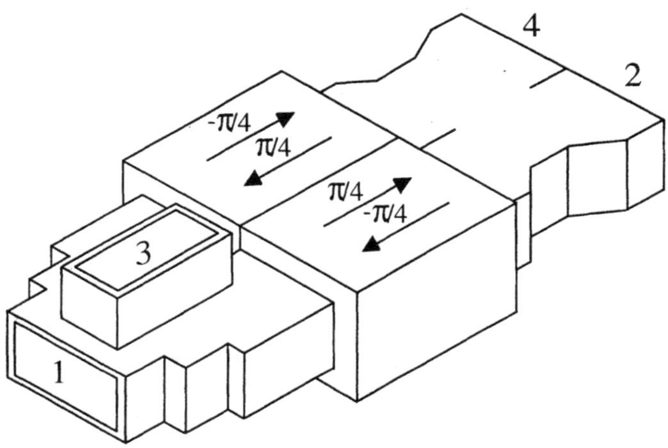

Figure 15.10 Schematic diagram of four-port differential phase shift circulator using H-plane folded magic-tee and 3 dB hybrid

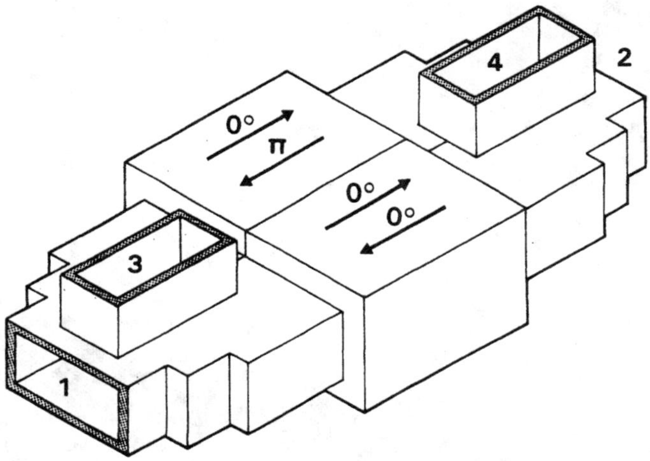

Figure 15.11 Schematic diagram of four-port differential phase circulator using H-plane folded magic-tees

Reprinted with permission, Hogan (1952)

therefore add. Transmission is therefore from port 1 to port 2. In a similar way transmission occurs from port 2 to port 3, and so on, in a cyclic manner.

Another geometry consists of two magic-tees between a dual waveguide section. Its topology is illustrated in Figure 15.11. The middle wave guide section is in this arrangement similar to that of the conventional structure using one hybrid and one magic-tee except that one of the two dual waveguides incorporates a 90° reciprocal dielectric phase shifter in order to cater for the phase difference between the two hybrids. The phase settings in the primary and secondary waveguides are given in the forward direction of propagation by

$$\phi_1^+ = \left(\frac{\pi}{4}\right) \text{ rad}$$

$$\phi_2^+ = \left(\frac{\pi}{2}\right) - \left(\frac{\pi}{4}\right) \text{ rad}$$

In the other direction of propagation

$$\phi_1^- = -\left(\frac{\pi}{4}\right) \text{ rad}$$

$$\phi_2^- = \left(\frac{\pi}{2}\right) + \left(\frac{\pi}{4}\right) \text{ rad}$$

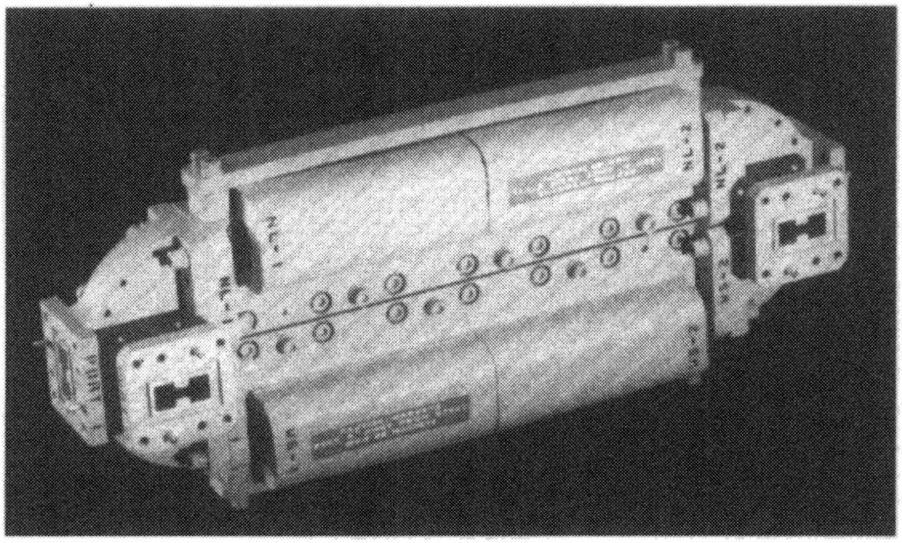

Figure 15.12 Photograph of ridge waveguide differential phase shift circulator

Courtesy: Raytheon Inc.

A wave incident on port 1, at the H arm of the magic-tee, divides equally in phase between the two waveguides. These two waves are in-phase at the output magic-tee after traversing the dual waveguide section. Transmission is therefore from port 1 to port 2. In a similar way transmission occurs from port 2 to port 3, and so on, in a cyclic manner. Figure 15.12 depicts an X-band commercial circulator.

Chapter 16

Finline waveguide

16.1 Introduction

A waveguide arrangement whose topology has many of the features of the ridge one is the finline structure. It consists of a dielectric sheet across the waveguide metallised on either one or both sides in a number of different ways. Four typical arrangements are the unilateral geometry, the bilateral one, the insulated finline and the antipodal structures. The bilateral arrangement may be visualised as a quasi-ridge waveguide. One benefit of this family of waveguides is that quite complicated circuits can be fabricated using planar techniques. A complete description of a typical finline waveguide includes its cut-off space, its impedance, its attenuation, its propagation constant and a description of its field pattern. It also requires an understanding of the location of the planes of circular polarisation in the waveguide. The solution of this class of waveguide has been the object of a host of numerical descriptions. The approach used here relies, in each instance, on an approximate closed form formulation of the field in each region of the structure. The effects of finite metallisation on both the propagation constant and the impedance in these sorts of waveguides are outside the remit of this paper. The chapter includes the description of one resonance isolator.

16.2 Finline waveguide topologies

A number of different configurations of the finline waveguide have by now been described. Figure 16.1 indicates some possibilities. Since a typical finline waveguide is an inhomogeneous structure its field is described in terms of LSE and LSM modes instead of the TE and TM ones met in connection with the conventional rectangular waveguide. The unilateral and bilateral

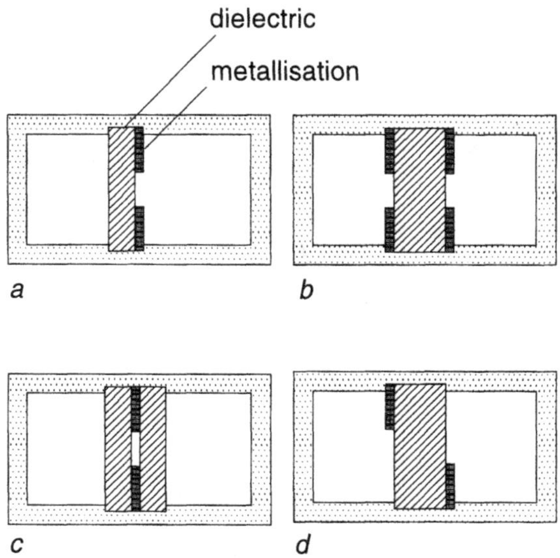

Figure 16.1 Schematic diagrams of unilateral, bilateral, insulated and antipodal finline waveguides

a Unilateral; *b* bilateral; *c* insulated; *d* antipodal

arrangements have, however, received the most attention. One construction technique is indicated in Figure 16.2. The relative dielectric constant (ε_d) employed in the construction of a finline waveguide is usually of the order 2.2. Some typical field patterns in these sorts of waveguides are illustrated in Figure 16.3.

16.3 Normalised wavelength and impedance in finline

A host of exact and approximate expressions for the wavelength and impedance of bilateral and unilateral finlines have been described in the literature. Since the finline waveguide is an inhomogeneous transmission line its guide wavelength must be calculated at each and every frequency. The purpose of this section is merely to give one typical result for each structure. The physical nomenclature met in connection with the important unilateral and bilateral lines are separately depicted in Figure 16.4. Some results are indicated in Figures 16.5 and 16.6. Scrutiny of these illustrations suggests that the main difference between the two configurations in the WR 28 waveguide is that the impedance of the bilateral arrangement is twice that of the unilateral geometry.

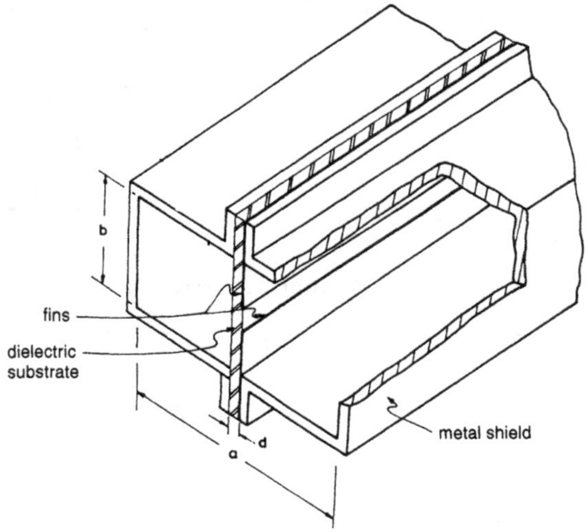

Figure 16.2 Fabrication of finline waveguide

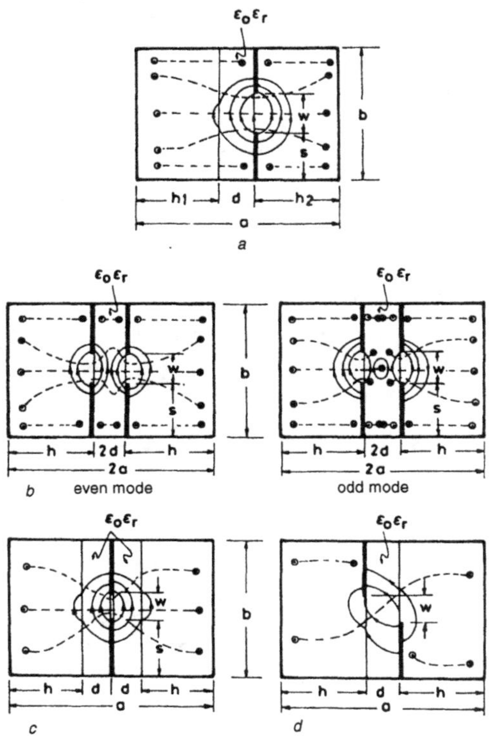

Figure 16.3 Electric fields in finline structures

a Unilateral finline; *b* bilateral finline; *c* insulated finline; *d* antipodal finline

Reprinted with permission, Bhat & Koul (1987)

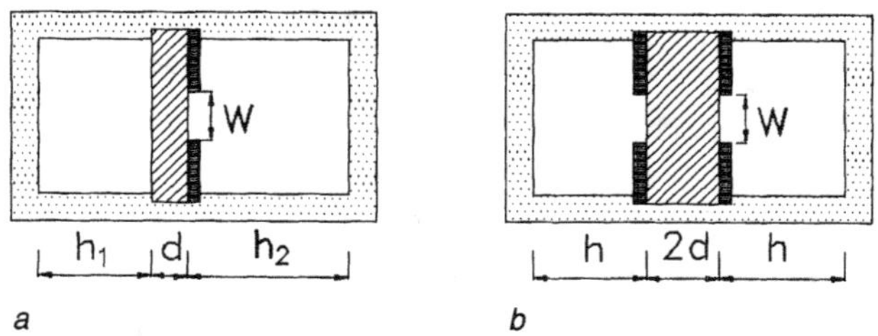

Figure 16.4 Physical details of unilateral and bilateral structures

a Unilateral; *b* bilateral

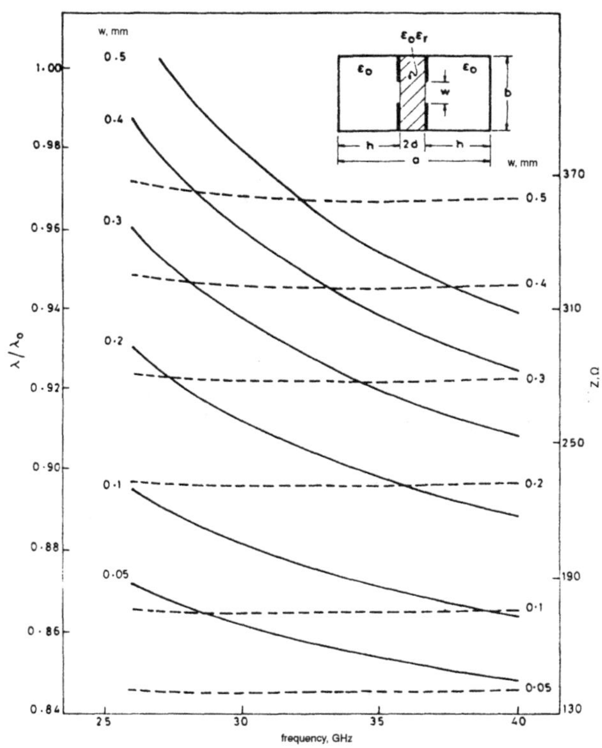

Figure 16.5 Normalised waveguide wavelength and impedance in bilateral WR 28 waveguide

$d = 0.127$ mm, $h = 3.429$ mm, $\varepsilon_r = 2.22$

Reprinted with permission, Bhat & Koul (1987)

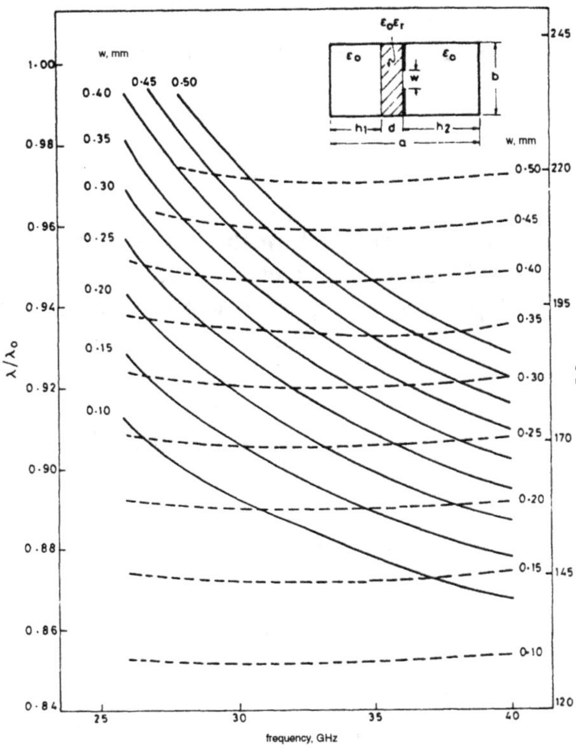

Figure 16.6 Normalised waveguide wavelength and impedance in unilateral WR 28 waveguide

$d = 0.254$ mm, $h_1 = 3.302$ mm, $h_2 = 3.556$ mm, $\varepsilon_r = 2.22$

Reprinted with permission, Bhat and Koul (1987)

16.4 Empirical expressions for propagation in bilateral and unilateral finline

The propagation constant in finline is sometimes characterised, for engineering purposes, in terms of an equivalent ridge waveguide of identical dimensions with an effective frequency dependent dielectric constant. This notion allows the usual relationships between λ_0, λ_c and λ_g to be retained:

$$\left(\frac{2\pi}{\lambda_g}\right)^2 = \left(\frac{2\pi}{\lambda_0}\right)^2 \varepsilon_{\text{eff}}(f_0) - \left(\frac{2\pi}{\lambda_{cr}}\right)\left(\frac{2\pi}{\lambda_{cr}}\right)^2 \tag{1}$$

where

$$\varepsilon_{\text{eff}}(f_0) = F(d/b, s/a, \lambda_0, \varepsilon_r)\varepsilon_{\text{eff}}(f_c) \tag{2}$$

and

$$\varepsilon_{\text{eff}}(f_c) = \left(\frac{\lambda_{\text{cf}}}{\lambda_{\text{cr}}}\right)^2 \tag{3}$$

λ_{cf} and λ_{cr} are the cut-off wavelength in the finline and in the ridge waveguides, respectively. The correction factor F entering into the description of the frequency dependent effective dielectric constant is empirically adjusted to cater for dispersion effects by using a rigorous calculation. Since the finline waveguide is an inhomogeneous transmission line its guide wavelength must be calculated at each and every frequency.

One empirical expression for the cut-off number of either a bilateral or unilateral finline is

$$\frac{b}{\lambda_{\text{cf}}} = A\left(\frac{d}{b}\right)^p\left(\frac{s}{a}\right)^q \tag{4}$$

provided

$$\frac{1}{16} \leqslant \frac{d}{b} \leqslant \frac{1}{4}$$

$$\frac{1}{32} \leqslant \frac{s}{a} \leqslant \frac{1}{4}$$

A, p and q are given for the unilateral finline with $\varepsilon_r = 2.22$ by

$$A = 0.1748$$

$$p = \begin{cases} 0.16(s/a)^{-0.07} & 1/32 \leqslant s/a \leqslant 1/20 \\ 0.16(s/a)^{-0.07} - 0.001 \ln[(s/a) - 1.32] & 1/20 \leqslant s/a \leqslant 1/4 \end{cases}$$

$$q = -0.0836$$

The arbitrary constants for the bilateral finline with $\varepsilon_r = 2.22$ are

$$A = 0.15$$

$$p = \begin{cases} 0.225(s/a)^{-0.042} & 1/32 \leqslant s/a \leqslant 1/10 \\ 0.149(s/a)^{-0.23} & 1/10 \leqslant s/a \leqslant 1/4 \end{cases}$$

$$q = -0.14$$

Calculations based on these expressions with b/a equal to 0.50 agree to within ±1% with more rigorous ones.

Once λ_{cr} and λ_{cf} are evaluated $\varepsilon_{eff}(f_c)$ may be calculated and $\varepsilon_{eff}(f_0)$ is obtained from a statement of the correction factor F.

The arbitrary constant F is for the unilateral finline with $\varepsilon_r = 2.22$ given by

$$F = \begin{cases} [1.0 + 0.43(s/a)](d/b)^{p_1} & 1/32 \leqslant s/a \leqslant 1/8 \\ [1.02 + 0.264(s/a)](d/b)^{p_1} & 1/8 \leqslant s/a \leqslant 1/4 \end{cases}$$

where

$$p_1 = 0.096(s/a) - 0.007$$

The corresponding quantities for the bilateral finline with $\varepsilon_r = 2.22$ are

$$F = \begin{cases} [0.78(s/a)^{-0.098}](d/b)^{0.109} & 1/32 \leqslant s/a \leqslant 1/8 \\ [1.04 - 0.2(s/a)](d/b)^{p_1} & 1/8 \leqslant s/a \leqslant 1/4 \end{cases}$$

where

$$p_1 = 0.152 - 0.256(s/a)$$

16.5 Fields in unilateral finline waveguide

While the finite element method and a host of other numerical techniques may again be employed to calculate the parameters of finline waveguides the solution summarised here relies on an approximate closed form formulation. The fields in these types of waveguides are denoted as either longitudinal sectional electric LSE or longitudinal sectional magnetic LSM according to whether E_x or H_x is equal to zero. A knowledge of the fields in the waveguide is sufficient for the calculation of its power flow and its impedance. The components of the electric field are summarised below for the unilateral line and in the next section for a bilateral structure.

The electric and magnetic fields in each region for the longitudinal section electric LSE modes unilateral line are:

$$E_x^{\mathrm{h}} = 0 \tag{5a}$$

$$E_y^{\mathrm{h}} = \sum_{n=0}^{N} (jA_n\beta)\phi_n(x)\cos\left(\frac{n\pi y}{b}\right) \tag{5b}$$

$$E_z^{\mathrm{h}} = \sum_{n=0}^{N} \left(\frac{-A_n n\pi}{b} \right) \phi_n(x) \sin\left(\frac{n\pi y}{b} \right) \tag{5c}$$

$$H_x^{\mathrm{h}} = \sum_{n=0}^{N} \left(\frac{-jA_n k_{cn}^2}{\omega\mu_0} \right) \phi_n(x) \cos\left(\frac{n\pi}{b} \right) \tag{5d}$$

$$H_y^{\mathrm{h}} = \sum_{n=0}^{N} \left(\frac{-A_n \alpha_n n\pi}{b\omega\mu_0} \right) \psi_n(x) \sin\left(\frac{n\pi}{b} \right) \tag{5e}$$

$$H_z^{\mathrm{h}} = \sum_{n=0}^{N} \left(\frac{-jA_n \alpha_n \beta}{\omega\mu_0} \right) \psi_n(x) \cos\left(\frac{n\pi}{b} \right) \tag{5f}$$

The corresponding electric and magnetic fields in each region for the LSM modes are:

$$E_x^{\mathrm{e}} = \sum_{n=0}^{N} \left(\frac{jB_n k_{cn}^2}{\alpha_n} \right) \psi_n(x) \sin\left(\frac{n\pi y}{b} \right) \tag{6a}$$

$$E_y^{\mathrm{e}} = \sum_{n=0}^{N} \left(\frac{B_n n\pi}{b} \right) \phi_n(x) \cos\left(\frac{n\pi y}{b} \right) \tag{6b}$$

$$E_z^{\mathrm{e}} = \sum_{n=0}^{N} (-jB_n \beta) \phi_n(x) \cos\left(\frac{n\pi y}{b} \right) \tag{6c}$$

$$H_x^{\mathrm{e}} = 0 \tag{6d}$$

$$H_y^{\mathrm{e}} = \sum_{n=0}^{N} \left(\frac{-jB_n \beta\omega\varepsilon_0\varepsilon_{\mathrm{r}}}{\alpha_n} \right) \psi_n(x) \sin\left(\frac{n\pi y}{b} \right) \tag{6e}$$

$$H_z^{\mathrm{e}} = \sum_{n=0}^{N} \left(\frac{-B_n \omega\varepsilon_0\varepsilon_{\mathrm{r}} n\pi}{b} \right) \psi_n(x) \cos\left(\frac{n\pi y}{b} \right) \tag{6f}$$

A_n, B_n, ϕ_n and ψ_n take on specific values in regions I, II and III for both the unilateral and bilateral structures.

In region I (unilateral finline),

$$\phi_n(x) = j\sin(\alpha_{1n}x) \tag{7}$$

$$\psi_n(x) = \cos(\alpha_{1n}x) \tag{8}$$

$$A_n = \frac{-j2\beta W_n}{\exp(j\alpha_{1n}\ell_1) - \exp(-j\alpha_{1n}\ell_1)} \tag{9}$$

$$B_n = \frac{\left(\frac{2n\pi}{b}\right) W_n}{\exp(j\alpha_{1n}\ell_1) - \exp(-j\alpha_{1n}\ell_1)} \tag{10}$$

$$W_n = \frac{\gamma_n}{n\pi k_{cn}^2}\left[\sin\left(\frac{n\pi}{b}(h+d)\right) - \sin\left(\frac{n\pi}{b}h\right)\right] \tag{11}$$

In region II (unilateral finline),

$$\phi_n(x) = S_n \exp(j\alpha_{2n}(x-\ell_1)) + \exp(-j\alpha_{2n}(x-\ell_1)) \tag{12}$$

$$\psi_n(x) = S_n \exp(j\alpha_{2n}(x-\ell_1)) - \exp(-j\alpha_{2n}(x-\ell_1)) \tag{13}$$

$$A_n = \frac{-j\beta W_n}{(1+S_n)} \tag{14}$$

$$B_n = \frac{n\pi W_n}{b(1+S_n)} \tag{15}$$

$$S_n = \frac{(1-C_n) - (1+C_n)\exp(-j2\alpha_{3n}\ell_3)}{(1+C_n) - (1-C_n)\exp(-j2\alpha_{3n}\ell_3)}\exp(-j2\alpha_{3n}\ell_3) \tag{16}$$

$$C_n = \frac{\alpha_{3n}^*}{\alpha_{2n}^*}, \quad \text{for LSE modes} \tag{17}$$

$$C_n = \frac{\alpha_{2n}^*}{\varepsilon_r \alpha_{3n}^*}, \quad \text{for LSM modes} \tag{18}$$

In region III (unilateral finline),

$$\phi_n(x) = j\sin\alpha_{3n}(x-a) \tag{19}$$

$$\psi_n(x) = \cos\alpha_{3n}(x-a) \tag{20}$$

$$A_n = -j2\beta W_n Y_n \tag{21}$$

$$B_n = \left(\frac{2n\pi}{b}\right) W_n Y_n \tag{22}$$

$$Y_n = \frac{\exp(-j(\alpha_{2n}\ell_2)) + S_n \exp(j(\alpha_{2n}\ell_2))}{(1 + S_n)(\exp(-j\alpha_{3n}\ell_3)) - (\exp(j\alpha_{3n}\ell_3))} \tag{23}$$

α_{in} in the preceding equations is given by

$$\alpha_{in}^2 = k_0^2\varepsilon_{ri} - k_{cn}^2 \tag{24}$$

where

$$k_{cn}^2 = \left(\frac{n\pi}{b}\right)^2 + \beta^2$$

Furthermore, $\gamma_n = 1$ for $n = 0$, $\gamma_n = 2$ for $n = 1, 2, 3$, etc.

The electric fields supported by these structures are separately illustrated in Figure 16.3.

16.6 Bilateral finline

Since the bilateral line is a symmetric configuration, it is only necessary to consider half its structure. The fields in region I of this waveguide are identical to those of the unilateral arrangement and need therefore not be repeated. In region II, however, A_n, B_n, ϕ_n and ψ_n have to be modified. The new quantities are defined by

$$\phi_n(x) = \cos\alpha_n\left(x - \frac{a}{2}\right) \tag{25}$$

$$\psi_n(x) = j\sin\alpha_n\left(x_n - \frac{a}{2}\right) \tag{26}$$

$$A_n = -j\beta W_n Y_n \tag{27}$$

$$B_n = \left(\frac{n\pi}{b}\right) W_n Y_n \tag{28}$$

$$Y_n = \frac{\exp(-j\alpha_{2n}\ell_2/2)}{1 + \exp(-j\alpha_{2n}\ell_2)} \tag{29}$$

16.7 Empirical formulation of impedance in bilateral finline waveguide

A calculation of impedance based on an equivalent waveguide model has also been described,

$$Z_0(\omega) = Z_0(\infty)\left(\frac{\lambda_0}{\lambda_g}\right) \tag{30}$$

where (λ_0/λ_g) is specified by equation (1) and $Z_0(\infty)$ is the impedance of the equivalent ridge waveguide with identical dimensions to those of the finline one. Some results are superimposed in Figures 16.5 and 16.6.

16.8 Circular polarisation in bilateral and unilateral finline waveguides

The field patterns in finline waveguides also support planes of circular polarisation. It corresponds to equal amplitude waves in time space quadrature. The existence of such planes may be established approximately by using the closed form descriptions of H_x and H_z. Figures 16.8 and 16.9 indicate two typical solutions, at the plane of the electric symmetry wall, for different gap dimensions in the case of a unilateral configuration. These sorts of waveguides may also therefore be utilised in the construction of nonreciprocal ferrite phase shifters and resonance isolators.

16.9 Finline isolator using hexagonal ferrite substrate

One important nonreciprocal device is the ferrite resonance isolator. It relies for its operation on the different values of dissipation associated with a magnetised ferrite material under the influence of clockwise and anti-clockwise polarised alternating magnetic fields. The schematic diagram of one possible finline isolator is illustrated in Figure 16.10. It consists of a unilateral finline on a carrier loaded on one side by alumina and hexagonal ferrite substrates, with the circuit metallisation deposited on the finline substrate. The gap dimension of the input and output finline circuits is flared on both sides to that of the finline isolator which may, but need not, coincide with the narrow dimensions of the waveguide; the alumina is also tapered at the input and output terminals of the device for matching purposes.

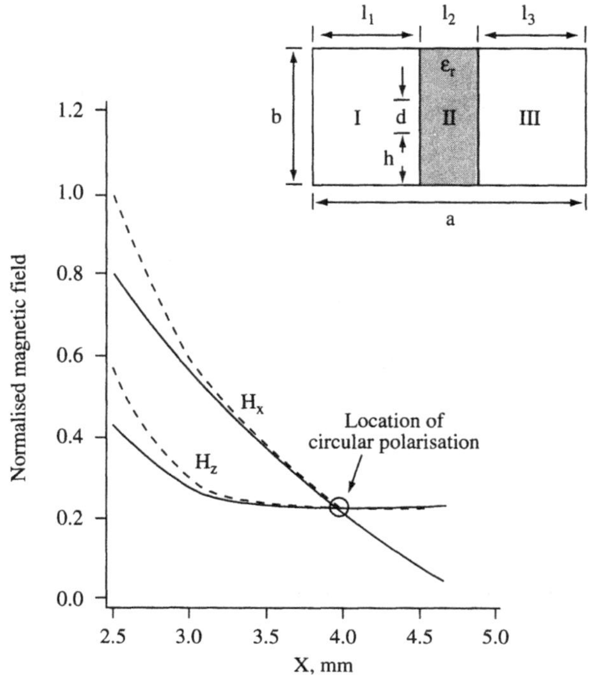

Figure 16.8 Approximate and hybrid mode descriptions of magnetic field in unilateral waveguide

$d/b = 0.20$, $a/b = 2$, $a = 4.775$ mm, $l_2 = 0.127$ mm, $l_1 = 2.387$ mm, $\varepsilon_r = 2.2$

Reprinted with permission, Mansour *et al.* (1988)

Contact problems between the wide dimensions of the waveguide and the finline circuit are avoided by incorporating suitable chokes in the assembly. A small external direct magnetic field is provided to orientate the internal field along the positive or negative C-axis of the uni-axial ferrite material and for tuning purposes.

The figure of merit for this class of device is defined as the ratio of its isolation and insertion loss at its midband frequency. It is in part determined by the power density in each region of the structure, the quality of the two hands of polarisation and the relative dielectric constants of the different regions, and it is in part dependent on the dielectric loss tangents of the ferrite and the other dielectric regions, and on the uniform linewidth of the ferrite material. The bandwidth is mainly fixed by the linewidth of the profile of the direct magnetic field. A full theoretical model, although perhaps a standard problem, must still deal, even with the finline gap

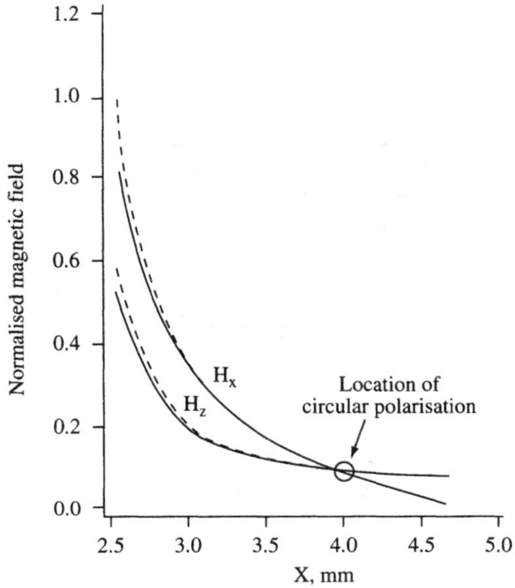

Figure 16.9 Approximate and hybrid mode descriptions of magnetic field in unilateral waveguide

$d/b = 0.25$, $a/b = 2$, $a = 4.775$ mm, $l_2 = 0.127$ mm, $l_1 = 2.387$ mm, $\varepsilon_r = 2.2$

Reprinted with permission, Mansour *et al.* (1988)

equal to that of the narrow dimension of the rectangular waveguide, with a six or more region boundary value problem. Furthermore, some of the regions exhibit tensor properties and magnetic and/or dielectric dissipation parameters – a situation in which the experimental approach is still second to none, provided the material resources are available.

One obvious parameter at hand in the experimental adjustment of this class of device is the finline gap. This was, in one instance, chosen to correspond with that of the input and output finline circuits, and in another case made equal to the narrow dimension of the rectangular waveguide. The circuit in the latter situation is, of course, akin to that of the standard E-plane isolator. The figure of merit of the first arrangement varied between 15 and 18 and that of the second one ranged between 22 and 24. Figure 16.11 indicates the insertion loss and isolation of one typical device in a WR 51 waveguide. This result was obtained with a residual direct magnetic field of about 1.8 kA/m. The dimensions of the layers W_1–W_6 in Figure 16.10 were chosen easily on the basis of experience; the length of the ferrite substrate

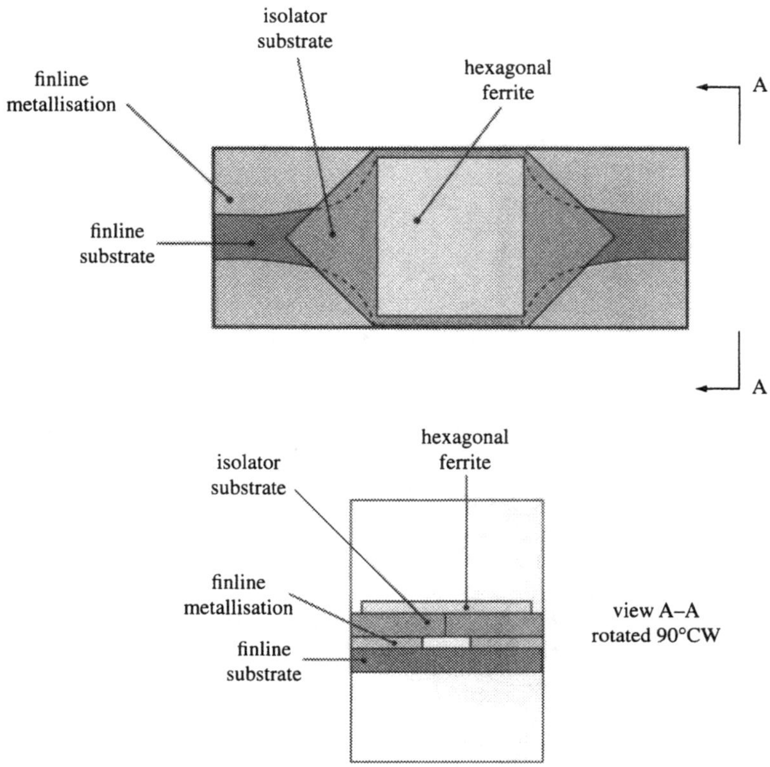

Figure 16.10 Schematic diagram of unilateral resonance isolator

Reprinted with permission, Helszajn & Thorpe (1985)

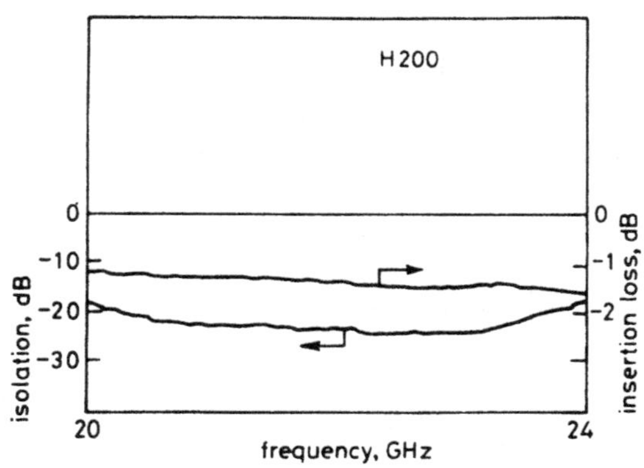

Figure 16.11 Frequency response of unilateral resonance isolator with finline gap equal to narrow dimension of WR 42 waveguide

Reprinted with permission, Helszajn & Thorpe (1985)

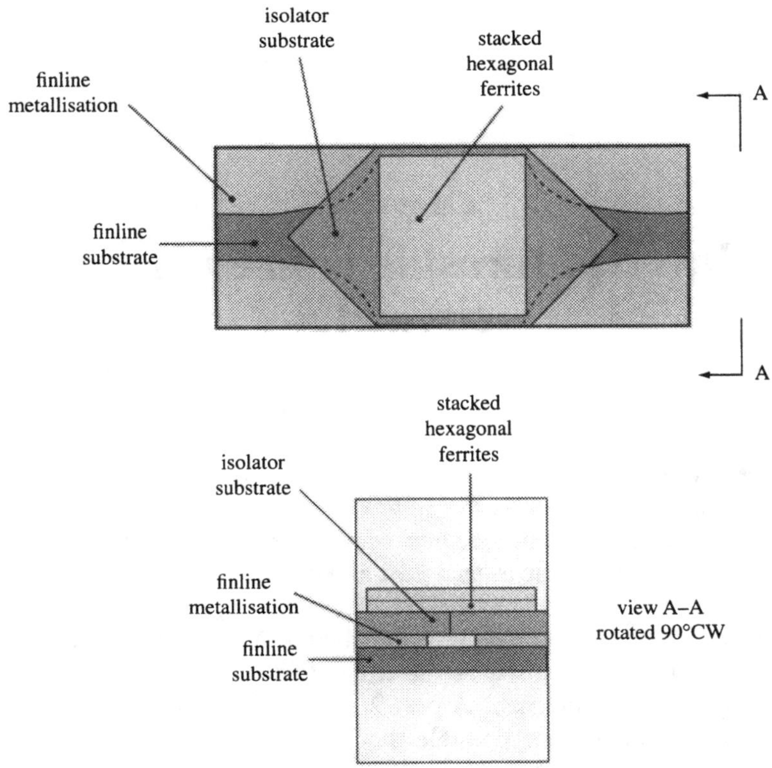

Figure 16.12 Topology of wideband unilateral resonance isolator

Reprinted with permission, Helszajn & Thorpe (1985)

was 45.7 mm. The overall length of the device including the tapers was of the order of six wavelengths. The return loss at each port (not shown) was better than 18 dB. The data in this illustration are not considered optimum in any sense of the word in that the dimensions W_1–W_6 were not varied. The topology of a full bandwidth waveguide arrangement is shown in Figure 16.12.

Chapter 17
Inverted turnstile finline junction circulator

17.1 Introduction

The 3-port circulator is a unique nonreciprocal symmetrical junction having one typical input port, one output port and one decoupled port. The fundamental definition of the junction circulator has its origin in energy conservation. It states that the only matched symmetrical 3-port junction corresponds to the definition of the circulator. In such a junction a wave incident at port 1 is emergent at port 2, one at port 2 at port 3, and so on in a cyclical manner. One possible model of a circulator is a magnetised ferrite or garnet gyromagnetic resonator having three-fold symmetry connected or coupled to three transmission lines or waveguides. The introduction of any such resonator at the junction of three E- or H-plane waveguides or finline circuits readily produces a degree-1 circulation solution. In practice the gyromagnetic resonator is embedded in a filter circuit to produce a degree-2 or -3 frequency response. Since a matched 3-port junction is a circulator by definition, matching such a magnetised resonator is both necessary and sufficient for design purposes. One possibility is to utilise one or two quarter-wave long resonators open-circuited at one end and short-circuited at the other. The operation of a 3-port junction circulator may be understood by using superposition, and this is the approach employed in this text.

17.2 Turnstile junction circulator

The original waveguide junction circulator, known as the turnstile circulator, was first described by Tor Schaug-Pattersen. It consists of a circular guide containing a longitudinally magnetised ferrite section at the junction of three rectangular waveguides. This arrangement relies for its

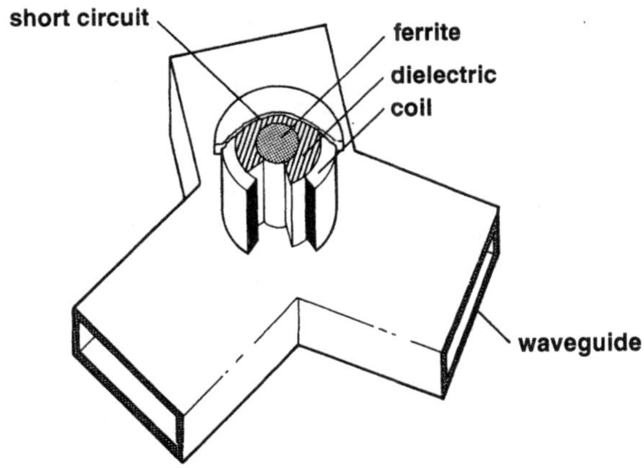

Figure 17.1 Schematic diagram of 3-port H-plane turnstile circulator using single-turnstile resonator

Reprinted with permission, Schaug-Patterson (1958)

operation on the well-known Faraday rotation principle along a magnetised ferrite loaded circular waveguide. Figure 17.1 depicts the original H-plane topology.

The operation of any circulator may be understood by using superposition. It starts by decomposing a single input wave at port 1 (say) into a linear combination of voltage settings at each port:

$$\begin{bmatrix} 1 \\ 0 \\ 0 \end{bmatrix} = \frac{1}{3}\begin{bmatrix} 1 \\ 1 \\ 1 \end{bmatrix} + \frac{1}{3}\begin{bmatrix} 1 \\ \alpha \\ \alpha^2 \end{bmatrix} + \frac{1}{3}\begin{bmatrix} 1 \\ \alpha^2 \\ \alpha \end{bmatrix} \qquad (1)$$

where

$$\alpha = \exp(j120)$$
$$\alpha^2 = \exp(j240)$$

A scrutiny of the first, so-called in-phase, generator setting, indicates that it produces an electric field along the axis of the circular waveguide which does not couple into it. The reflected waves at the three ports of the junction are therefore in this instance unaffected by the details of the gyromagnetic waveguide. A scrutiny of the second and third, so-called counter-rotating, generator settings, indicates, however, that these establish counter-rotating circularly polarised alternating magnetic fields at the open face of the circular gyromagnetic waveguide which readily propagate. The fields produced at the axis of the junction by each of these three possible generator settings are illustrated in Figure 17.2. Since a characteristic of such a waveguide is that it

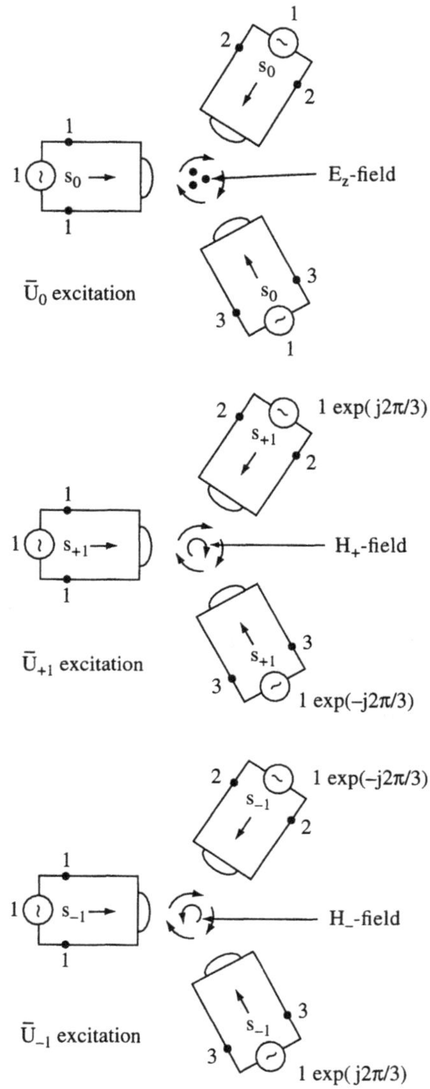

Figure 17.2 Eigenvalue adjustment of junction circulator

has different scalar permeabilities under the two arrangements, it provides one practical means of removing the degeneracy between the reflected waves associated with these two generator settings.

A typical reflected wave at any port is constructed by adding the individual ones due to each possible generator setting. A typical term is realised by taking the product of a typical incident wave and a typical reflection coefficient:

$$\begin{bmatrix} b_1 \\ b_2 \\ b_3 \end{bmatrix} = \frac{\rho_0}{3}\begin{bmatrix} 1 \\ 1 \\ 1 \end{bmatrix} + \frac{\rho_-}{3}\begin{bmatrix} 1 \\ \alpha \\ \alpha^2 \end{bmatrix} + \frac{\rho_+}{3}\begin{bmatrix} 1 \\ \alpha^2 \\ \alpha \end{bmatrix} \tag{2}$$

An ideal circulator is now defined as

$$\frac{\rho_0 + \rho_- + \rho_+}{3} = 0 \tag{3a}$$

$$\frac{\rho_0 + \alpha\rho_- + \alpha^2\rho_+}{3} = -1 \tag{3b}$$

$$\frac{\rho_0 + \alpha^2\rho_- + \alpha\rho_+}{3} = 0 \tag{3c}$$

To adjust this, and other circulators, requires a 120° phase difference between the reflection coefficients of the three different ways it is possible to excite the three rectangular waveguides. One solution is

$$\rho_+ = \exp\left[-j2\left(\theta_1 + \theta_+ + \frac{\pi}{2}\right)\right] \tag{4a}$$

$$\rho_- = \exp\left[-j2\left(\theta_1 + \theta_- + \frac{\pi}{2}\right)\right] \tag{4b}$$

$$\rho_0 = \exp[-j(2\theta_0)] \tag{4c}$$

provided that

$$\theta_1 = \theta_0 = \frac{\pi}{2} \tag{5a}$$

$$\theta_+ = -\theta_- = -\frac{\pi}{6} \tag{5b}$$

The required phase angles of the three reflection coefficients are established by adjusting the length of the demagnetised ferrite section so that the angle between the in-phase and counter-rotating reflection coefficients is initially 180°. The degenerate phase angles of the counter-rotating reflection coefficient are then separated by 120° by magnetising the ferrite region, thereby producing the ideal phase angles of the circulator. These two steps represent the necessary and sufficient conditions for the adjustment of this class of circulator.

Since the relationship between the incident and reflected waves at the terminals of a network or junction is often described in terms of a scattering matrix it is appropriate to reduce the result established here to that notation. Figure 17.3 indicates the nomenclature entering into the definition of this

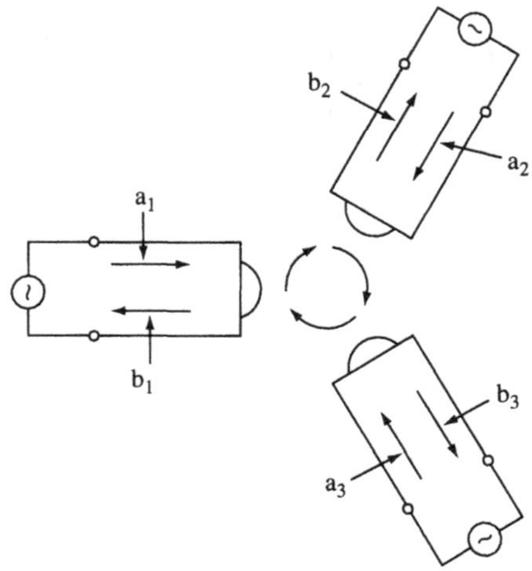

$$\begin{bmatrix} b_1 \\ b_2 \\ b_3 \end{bmatrix} = \begin{bmatrix} S_{11} & S_{21} & S_{31} \\ S_{31} & S_{11} & S_{21} \\ S_{21} & S_{31} & S_{11} \end{bmatrix} \begin{bmatrix} a_1 \\ a_2 \\ a_3 \end{bmatrix}$$

Figure 17.3 Definition of scattering parameters of 3-port junction circulator

matrix in the case of a 3-port network with three-fold symmetry. Its entries relate incident and reflected waves at suitable terminal planes of the circuit:

$$S_{11} = \frac{b_1}{a_1} \qquad a_2 = a_3 = 0 \tag{6a}$$

$$S_{21} = \frac{b_2}{a_1} \qquad a_2 = a_3 = 0 \tag{6b}$$

$$S_{31} = \frac{b_3}{a_1} \qquad a_2 = a_3 = 0 \tag{6c}$$

A scrutiny of these definitions indicates that the entries of the scattering matrix may be readily evaluated once the reflected waves at all the ports due to an incident wave at a typical port are established. Taking a_1 as unity and making use of the results for b_1, b_2 and b_3 gives the required parameters easily:

$$S_{11} = \frac{\rho_0 + \rho_- + \rho_+}{3} \tag{7a}$$

$$S_{21} = \frac{\rho_0 + \alpha\rho_- + \alpha^2\rho_+}{3} \tag{7b}$$

$$S_{31} = \frac{\rho_0 + \alpha^2\rho_- + \alpha\rho_+}{3} \tag{7c}$$

The entries of the scattering matrix are therefore linear combinations of the reflection variables at any port associated with each possible family of generator settings. One definition of an ideal circulator which is in keeping with the description of the turnstile junction circulator is therefore

$$S_{11} = 0 \tag{8a}$$

$$S_{21} = -1 \tag{8b}$$

$$S_{31} = 0 \tag{8c}$$

This solution may be separately established by using the unitary condition and may therefore be taken as a universal definition of a 3-port lossless junction circulator.

17.3 Re-entrant H-plane waveguide circulator

While the operation of the turnstile circulator may be readily visualised, it is not the only arrangement. Another possibility is to use one or two quarter-wave long open triangular or circular resonators open-circuited at one end and short-circuited at the other at the junction of either three E- or H-plane ridge waveguides. In the H-plane arrangement (the usual rectangular waveguide structure) one or two resonators are mounted on the ridges on the axis of the waveguide. In this instance a quarter-wave long magnetised ferrite resonator short-circuited at one end, and open-circuited or loaded by an image wall at the other, is a suitable prototype for the construction of this class of device. The operating frequency of this junction corresponds to the odd solution of two coupled open dielectric resonators constructed from sections of a dielectric waveguide propagating the hybrid HE_{11} mode. The single turnstile geometry has its origin in the turnstile structure indicated in Figure 17.1.

The re-entrant H-plane junction circulator using a single resonator can be visualised as a 5-port network consisting of a 3-port H-plane junction symmetrically coupled along its axis to a cylindrical gyromagnetic waveguide supporting two orthogonal ports. The network is reduced to a 3-port one by closing the gyromagnetic waveguide by a short-circuit piston. The duality between the two arrangements may be achieved by replacing the electric wall in the circular waveguides of the original turnstile junction by open or magnetic surfaces. The first circulation condition in this sort of circulator coincides with that for which the in-phase and degenerate counter-rotating reflection coefficients are in anti-phase. The second one is established by splitting the degeneracy between the counter-rotating ones by a suitable direct magnetic field. No H-plane finline junction circulator has been realised to date.

17.4 Re-entrant E-plane waveguide circulator

Another important waveguide circulator topology is the E-plane one. It is realised by locating quarter-wave long open resonators at each side-wall of the junction of three E-plane waveguides. The operating frequency of this type of junction corresponds to the even eigensolution of two coupled open dielectric resonators. Some appreciation of the operation of the symmetrical configuration employing H-plane turnstile resonators may be gained by recognising that it is essentially a 7-port circuit comprising three E-plane rectangular waveguide or finline ports and two orthogonal ports for each round re-entrant turnstile open waveguide. Since the output terminals of the round waveguide are short-circuited, its overall matrix description reduces to the necessary symmetrical 3-port network. In the E-plane device, with the direction of propagation taken along the z-axis, H_z rather than H_x is perpendicular to the symmetry axis of the device. The magnetic fields corresponding to the counter-rotating eigenvectors are therefore circularly polarised at the side instead of the top and bottom walls of the waveguide. The E-plane topology is readily realised in a finline waveguide. Figure 17.4 depicts one arrangement using a single gyromagnetic waveguide.

17.5 Closed gyromagnetic resonator

The split frequencies of gyromagnetic resonators are important quantities in the description of weakly magnetised resonators. The simplest topology met in connection with this sort of problem region is illustrated in Figure 17.5. It consists of a quarter-wave long demagnetised or magnetised ferrite waveguide with an ideal or an open magnetic wall open-circuited at one end and short-circuited at the other. The characteristic equations associated with the two situations are

$$\cot(\beta_0 L_0) = 0 \tag{9}$$

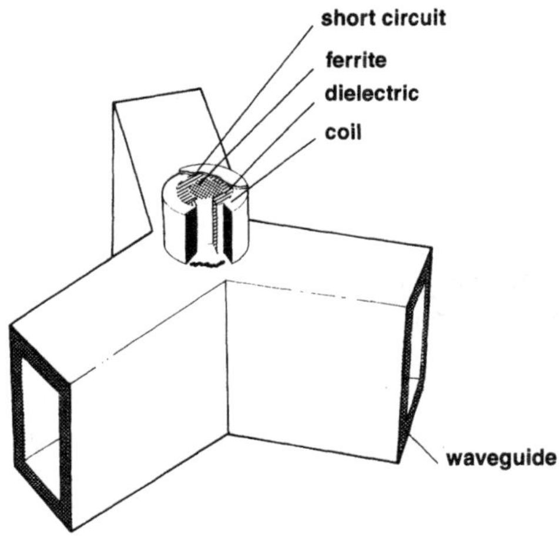

Figure 17.4 Schematic diagram of 3-port E-plane junction using single-turnstile resonator

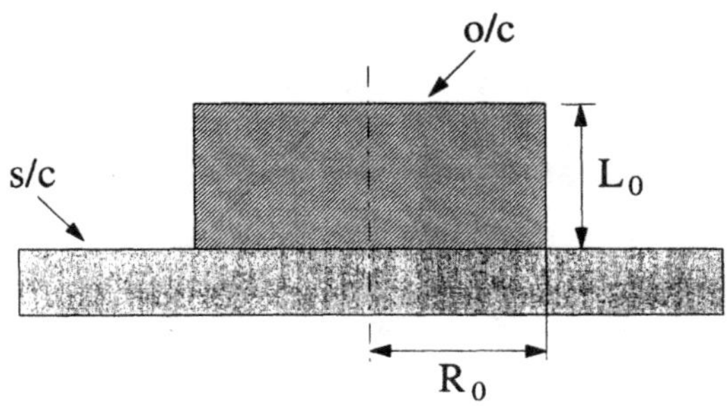

Figure 17.5 Schematic diagram of quarter-wave long dielectric resonator

$$\cot(\beta_{\pm}L_0) = 0 \tag{10}$$

respectively.

The first of these two equations fixes the length of the resonator from a knowledge of k_0R and frequency. The first root of its characteristic equation is

$$\beta_0 L_0 = \frac{\pi}{2} \tag{11}$$

where

$$\beta_0^2 = k_0^2 \mu_{\text{eff}} \varepsilon_{\text{eff}} - \left(\frac{1.84}{R}\right)^2 \tag{12}$$

The second characteristic equation may be solved for the relationship between the split frequencies of the resonator and the magnetic variables in the neighbourhood of the demagnetised one once the characteristic equation for $\beta_{\pm}$ is at hand,

$$\beta_{\pm} L_0 = \frac{\pi}{2} \tag{13}$$

The simplest calculation met in this sort of circuit is the closed gyromagnetic resonator with an ideal magnetic wall. The exact open problem may be derived from that of the related problem region of a partially filled circular waveguide with an electric side-wall by allowing the outside radius to approach infinity, but its solution is outside the remit of this text.

17.6 Perturbation theory of closed cylindrical gyromagnetic resonator

If the only engineering interest is a calculation of the split frequencies of the gyromagnetic resonator then it is sufficient to use a description of the split propagation constants based on perturbation theory. The required result for the split phase constants of a gyromagnetic waveguide with an ideal magnetic wall is

$$\beta_{\pm}^2 \approx k_0^2 \varepsilon_{\mathrm{f}} (\mu \mp C_{11} \kappa) - \left(\frac{1.84}{R}\right)^2 \tag{14}$$

where

$$C_{11} = \frac{2}{(1.84)^2 - 1} \tag{15}$$

$\varepsilon_{\mathrm{eff}}$ is an effective dielectric constant; μ and κ are the off-diagonal elements of the tensor permeability.

This solution correctly displays both the split phase-constants and split cut-off numbers of this type of waveguide.

If a quarter-wave long cavity open-circuited at one flat face and short-circuited at the other is formed from such a waveguide then

$$(k_+ R)^2 = \frac{\left(\frac{\pi}{2}\right)^2 \left(\frac{R}{L}\right)^2 + 1.84^2}{(\mu - C_{11} \kappa) \varepsilon_{\mathrm{f}}} \tag{16a}$$

$$(k_-R)^2 = \frac{\left(\frac{\pi}{2}\right)^2\left(\frac{R}{L}\right)^2 + 1.84^2}{(\mu + C_{11}\kappa)\varepsilon_f} \tag{16b}$$

$$(k_0R)^2 = \frac{\left(\frac{\pi}{2}\right)^2\left(\frac{R}{L}\right)^2 + 1.84^2}{\mu_{eff}\varepsilon_f} \tag{16c}$$

where

$$\mu_{eff} = \frac{\mu^2 - (C_{11}\kappa)^2}{\mu} \tag{17}$$

The split frequencies in such a gyromagnetic resonator are therefore described by

$$\frac{\omega_+ - \omega_-}{\omega_0} = C_{11}\kappa \tag{18}$$

The upper bound on κ is, of course, fixed by the cut-off condition in equation (14). C_{11} is unity for an anisotropic waveguide.

Figure 17.6 depicts one result based on the perturbation formulation of the gyromagnetic resonator.

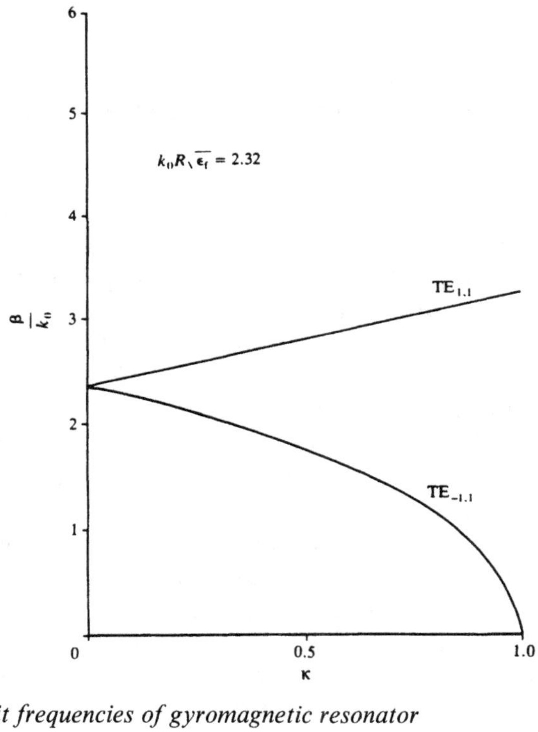

Figure 17.6 Split frequencies of gyromagnetic resonator

17.7 Quality factor of closed gyromagnetic resonator

One of the most important quantities in the theory of quarter-wave coupled circulators is the loaded quality factor (Q_L). This parameter is related to the split frequencies of the magnetised resonator in a particularly simple fashion by

$$Q_L = \left[\sqrt{3}\left(\frac{\omega_+ - \omega_-}{\omega_0}\right)\right]^{-1} \tag{19}$$

Although the calculations of the split frequencies are exact, that of the quality factor neglects the influence of higher order modes (in a strongly magnetised resonator) on the description of the gyrator circuit. It separately assumes that the frequency variation of the in-phase eigen-network can be disregarded compared to those of the split counter-rotating ones. It also assumes that only the first pair of counter-rotating modes need to be catered for in forming the complex gyrator circuit of the device. A knowledge of the onset of the first higher order split pair of modes is therefore desirable. Its description therefore applies to a weakly magnetised resonator only. For the purpose of this work, this condition is satisfied provided the quality factor has a lower bound equal to approximately two. Such a value of loaded Q-factor is compatible with the performance of many commercial devices.

17.8 E-plane finline circulator using coupled H-plane turnstile resonators

One important practical arrangement which is akin to a ridge circulator is the E-plane finline geometry. It consists of a junction of three finlines symmetrically loaded in the H-plane by two open quarter-wave long ferrite resonators. The structure is therefore a seven-port circuit, comprising three symmetrical finline E-plane ports and two orthogonal ports for each round open waveguide port in the H-plane. Figures 17.7*a* and *b* depict arrangements using bilateral and unilateral circuits. Its geometry is fixed by its overall axial dimensions ($2H$), and the radius (R) and length (L) of each ferrite or garnet resonator. Figure 17.8 illustrates one typical finline circuit.

The operating frequency of this sort of device corresponds to the even solution of two coupled $TM_{11\frac{1}{2}}$ resonators with quasi-magnetic walls. Its counter-rotating eigen-networks may therefore be represented in terms of quarter-wave long open resonators, with one flat face open-circuited and the other short-circuited. The details of the junction are separately adjusted so that it is cut-off for the in-phase eigen-network. Its degenerate eigen-networks may therefore be approximately idealised by an electric wall at the

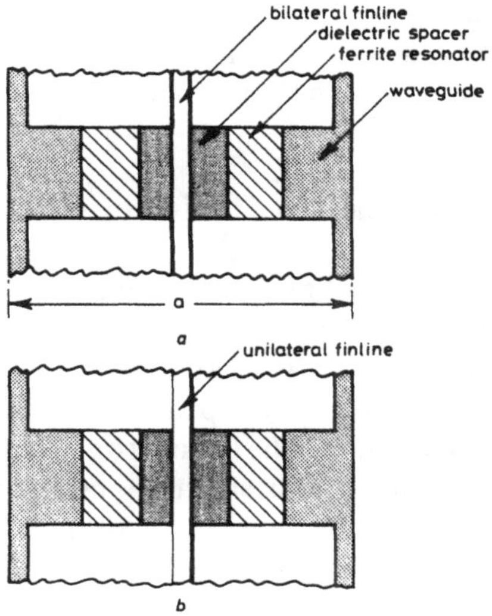

Figure 17.7 Schematic diagrams of E-plane finline circulator using H-plane turnstile resonators

terminals of the junction. Symmetry would otherwise require that it exhibits a magnetic wall there. This arrangement separately ensures that the overall device acts as a bandpass circuit in its demagnetised state in keeping with experiment. The frequency variation of the in-phase eigenvalue in this sort of arrangement is often neglected, compared to those of the split or degenerate counter-rotating ones. The coupling between the terminals of

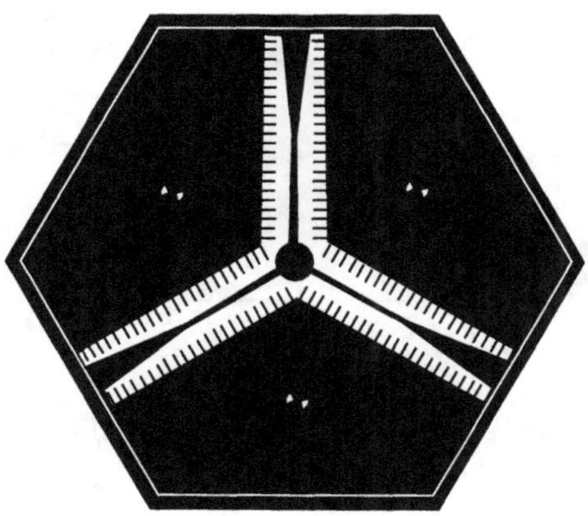

Figure 17.8 Bilateral finline E-plane junction

Reprinted with permission, Helszajn *et al.* (1988)

the eigen-resonators and those of the eigen-networks is less well understood. It is usually described by an ideal transformer; the dielectric spacer provides one independent means of varying this quantity.

17.9 Experimental adjustment of finline turnstile circulator

The relationship between the filling factor L/H and the radial wavenumber k_0R of one experimental device is depicted in Figure 17.9 for four different values of the aspect ratio R/L of the resonator. The even-mode solutions of two coupled $TM_{11\frac{1}{2}}$ resonators, using open and closed walls, are separately superimposed on this illustration. The open resonator model adopted in this calculation consists of a pair of quarter-wave long resonators, with an ideal magnetic side-wall and an effective dielectric constant to cater for the open wall condition, separated by a contiguous section of cut-off waveguide with a magnetic side-wall. The discrepancy between theory and experiment is, in part, due to the fact that the operating frequency of circulators for which the in-phase eigen-network has not been idealised coincides with the frequency

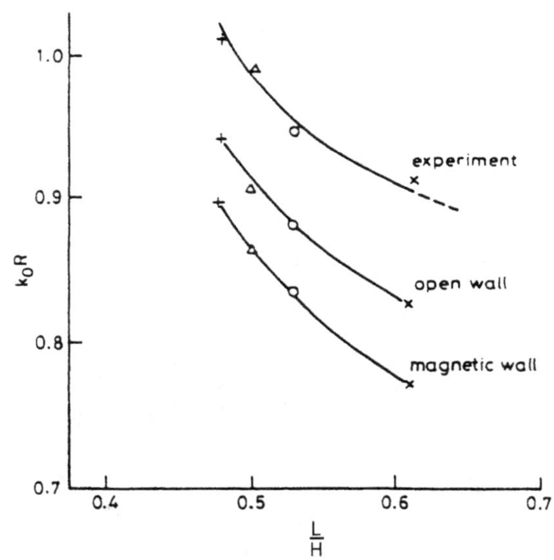

Figure 17.9 Relationship between filling factor and radial wavenumber for different aspect ratios of the resonator

R/L: + 1.96; △ 1.85; ⊙ 1.75; × 1.54

Reprinted with permission, Helszajn, McDonald and Sutherland (1988)

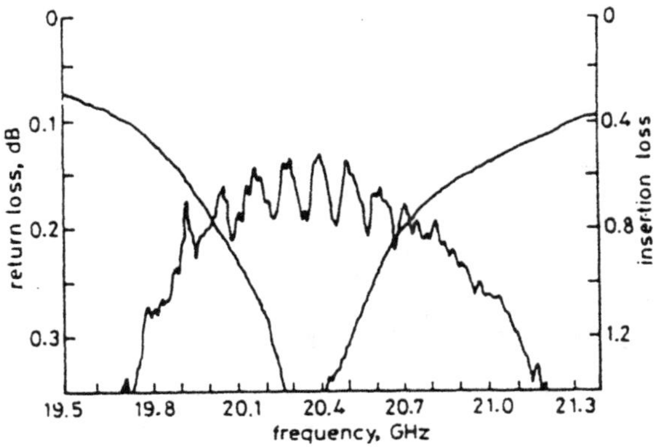

Figure 17.10 Frequency response of directly coupled E-plane finline circulator using H-plane turnstile resonators

Reprinted with permission, Helszajn *et al.* (1988)

at which the phase angles of the counter-rotating and in-phase eigen-networks are in antiphase, rather than at that for which these are commensurate.

The radius (R) of the junction employed in obtaining the results under discussion was fixed at 2.15 mm and its overall axial dimension ($2H$) was kept constant at 4.6 mm. This radius corresponds to that of the junction of the three E-plane WR 42 waveguides employed to form the device. The length (L) of each resonator was separately varied between 1.10 and 1.40 mm; each experimental point in this illustration is therefore associated with a resonator with a different aspect ratio (R/L). The material used in this work was a lithium ferrite with a magnetisation ($\mu_0 M_0$) of 0.4100 T and a relative permittivity (ε_f) of 14.8.

Figure 17.10 depicts the insertion and return losses of one directly coupled circulator. The insertion loss of the device includes that of the finline circuits at each port. It is of note that its susceptance slope parameter is typically of the order encountered in the related E-plane problem region.

Chapter 18

Semi-tracking ridge circulator

18.1 Introduction

Ridge or finline circulators may be realised by introducing a suitable gyromagnetic resonator at the junction of three E- or H-plane waveguides. The arrangement met in connection with the finline waveguide is an E-plane junction embodying one or two re-entrant quarter-wave long gyromagnetic resonators open-circuited at one end and short-circuited at the other. The purpose of this chapter is to describe an H-plane ridge structure which relies for its operation on a planar resonator. Its topology is completely described by its radius, its gyrotropy and the angle that the ridge subtends at a typical resonator terminal. This geometry supports a host of solutions, which may be labelled according to whether the effective permeability is positive or negative. In the former case the gyrotropy of the gyromagnetic resonator may be catalogued according to whether it is very weakly, weakly, moderately or strongly magnetised. A useful rule in choosing any particular solution is to recognise that the gain-bandwidth of this sort of device is determined by the intensity of the gyrotropy. Since the ridge waveguide is essentially a wideband transmission line the strongly magnetised solution is in this instance appropriate for design. The gyromagnetic resonator may also, in some cases, be very strongly magnetised. This situation is associated with edge mode or field displacement effects. It is outside the remit of this work. While propagation in the ridge waveguide is not exactly TEM it will be assumed to be so for the purpose of this work. In practice the admittance level of the complex gyrator circuit of the circulator does not usually coincide with that of the characteristic admittance of the ridge waveguide, so some sort of matching circuit is necessary. The usual topology consists of a cascade of quarter-wave impedance transformers. Its solution is a classic problem in the filter literature but is outside the remit of this work. Its synthesis may therefore proceed in a similar way to that met in connection with the design of stripline devices. A description of the impedance and propagation constant of the dielectric loaded ridge waveguide is also necessary.

18.2 Phenomenological adjustment

In its simplest form the ridge waveguide circulator consists of a gyromagnetic resonator with top and bottom electric walls at the junction of three ridge waveguides. Figure 18.1 illustrates its topology. The phenomenological adjustment of this class of device involves, under some simplifying conditions, the removal of the degeneracy of a pair of counter-rotating modes under the influence of a direct magnetic field and the rotation of a figure of eight standing wave pattern. This is done in such a way as to locate a null in its pattern at one of the three ports of the circulator. The rotation of the standing wave pattern under the application of a direct magnetic field may be understood by decomposing the linearly polarised radio frequency magnetic field on the axis of the resonator into counter-rotating ones, which are then split by its gyrotropy. Figure 18.2 illustrates this situation. In the tracking or semi-tracking solution, to be dealt with here, the split eigen-networks are associated with the $n = +1$ and $n = -2$ split branches of the $n = \pm 1$ and $n = \pm 2$ degenerate modes.

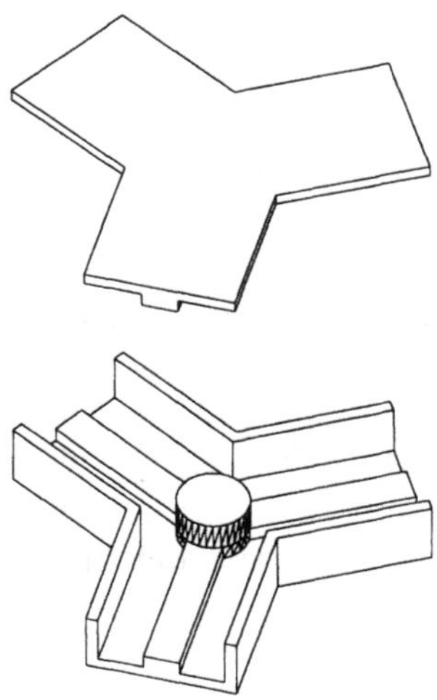

Figure 18.1 Schematic diagram of a ridge circulator

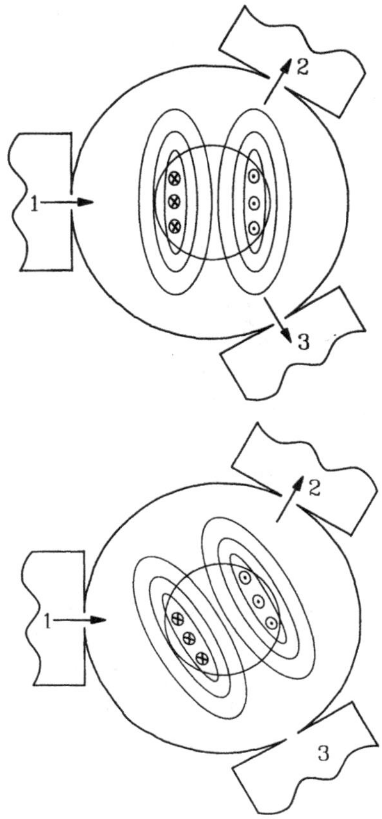

Figure 18.2 Phenomenological operation of waveguide circulator

18.3 Impedance matrix

There are in general six different ways in which circulator boundary conditions can be applied. This may be done in terms of the scattering, impedance and admittance matrices of an ideal circulator and also in terms of the corresponding triplets of eigenvalues. The actual choice is usually determined by the physical problem. It is recalled that although it is always possible to form the scattering matrix of a junction it is not always possible to construct an impedance or admittance one.

The description of a 3-port junction circulator often starts by establishing its impedance matrix,

$$\mathbf{Z} = \begin{bmatrix} Z_{11} & Z_{12} & Z_{13} \\ Z_{13} & Z_{11} & Z_{12} \\ Z_{12} & Z_{13} & Z_{11} \end{bmatrix} \tag{1}$$

The open-circuit parameters of the junction are completely described once its in-phase (Z^0) and counter-rotating (Z^+ and Z^-) impedance eigenvalues are available. The required relationships have the nature met in connection with its scattering variables in Chapter 17:

$$Z_{11} = \frac{Z^0 + Z^+ + Z^-}{3} \tag{2a}$$

$$Z_{12} = \frac{Z^0 + Z^+ \exp(j2\pi/3) + Z^- \exp(-j2\pi/3)}{3} \tag{2b}$$

$$Z_{13} = \frac{Z^0 + Z^+ \exp(-j2\pi/3) + Z^- \exp(j2\pi/3)}{3} \tag{2c}$$

Each of the three 1-port reactances Z^0, Z^+ and Z^- appearing in the descriptions of the open-circuit parameters of a junction circulator may be expanded in terms of its poles in a First Foster or partial fractions form in the manner indicated in Figure 18.3. The symmetric poles and those with the threefold symmetry of the junction are identified with the in-phase eigen-network and the other split poles with the two counter-rotating eigen-networks. Adopting this nomenclature allows the desired 1-port reactances to be defined by

$$Z^0 = \sum Z_n, \qquad n = 0, \pm 3, \pm 6, \pm 9, \ldots \tag{3a}$$

$$Z^+ = \sum Z_n, \qquad n = +1, -2, +4, -5, \ldots \tag{3b}$$

$$Z^- = \sum Z_n, \qquad n = -1, +2, -4, +5, \ldots \tag{3c}$$

The exact nature of a typical pole in the definitions of the eigenvalues of the junction requires a knowledge of the electromagnetic problem. The usual boundary conditions adopted in the description of a planar circuit is that the magnetic field is a constant over the width of each ridge and zero elsewhere:

$$\begin{aligned} &-\psi < \phi < \psi, && H_\phi = H_1 \\ &-120^\circ - \psi < \phi < \psi - 120^\circ, && H_\phi = H_2 \\ &120^\circ - \psi < \phi < \psi + 120^\circ, && H_\phi = H_3 \\ &\text{elsewhere}, && H_\phi = 0 \end{aligned}$$

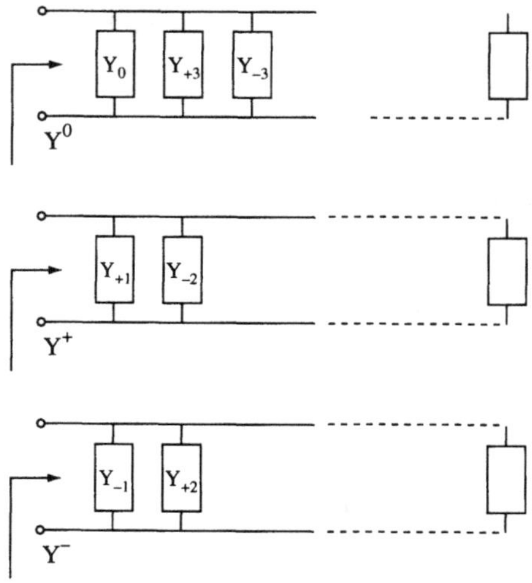

a

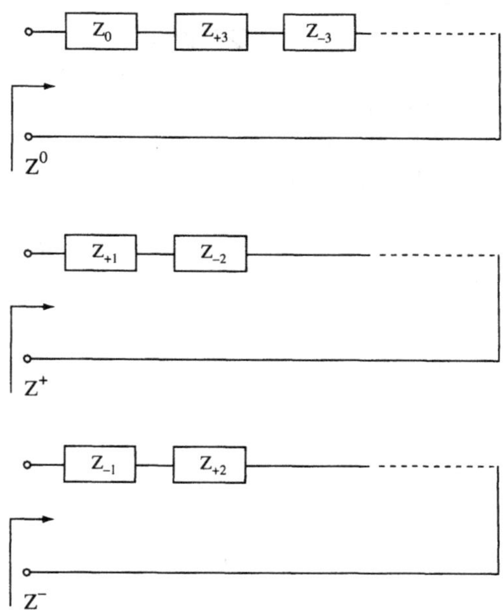

b

Figure 18.3 First Foster Form expansion of in-phase and counter-rotating eigen-networks

The coupling angle (ψ) is defined by

$$\sin\psi = \frac{S}{2R} \tag{4}$$

S is the width of the ridge waveguide at the terminals of the junction. The physical variables of the structure in question is illustrated in Figure 18.4.

A detailed solution of this problem indicates that a typical pole is described by

$$Z_n = \frac{j3Z_e(\omega)\sin^2(n\psi)J_n(k_eR)}{n^2\pi\psi\left[J_n'(k_eR) - \left(\frac{\kappa}{\mu}\right)\frac{nJ_n(k_eR)}{k_eR}\right]} \tag{5}$$

$Z_e(\omega)$ is related to the characteristic impedance of the ridge waveguide $Z_r(\omega)$ at the resonator terminals:

$$Z_e(\omega) = \sqrt{\frac{\mu_e}{\varepsilon_f}}Z_r(\omega) \tag{6}$$

$Z_r(\omega)$ need not be equal to $Z_0(\omega)$.

k_e is given by

$$k_e = k_0\sqrt{\mu_e\varepsilon_f}$$

and

$$\mu_e = \frac{\mu^2 - \kappa^2}{\mu} \tag{7}$$

$\mu, \kappa, k_eR, Z_0(\omega)$ and $Z_r(\omega)$ are all in practice frequency dependent.

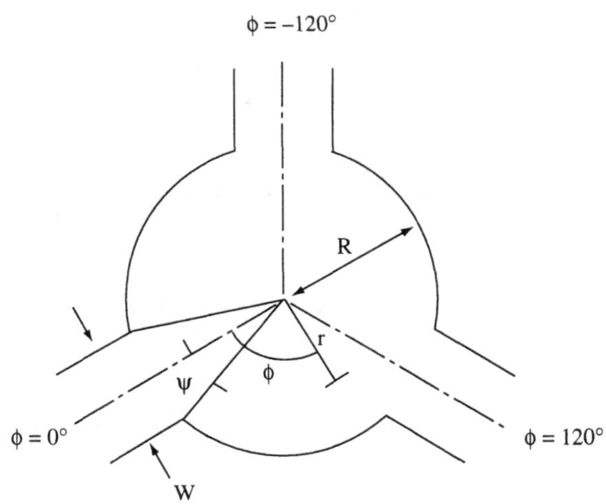

Figure 18.4 Physical variables of ridge circulator

μ and κ are the diagonal and off-diagonal entries of the tensor permeability. $J_n(x)$ is the Bessel function of order n and argument x; $J_n'(x)$ is its derivative.

The input impedance or complex gyrator circuit of a symmetric junction circulator is defined in terms of its open-circuit parameters by setting $V_3 = I_3 = 0$. The solution is a classic result in the literature

$$Z_{in} = Z_{11} + \frac{Z_{12}^2}{Z_{13}} \tag{8}$$

Since the 1-port equivalent circuit of a junction circulator using a gyromagnetic resonator with magnetic side-walls is normally a shunt resonator in parallel with the gyrator conductance of the device. It is therefore usual to express its complex gyrator in terms of Y_{in} instead of Z_{in}:

$$Y_{in} = \frac{1}{Z_{in}} \tag{9}$$

The boundary conditions of a classic circulator are then given by

$$\mathrm{Im}(Y_{in}) = 0 \tag{10a}$$

$$\mathrm{Re}(Y_{in}) = G \tag{10b}$$

The first of these equations determines the centre frequency of the junction, whereas the second fixes its gyrator conductance.

A knowledge of the series form for $J_0(x)$ and $J_1(x)$,

$$\begin{aligned} J_0(x) = 1 &- 2.2499997\left(\frac{x}{3}\right)^2 + 1.2656208\left(\frac{x}{3}\right)^4 \\ &- 0.3163866\left(\frac{x}{3}\right)^6 + 0.0444479\left(\frac{x}{3}\right)^8 \\ &- 0.0039444\left(\frac{x}{3}\right)^{10} + 0.0002100\left(\frac{x}{3}\right)^{12} \end{aligned}$$

$$\begin{aligned} J_1(x) = x\Bigg[0.50 &- 0.56249985\left(\frac{x}{3}\right)^2 + 0.21093573\left(\frac{x}{3}\right)^4 \\ &- 0.03954289\left(\frac{x}{3}\right)^6 + 0.00443319\left(\frac{x}{3}\right)^8 \\ &- 0.00031761\left(\frac{x}{3}\right)^{10} + 0.00001109\left(\frac{x}{3}\right)^{12}\Bigg] \end{aligned}$$

and the recurrence formulae

$$J_{n+1}(x) = J_{n-1}(x) + \frac{2n}{x} J_n(x)$$

$$J_{-n}(x) = (-1)^n J_n(x)$$

is sufficient for computation.

18.4 Complex gyrator circuit

The nature of the complex gyrator circuit of a junction circulator may be constructed by taking the real and imaginary parts of Y_{in}. This gives

$$G_{in} = \frac{R_{in}}{R_{in}^2 + X_{in}^2} \tag{11a}$$

$$B_{in} = \frac{-jX_{in}}{R_{in}^2 + X_{in}^2} \tag{11b}$$

The two circulation conditions at a single frequency are now deduced by satisfying equations (10a) and (10b). The first condition defines the planar circuit in terms of ψ, κ/μ and $k_e R$, and the second fixes the absolute conductance level in terms of the admittance (Y_r) of the ridge waveguide at the resonator terminals. The quality of the frequency response and the gain-bandwidth of a circulator is usually expressed in terms of the susceptance slope parameter (B') or the quality factor (Q_L) of its complex gyrator circuit.

The susceptance slope parameter is determined from a knowledge of the imaginary part of Y_{in} in the vicinity of the first circulation condition,

$$B' = \frac{k_0 R}{2}\left[\frac{B_{in}(\delta) - B_{in}(-\delta)}{k_0 R(1+\delta) - k_0 R(1-\delta)}\right] = \frac{B_{in}(\delta) - B_{in}(-\delta)}{4\delta} \tag{12}$$

where

$$\delta = \frac{\omega - \omega_0}{\omega_0} \tag{13}$$

The quality factor (Q_L) of the circuit is calculated by forming

$$Q_L = \frac{B'}{G_0} \tag{14}$$

The two most important parameters in the synthesis of any circulator are the loaded Q-factor of the junction and the frequency interval over which the complex gyrator circuit has a nearly frequency independent conductance and a nearly constant susceptance slope. Once these quantities are fixed (by the coupling angle of the resonator and its gyrotropy), it is merely necessary to fix the absolute levels of the real and imaginary parts of the junction by adjusting the gap between the ridges to meet the required specification. A description of the junction in terms of its complex gyrator admittance and loaded Q-factor is therefore necessary and sufficient. Although no closed form solution is in general available it may be derived graphically from a knowledge of the frequency response of the device. A detailed investigation of this problem indicates that a host of semi-tracking solutions may be realised in the vicinity of the tracking one by properly adjusting the details of the junction. One possibility is the design of a degree-3 equal ripple frequency response over one octave.

18.5 Semi-tracking complex gyrator circuit

The classic junction circulator using a planar disk resonator exhibits a host of solutions, some of which are well behaved and others which are not. A particularly attractive one is the so-called semi-tracking one for which the gyrator circuit is a nearly frequency independent conductance over approximately an octave frequency band. This circulator relies for its operation on a standing wave solution established by split counter-rotating eigen-networks whose degenerate eigenvalues do not have common poles. Figure 18.5 depicts the mode chart of a gyromagnetic resonator. The semi-tracking solution is approximately defined with the gyrotropy between $0.50 < \kappa < 1.0$ and $\mu = 1.0$.

Table 18.1 depicts semi-tracking solutions with κ/μ and ψ in the neighbourhood of the tracking solution. Such solutions are particularly attractive for the design of octave band devices.

Intermediate values of ψ and Q_L may be approximated in terms of the first two terms of the Taylor expansion or by some more elaborate interpolation procedure,

$$\psi_i = \psi_{i-1} + \frac{\psi_{i-1} - \psi_{i+1}}{Q_{i-1} - Q_{i+1}}(Q_i - Q_{i-1}) \tag{15}$$

i indicates the required quantities; $i-1$ and $i+1$ refer to the known quantities.

The results in these tables marked by an asterisk are not suitable for the design of such devices in that the solutions defined by these boundary conditions exhibit a reversal in the direction of the circulation at the high frequency end of the band.

While a knowledge of the quality factor is mandatory it is in itself not sufficient in that it is also necessary to ensure that the complex gyrator circuit is well behaved over the frequency interval of interest. Figure 18.6 illustrates the frequency response of one typical semi-tracking solution. It indicates

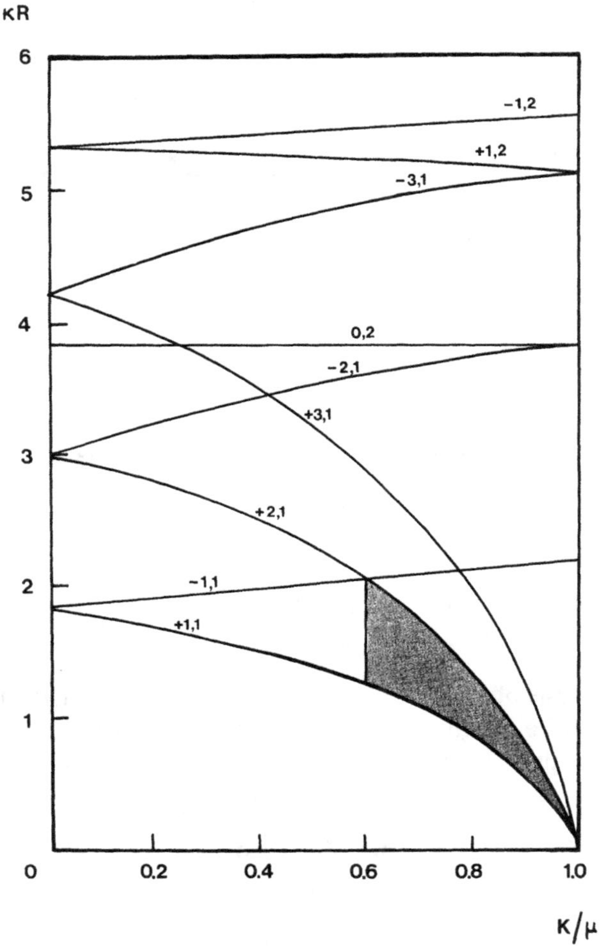

Figure 18.5 Split mode chart of gyromagnetic post-resonator

that the frequency characteristics of semi-tracking solutions using disk resonators are exceptionally well behaved and are indeed appropriate for the design of octave-band devices. The complex gyrator of this solution is fixed by

$$G = 0.059\,\mathrm{S}$$

$$B' = 0.030$$

$$Q_{\mathrm{L}} = 0.515$$

$$2\delta_0 = 0.65$$

$$Z_{\mathrm{r}} = 50\Omega$$

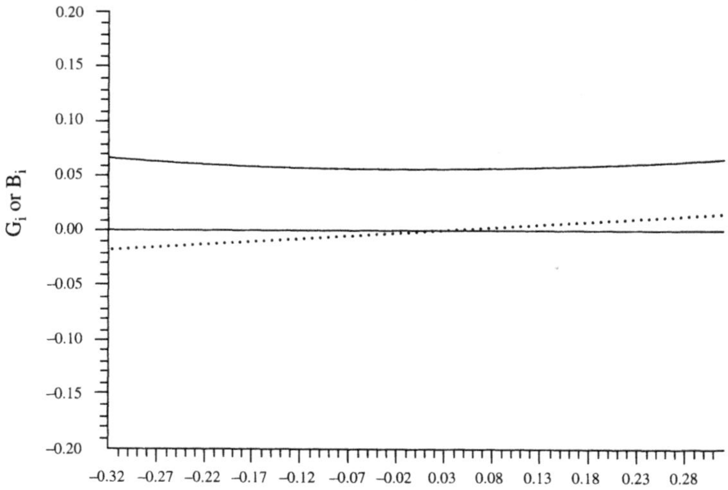

Figure 18.6 The real and imaginary parts of the complex gyrator circuit with $\mu = 1$, $\kappa = 0.65$, $\psi = 0.65$ *rad*, $Z_r = 50\ \Omega$, $kR = 1.291$

Reprinted with permission, Helszajn (1995)

While the synthesis of extended octave-band devices is outside the scope of this work it may be worthwhile to observe that well behaved semi-tracking gyrator circuits may also be realised. Figure 18.7 illustrates one such solution. Its complex gyrator is characterised by

$$G = 0.0456\,\mathrm{S}$$

$$B' = 0.2605$$

$$Q_L = 0.626$$

$$2\delta_0 = 0.70$$

$$Z_r = 60\Omega$$

The immittance levels in these two examples apply to stripline calculators with 50 Ohm terminations. The immittance levels in these solutions must therefore be renormalised to the midband impedance of ridge waveguide $(Z_0(\omega))$ in order to cater for the ridge geometry

$$Z_0(\omega) = Z_0(\infty)\left(\frac{\lambda g}{\lambda_0}\right)$$

A perusal of Table 18.1 indicates that the quality factor of the gyrator circuit is not unique in that some trade-off is possible between the gyrotropy and the coupling angle. In general, the use of wide coupling angles and large

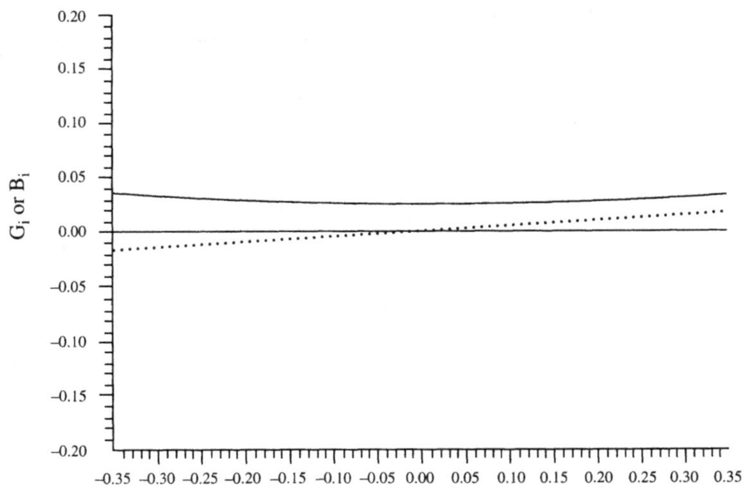

Figure 18.7 The real and imaginary parts of the complex gyrator circuit with $\mu = 1$, $\kappa = 0.60$, $\psi = 0.70$ *rad,* $Z_r = 60\ \Omega$, $kR = 1.350$

Reprinted with permission, Helszajn (1995)

values of magnetisation moves the frequency at which the direction of circulation reverses in this type of junction outside the required passband of the specification. It also produces, loosely speaking, the more optimum frequency responses. The synthesis of devices with very low ripple levels is, of course, quite demanding on the quality of the equivalent circuit. It, by and large, appears to be good enough for the realisation of degree-3 equiripple octave-band devices, with a typical value of VSWR $\leqslant$ 1.20 at the bandedges.

18.6 Direct magnetic field and magnetisation of semi-tracking circulators

To proceed with the design of circulators and other nonreciprocal ferrite devices, it is necessary to have some appreciation of the magnetic variables entering into the description of the gyromagnetic resonator. The two quantities entering into its description are the gyrotropy (κ/μ) and the effective permeability (μ_{eff}). One essential material requirement for this type of

Table 18.1 Complex gyrator variables of semi-tracking circulators

ψ	kR	G	B'	Q_L
0.543*	1.6610	$0.8880\,Y_f$	$0.4333\,Y_f$	0.4880
0.550*	1.6453	$0.8727\,Y_f$	$0.4480\,Y_f$	0.5138
0.575*	1.6018	$0.8238\,Y_f$	$0.4855\,Y_f$	0.5894
0.600*	1.5685	$0.7841\,Y_f$	$0.5074\,Y_f$	0.6472
0.625	1.5413	$0.7506\,Y_f$	$0.5197\,Y_f$	0.6924
0.650	1.5183	$0.7219\,Y_f$	$0.5253\,Y_f$	0.7275
0.675	1.4985	$0.6971\,Y_f$	$0.5267\,Y_f$	0.7543
0.700	1.4810	$0.6763\,Y_f$	$0.5233\,Y_f$	0.7739

* See comment below eqn (15)
$\kappa = 0.525$, $\mu = 1$

ψ	kR	G	B'	Q_L
0.538*	1.6062	$0.9302\,Y_f$	$0.3486\,Y_f$	0.3486
0.550	1.5727	$0.9011\,Y_f$	$0.3628\,Y_f$	0.4021
0.575	1.5241	$0.8507\,Y_f$	$0.4118\,Y_f$	0.4842
0.600	1.4879	$0.8098\,Y_f$	$0.5446\,Y_f$	0.5446
0.625	1.4589	$0.7757\,Y_f$	$0.5905\,Y_f$	0.5905
0.650	1.4347	$0.7466\,Y_f$	$0.6254\,Y_f$	0.6254
0.675	1.4139	$0.7216\,Y_f$	$0.4670\,Y_f$	0.6513
0.700	1.3956	$0.7002\,Y_f$	$0.4689\,Y_f$	0.6697

* See comment below eqn (15)
$\kappa = 0.575$, $\mu = 1$

ψ	kR	G	B'	Q_L
0.535*	1.5740	$0.947\,Y_f$	$0.2772\,Y_f$	0.2927
0.550*	1.5290	$0.9093\,Y_f$	$0.3305\,Y_f$	0.3634
0.575*	1.4794	$0.8598\,Y_f$	$0.3834\,Y_f$	0.4453
0.600	1.4428	$0.8192\,Y_f$	$0.4138\,Y_f$	0.5051
0.625	1.4136	$0.7852\,Y_f$	$0.4319\,Y_f$	0.5500
0.650	1.3892	$0.7561\,Y_f$	$0.4416\,Y_f$	0.5839
0.675	1.3682	$0.7313\,Y_f$	$0.4452\,Y_f$	0.6089
0.700	1.3498	$0.7098\,Y_f$	$0.4445\,Y_f$	0.6263

* See comment below eqn (15)
$\kappa = 0.600$, $\mu = 1$

continued on next page

Table 18.1 continued

ψ	kR	G	B'	Q_L
0.531*	1.5384	$0.9537 Y_f$	$0.2433 Y_f$	0.2552
0.550	1.4806	$0.9151 Y_f$	$0.3053 Y_f$	0.3336
0.575	1.4313	$0.8666 Y_f$	$0.3581 Y_f$	0.4133
0.600	1.3949	$0.8267 Y_f$	$0.3897 Y_f$	0.4714
0.625	1.3658	$0.7930 Y_f$	$0.4083 Y_f$	0.5150
0.650	1.3414	$0.7642 Y_f$	$0.4185 Y_f$	0.5475
0.675	1.3205	$0.7396 Y_f$	$0.4224 Y_f$	0.5712
0.700	1.3021	$0.7183 Y_f$	$0.4220 Y_f$	0.5876

* See comment below eqn (15)
$\kappa = 0.625$, $\mu = 1$

ψ	kR	G	B'	Q_L
0.550	1.4281	$0.9818 Y_f$	$0.2853 Y_f$	0.3107
0.575	1.3799	$0.8715 Y_f$	$0.3372 Y_f$	0.3870
0.600	1.3442	$0.8324 Y_f$	$0.3685 Y_f$	0.4427
0.625	1.3155	$0.7993 Y_f$	$0\ 3872 Y_f$	0.4844
0.650	1.2915	$0.7708 Y_f$	$0.3972 Y_f$	0.5153
0.675	1.2708	$0.7465 Y_f$	$0.4013 Y_f$	0.5376
0.700	1.2525	$0.7255 Y_f$	$0.4009 Y_f$	0.5527

* See comment below eqn (15)
$\kappa = 0.650$, $\mu = 1$

ψ	kR	G	B'	Q_L
0.522	1.4650	$0.9830 Y_f$	$0.1735 Y_f$	0.1765
0.550	1.3834	$0.9200 Y_f$	$0.2726 Y_f$	0.2963
0.575	1.3367	$0.8744 Y_f$	$0.3228 Y_f$	0.3692
0.600	1.3017	$0.8359 Y_f$	$0.3533 Y_f$	0.4227
0.625	1.2737	$0.8034 Y_f$	$0.3717 Y_f$	0.4626
0.650	1.2501	$0.7755 Y_f$	$0.3816 Y_f$	0.4921
0.675	1.2296	$0.7513 Y_f$	$0.3855 Y_f$	0.5131
0.700	1.2115	$0.7303 Y_f$	$0.3850 Y_f$	0.5272

* See comment below eqn (15)
$\kappa = 0.670$, $\mu = 1$

Reprinted with permission, Helszajn (1995)

circulator is that the ferrite should be saturated. The diagonal and off-diagonal elements of the tensor permeability are given in this situation by

$$|\kappa| = \frac{\omega_{\mathrm{m}}}{\omega}$$

$$\mu = 1$$

The effective permeability is

$$\mu_{\mathrm{eff}} = 1 - \left(\frac{\omega_{\mathrm{m}}}{\omega}\right)^2$$

where

$$\omega_{\mathrm{m}} = \gamma M_0$$

γ is the gyromagnetic ratio (2.21×10^5 rad/s per A/m), M_0 is the magnetisation (A/m), μ_0 is the free space permeability ($4\pi \times 10^{-7}$ H/m), and ω is the radian frequency (rad/s). μ and κ are the diagonal and off-diagonal entries of the tensor permeability.

The internal direct magnetic field (H_i) differs from that of the external field (H_0). The two quantities are related by the direct magnetisation (M_0) and the shape demagnetising factor (N_z) by

$$H_i = H_0 - N_z M_0$$

where

$$N_z \approx 1 - \left(\frac{H}{2R}\right)\left\{1 + \left(\frac{H}{2R}\right)^2\right\}^{-\frac{1}{2}}$$

H is the thickness of the ferrite disk (m), R is its radius (m), N_z is the demagnetising factor, and

$$H_0 = N_z M_0$$

The second essential material requirement in the design of any circulator is the value of the gyrotropy at the centre frequency. One value met in connection with the design of the tracking circulator is

$$\left|\frac{\kappa}{\mu}\right| = 0.67$$

This choice of material ensures that the gyrotropy of the resonator varies in the region

$$0.50 < \left|\frac{\kappa}{\mu}\right| < 1$$

for

$$0.50 < \left|\frac{\omega_{\mathrm{m}}}{\omega}\right| < 1$$

18.7 Physical variables of semi-tracking circulators

The physical and magnetic variables in the tables may be understood by constructing an example. One possibility is defined by

$$\kappa = \frac{\omega_m}{\omega} \approx 0.67$$

$$\mu = 1$$

$$\psi \approx 0.522$$

$$k_e R \approx 1.465$$

In the preceding equations kR and ψ are defined in the standard way:

$$k_e R = \frac{2\pi\sqrt{\mu_e \varepsilon_f}}{\lambda_0} R \tag{16}$$

$$\sin\psi = \frac{S}{2R} \tag{17}$$

μ_e and ω_m have the usual meanings and ε_f is the relative dielectric constant of the ferrite material. S, R and ψ are defined in Figure 18.2; H is the thickness of the resonator.

The absolute gyrator conductance realised by the above variables is given by

$$G = 0.9822\,Y_f \tag{18}$$

where

$$Y_f = Y_r(\omega)\sqrt{\varepsilon_f} \tag{19}$$

$Y_r(\omega)$ is the free space conductance at finite frequency of the ridge waveguide at a typical port. It is fixed by the coupling angle and the gap between the ridges. The dielectric constant ε_t of the dielectric region adjacent to the junction may then be employed to set the admittance Y_{0n-1} of the first quarter-wave transformer adjacent to the junction:

$$Y_{0n-1} = Y_r(\omega)\sqrt{\varepsilon_t} \tag{20}$$

18.8 Network problem

A 1-port G-STUB load for which the real part (conductance) is nearly frequency independent and the imaginary part (susceptance) has a small but nonzero value is compatible with the synthesis of a degree-3 equal ripple frequency specification. The topology of this network and its

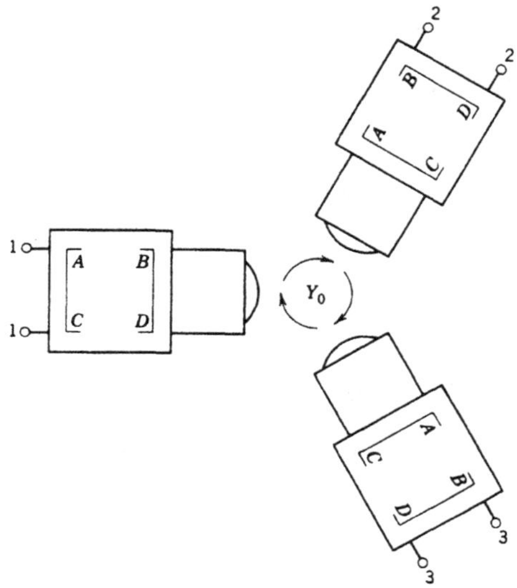

Figure 18.8 Topology of 3-port circulator with 2-port matching networks

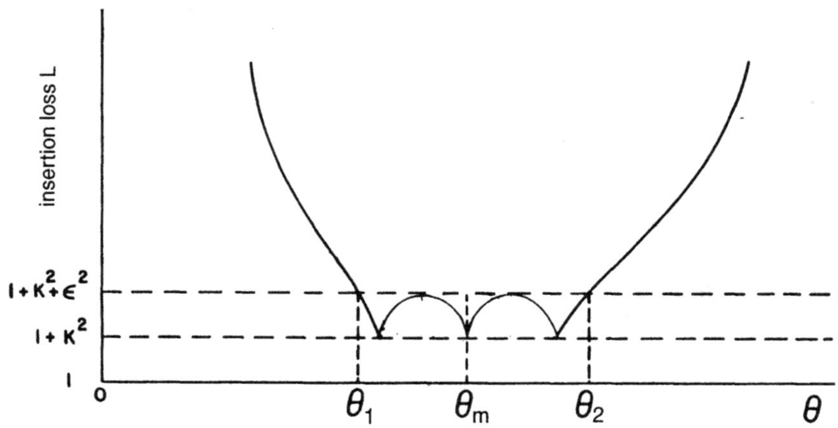

Figure 18.9 Frequency response of n = 3 network with nonzero minima in the reflection coefficient

$$S(\max) = (\sqrt{1 + k^2 + \varepsilon^2} + \sqrt{k^2 + \varepsilon^2})^2$$

$$S(\min) = (\sqrt{1 + k^2} + k)^2$$

$$w = \frac{2(\theta_2 - \theta_1)}{\theta_2 + \theta_1} = \frac{\theta_2 - \theta_1}{\theta_m}$$

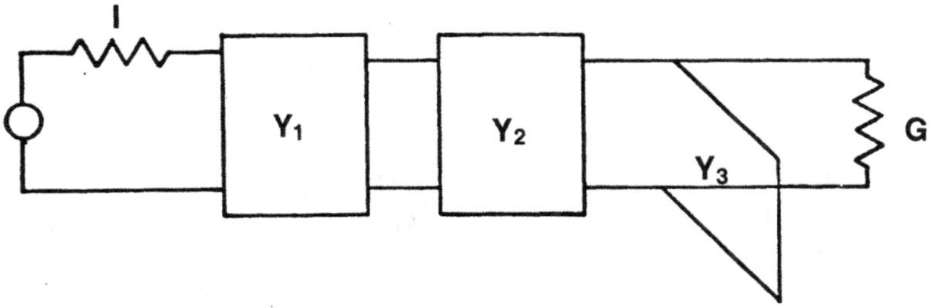

Figure 18.10 Topology of quarter-wave coupled complex gyrator circuit

frequency response are illustrated in Figures 18.8 and 18.9. The topology of the 1-port complex gyrator circuit is specified in Figure 18.10. The relationship between the frequency response of the network and the circuit elements is a standard problem in the literature. Table 18.2 relates the specifications (VSWR(max), VSWR(min)) and bandwidth to the elements of the gyrator circuit (normalised susceptance slope parameter b', normalised conductance g, loaded Q-factor Q_L). It also defines the circuit admittances (y_1 and y_2).

A scrutiny of the network problem suggests that a precise realisation of the susceptance slope parameter may not be as critical as is historically supposed provided that the minima in the reflection coefficient of the device are not forced to pass through zero. This result suggests that the network problem can accommodate some uncertainty in the definition of the coupling angle.

18.9 Frequency response

The frequency response of a quarter-wave coupled circulator may be traced by forming the reflection coefficient at the input terminals of the 1-port complex gyrator network. The topology in question is illustrated in Figure 18.10. This allows the frequency response of the quarter-wave coupled junction to be displayed without difficulty:

$$|\Gamma_{in}|^2 = \frac{[AR_{in} - Z_0(D - CX_{in})]^2 + [B + AX_{in} - Z_0CR_{in}]^2}{[AR_{in} + Z_0(D - CX_{in})]^2 + [B + AX_{in} + Z_0CR_{in}]^2}$$

A, B, C and D are the parameters of the overall region. The ABCD parameters of a typical transformer region are

Table 18.2 Network variables for degree-3 circulators

Degree N = 3 S(max) = 1.15 S(min) = 1

W	*G*	*B′*	*Q*	*Y1*	*Y2*
0.200	969.349	3486.515	3.597	4.732	147.312
0.250	399.553	1138.049	2.848	3.818	76.309
0.300	194.299	455.324	2.343	3.214	44.805
0.350	106.011	209.635	1.977	2.789	28.712
0.400	62.989	106.970	1.698	2.474	19.635
0.450	39.988	59.049	1.477	2.233	14.123
0.500	26.777	34.687	1.295	2.044	10.579
0.550	18.741	21.430	1.143	1.893	8.196
0.600	13.620	13.805	1.014	1.770	6.533
0.667	9.356	8.106	0.866	1.639	5.013
0.700	7.902	6.336	0.802	1.584	4.454
0.750	6.265	4.472	0.714	1.514	3.789
0.800	5.083	3.229	0.635	1.454	3.277
0.850	4.211	2.378	0.565	1.403	2.878
0.900	3.555	1.782	0.501	1.359	2.562
0.950	3.054	1.356	0.444	1.321	2.309
1.000	2.665	1.045	0.392	1.289	2.104

Degree N = 3 S(max) = 1.15 S(min) = 1.02

W	*G*	*B′*	*Q*	*Y1*	*Y2*
0.200	893.332	3313.304	3.709	4.559	137.630
0.250	368.379	1081.745	2.936	3.681	71.355
0.300	179.240	432.918	2.415	3.102	41.940
0.350	97.862	199.388	2.037	2.693	26.909
0.400	58.196	101.785	1.749	2.392	18.428
0.450	36.983	56.215	1.520	2.161	13.275
0.500	24.794	33.042	1.333	1.981	9.961
0.550	17.378	20.428	1.176	1.836	7.731
0.600	12.649	13.170	1.041	1.719	6.175
0.667	8.709	7.742	0.889	1.594	4.752
0.700	7.366	6.055	0.822	1.542	4.228
0.750	5.852	4.279	0.731	1.475	3.604
0.800	4.758	3.093	0.650	1.418	3.125
0.850	3.951	2.280	0.577	1.370	2.751
0.900	3.344	1.711	0.512	1.329	2.455
0.950	2.880	1.303	0.453	1.294	2.217
1.000	2.520	1.006	0.399	1.263	2.025

continued on next page

Table 18.2 continued

Degree N = 3 S(max) = 1.15 S(min) = 1.04

W	*G*	*B'*	*Q*	*Y1*	*Y2*
0.200	801.796	3043.685	3.796	4.363	125.987
0.250	330.850	994.045	3.005	3.525	65.394
0.300	161.116	397.989	2.470	2.974	38.491
0.350	88.060	183.398	2.083	2.585	24.738
0.400	52.436	93.683	1.787	2.298	16.973
0.450	33.374	51.781	1.552	2.080	12.253
0.500	22.416	30.463	1.359	1.909	9.216
0.550	15.744	18.854	1.198	1.772	7.171
0.600	11.488	12.170	1.059	1.662	5.743
0.667	7.939	7.168	0.903	1.544	4.437
0.700	6.728	5.612	0.834	1.495	3.956
0.750	5.362	3.972	0.741	1.432	3.383
0.800	4.375	2.876	0.657	1.379	2.942
0.850	3.646	2.124	0.583	1.334	2.598
0.900	3.098	1.597	0.515	1.296	2.326
0.950	2.678	1.219	0.455	1.263	2.108
1.000	2.353	0.943	0.401	1.235	1.932

Degree N = 3 S(max) = 1.15 S(min) = 1.06

W	*G*	*B'*	*Q*	*Y1*	*Y2*
0.200	697.205	2688.010	3.855	4.139	112.522
0.250	287.967	878.302	3.050	3.348	58.498
0.300	140.406	351.866	2.506	2.828	34.498
0.350	76.861	162.270	2.111	2.462	22.222
0.400	45.854	82.970	1.809	2.193	15.286
0.450	29.252	45.912	1.570	1.988	11.067
0.500	19.700	27.047	1.373	1.827	8.351
0.550	13.880	16.766	1.208	1.700	6.521
0.600	10.163	10.841	1.067	1.597	5.242
0.667	7.062	6.402	0.907	1.488	4.071
0.700	6.002	5.020	0.836	1.443	3.640
0.750	4.806	3.561	0.741	1.385	3.126
0.800	3.941	2.585	0.656	1.336	2.731
0.850	3.302	1.914	0.580	1.295	2.423
0.900	2.820	1.443	0.512	1.260	2.178
0.950	2.451	1.104	0.450	1.230	1.983
1.000	2.165	0.856	0.396	1.205	1.825

continued on next page

Table 18.2 continued

Degree N = 3 S(max) = 1.15 S(min) = 1.08

W	*G*	*B'*	*Q*	*Y1*	*Y2*
0.200	580.786	2253.269	3.880	3.881	97.189
0.250	240.220	736.770	3.067	3.144	50.639
0.300	117.341	295.436	2.518	2.660	29.945
0.350	64.382	136.404	2.119	2.321	19.351
0.400	38.519	69.843	1.813	2.071	13.360
0.450	24.657	38.714	1.570	1.882	9.713
0.500	16.672	22.852	1.371	1.735	7.362
0.550	11.801	14.198	1.203	1.618	5.777
0.600	8.687	9.205	1.060	1.524	4.668
0.667	6.083	5.458	0.897	1.425	3.653
0.700	5.193	4.288	0.826	1.384	3.279
0.750	4.187	3.052	0.729	1.332	2.832
0.800	3.458	2.224	0.643	1.288	2.489
0.850	2.918	1.653	0.566	1.251	2.222
0.900	2.511	1.250	0.498	1.220	2.010
0.950	2.200	0.961	0.437	1.194	1.840
1.000	1.957	0.748	0.382	1.171	1.703

Degree N = 3 S(max) = 1.15 S(min) = 1.1

W	*G*	*B'*	*Q*	*Y1*	*Y2*
0.200	452.332	1742.610	3.853	3.572	79.668
0.250	187.508	570.432	3.042	2.900	41.652
0.300	91.859	229.069	2.494	2.460	24.733
0.350	50.587	105.957	2.095	2.153	16.060
0.400	30.402	54.376	1.789	1.928	11.150
0.450	19.567	30.223	1.545	1.758	8.157
0.500	13.315	17.897	1.344	1.627	6.225
0.550	9.494	11.160	1.175	1.523	4.921
0.600	7.047	7.266	1.031	1.440	4.008
0.667	4.996	4.334	0.868	1.353	3.171
0.700	4.293	3.417	0.796	1.317	2.862
0.750	3.497	2.445	0.699	1.272	2.494
0.800	2.920	1.791	0.613	1.234	2.212
0.850	2.492	1.338	0.537	1.203	1.991
0.900	2.168	1.018	0.470	1.176	1.816
0.950	1.920	0.787	0.410	1.154	1.677
1.000	1.727	0.616	0.356	1.135	1.564

continued on next page

Table 18.2 continued

Degree N = 3 S(max) = 1.15 S(min) = 1.12

W	*G*	*B'*	*Q*	*Y1*	*Y2*
0.200	309.011	1152.644	3.730	3.173	59.027
0.250	128.630	378.096	2.939	2.587	31.049
0.300	63.360	152.247	2.403	2.205	18.573
0.350	35.134	70.666	2.011	1.939	12.165
0.400	21.296	36.419	1.710	1.746	8.529
0.450	13.846	20.345	1.469	1.602	6.308
0.500	9.535	12.120	1.271	1.491	4.872
0.550	6.891	7.609	1.104	1.404	3.901
0.600	5.192	4.991	0.961	1.335	3.220
0.667	3.763	3.010	0.800	1.264	2.596
0.700	3.272	2.387	0.730	1.236	2.366
0.750	2.714	1.724	0.635	1.199	2.091
0.800	2.308	1.274	0.552	1.170	1.881
0.850	2.006	0.961	0.479	1.145	1.717
0.900	1.778	0.738	0.415	1.125	1.587
0.950	1.602	0.575	0.359	1.108	1.484
1.000	1.466	0.454	0.310	1.093	1.401

Degree N = 3 S(max) = 1.15 S(min) = 1.14

W	*G*	*B'*	*Q*	*Y1*	*Y2*
0.200	138.408	458.742	3.314	2.528	31.758
0.250	58.354	151.441	2.595	2.084	16.998
0.300	29.231	61.496	2.104	1.799	10.384
0.350	16.559	28.850	1.742	1.604	6.968
0.400	10.302	15.064	1.462	1.465	5.020
0.450	6.908	8.546	1.237	1.363	3.825
0.500	4.928	5.182	1.051	1.287	3.050
0.550	3.704	3.317	0.896	1.229	2.525
0.600	2.911	2.222	0.763	1.184	2.157
0.667	2.238	1.381	0.617	1.139	1.820
0.700	2.005	1.111	0.554	1.122	1.696
0.750	1.739	0.820	0.472	1.100	1.549
0.800	1.545	0.620	0.401	1.083	1.437
0.850	1.400	0.477	0.341	1.069	1.351
0.900	1.290	0.374	0.290	1.058	1.283
0.950	1.523	0.410	0.270	1.084	1.428
1.000	1.738	0.446	0.257	1.105	1.556

Degree-3 network variables (reproduced with permission, Levy, R. & Helszajn, J. (1982) specific equations for one and two section quarter-wave matching networks for stub-resistor loads, IEEE Trans. Microwave Theory Mech., MTT-30, 55–62)

Reproduced with permission, Levy & Helszajn (1982)

$$A = \cos\theta$$

$$B = \sin\theta / Z_t$$

$$C = Z_t \sin\theta$$

$$D = \cos\theta$$

θ is the electrical length of the transformer:

$$\theta = \frac{\pi}{2}(1 + \delta)$$

δ is the normalised radian frequency variable:

$$\delta = \frac{\omega - \omega_0}{\omega_0}$$

Z_r is the ridge characteristic impedance of the generator circuit and Z_t is that of the quarter-wave-long ridge transformer:

$$Z_t = \frac{Z_r}{\sqrt{\varepsilon_t}}$$

X_{in} and R_{in} are the imaginary and real parts of the complex gyrator impedance and ε_t is the relative permittivity of the transformer region. The admittances appearing in these relationships are, unlike the stripline situation, frequency dependent.

The frequency response of the $n = 3$ network is obtained with

$$A = \cos^2\theta - \left(\frac{y_{02}}{y_{01}}\right)\sin^2\theta$$

$$B = \left(\frac{1}{y_{01}} + \frac{1}{y_{02}}\right)\sin\theta\cos\theta$$

$$C = (y_{01} + y_{02})\sin\theta\cos\theta$$

$$D = \left(\frac{-y_{01}}{y_{02}}\right)\sin^2\theta + \cos^2\theta$$

Figures 18.11 and 18.12 depict two typical frequency responses.

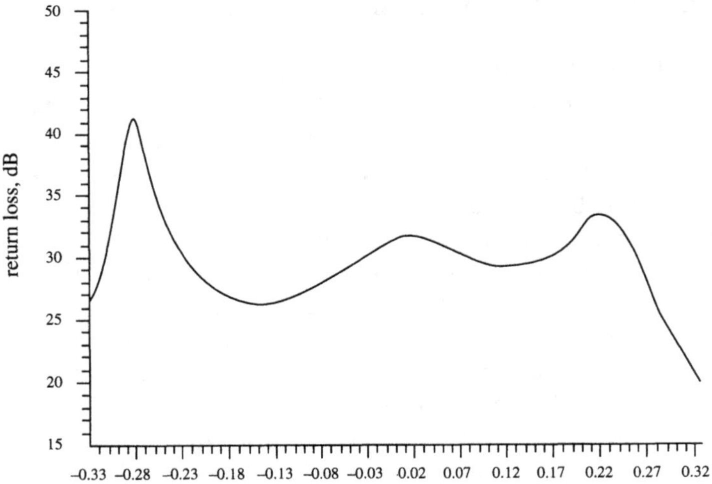

Figure 18.11 Theoretical frequency response of two section quarter-wave coupled semi-tracking junction with $\mu = 1$, $\kappa = 0.60$, $\psi = 0.65$, $G = 0.0780$, $B' = 0.0467$, $Q_L = 0.584$, $Z_r = 36.25\Omega$, $\varepsilon_d = 3.96$

Reprinted with permission, Helszajn (1995)

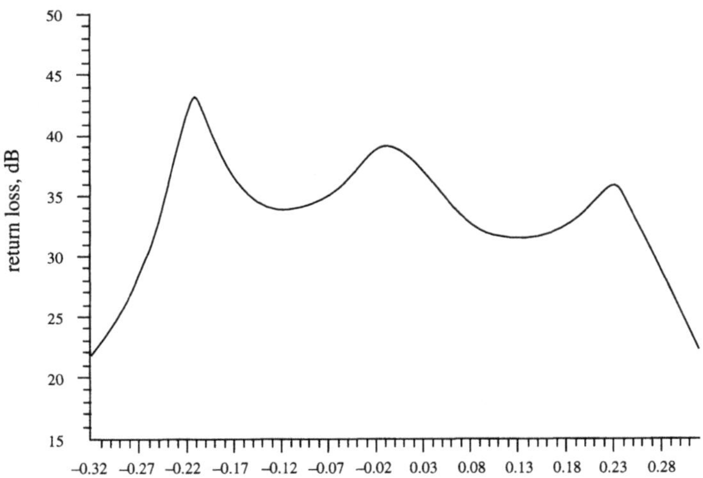

Figure 18.12 Theoretical frequency response of two section quarter-wave coupled semi-tracking junction with $\mu = 1$, $\kappa = 0.67$, $\psi = 0.70$, $G = 0.077$, $B' = 0.0404$, $Q_L = 0.527$, $Z_r = 36.50\Omega$, $\varepsilon_d = 3.50$

Reprinted with permission, Helszajn (1995)

18.10 Design of octave-band semi-tracking circulators

Figures 18.13 and 18.14 illustrate the details of quarter-wave coupled ridge circulators using ferrite posts and dielectric transformer sections.

One feature of the ideal semi-tracking solution of circulators outlined here is that the dielectric constant of the transformer region adjacent to the resonator has a value between 3 and 6. The use of such low values of dielectric constant adjacent to the resonator readily reproduces the assumed magnetic wall boundary condition between the ports of the junction. It therefore more readily ensures correlation between practice and theory than would otherwise be the case. Failure to accurately reproduce the boundary conditions between the ports of the junction leads to some uncertainty in the definition of the effective coupling angle of the junction (defined by the transmission lines and the resonator circuit) and of the radius of the resonator. There is also some corresponding modification in the susceptance slope parameter and to a lesser extent in the conductance of the complex gyrator circuit. Fortunately, the field of solutions of the semi-tracking subspace outlined here permits some laxity in the definition of the former parameters, and the network problem can accommodate some uncertainty in the latter quantities if the minima in the reflection coefficient are not forced to pass through zero.

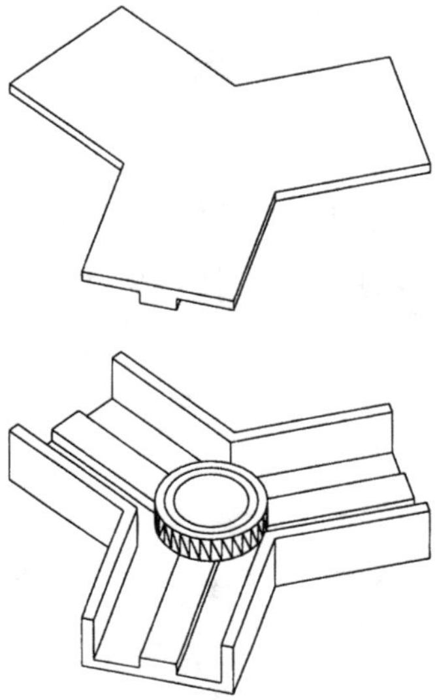

Figure 18.13 Schematic diagram of ridge circulator using dielectric transformers

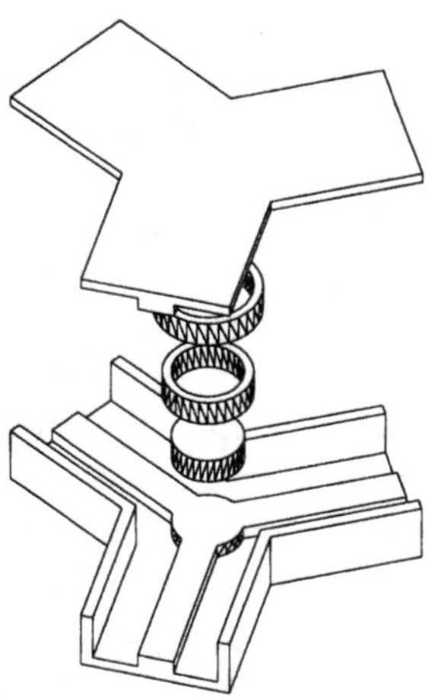

Figure 18.14 Details of ridge circulator using post-resonator and dielectric transformers

Chapter 19
Variational calculus, functionals and the Rayleigh-Ritz procedure

19.1 Introduction

A number of the calculations on the ridge waveguide described in this book amount to obtaining the eigenvalues and eigenvectors of the related planar problem region by using the finite element method. It is therefore appropriate to include a chapter on this technique. A property of a typical eigensolution of any problem region is that it must satisfy both the wave equation and the boundary conditions of the region. Solutions of the wave equation based on a separation of variables technique, however, only exist for regular geometries such as rectangular and circular structures. For irregular structures, a variational approach based on the fact that the stationary values of the energy functional of the problem region also satisfy the related scalar homogeneous Helmholtz differential equation must be employed. The stored energy of the circuit when integrated over the problem region is known as the functional of the problem region. If the region consists of top and bottom electrical walls then this quantity automatically satisfies a magnetic boundary condition on the side-wall of the problem region. If the problem region has top and bottom magnetic walls then it automatically satisfies an electric wall at its side-walls. If the former structure contains an electric wall or segments of such walls then these have to be separately catered for. A dual statement applies to the latter problem region. The process of obtaining a solution to a Helmholtz differential equation by extremising a functional is called a variational method. The use of the word 'functional' in this context serves as a reminder that it is not in itself a function but rather a function of functions. The stationary values obtained in this way satisfy, as will be demonstrated, the homogeneous Helmholtz differential equation. One means of extremising this sort of quantity is to separately vary each function inside the functional. One means of extremising a functional is the Rayleigh-Ritz one. It consists of introducing some polynomial or other approximation for the true solution into the functional prior to reducing it to a matrix eigenvalue problem. The problem may then be solved using standard numerical techniques. In the finite element approximation the problem region is discretised into triangular elements inside of which some suitable polynomial representation of the fields is introduced. The eigenvalues of the problem represent the cut-off frequencies of the circuit and the eigenvectors represent the fields at the nodes. The stationary values of this problem are then the required eigensolutions. The other field components are obtained by using Maxwell's equations in the usual way.

19.2 Stationary value of functional

One means of solving a homogeneous Helmholtz differential equation with a homogeneous boundary condition is to obtain the stationary values of the functional of the problem region. The main purpose of this chapter is the formulation of a suitable functional based on the stored energy in the circuit. An important property of a problem region with top and bottom electric walls is that its functional has a magnetic wall as a natural boundary condition of the problem. A dual remark is applicable to one with top and bottom magnetic walls.

Before attempting to construct an energy functional it is desirable to verify whether its extremisation satisfies the wave equation. The demonstration starts by writing the wave equation in terms of an operator L:

$$LE_z = 0 \tag{1}$$

where

$$L = \nabla_t^2 + k_e^2 \tag{2}$$

A property of the operator (**L**) is that it is self-adjoint. Such a matrix satisfies the following relationship

$$\mathbf{ALB} = \mathbf{BLA}$$

It continues by defining a functional by premultiplying the wave equation by the complex conjugate of the field variable prior to integrating the ensuing equation over the problem region,

$$I = \iint_S [E_z^* LE_z] \, \mathrm{d}S \tag{3}$$

The derivation of the required result now continues by assuming a trial function (ϕ) in the vicinity of the exact value (E_z) of the form

$$\phi = E_z + \alpha\theta \tag{4}$$

Introducing this approximation into the definition of the functional readily gives

$$I = \iint_S [(E_z^* + \alpha\theta^*)L(E_z + \alpha\theta)]\,\mathrm{d}S \tag{5}$$

A typical stationary condition for the functional is now defined by

$$\left.\frac{\mathrm{d}I}{\mathrm{d}\alpha}\right|_{\alpha=0} = 0 \tag{6}$$

If this condition is evaluated for the assumed form for I, and if terms in α^2 are neglected, it is then readily seen that

$$\iint_S (E_z^* L\theta + \theta^* LE_z)\,\mathrm{d}S = 0 \tag{7}$$

The derivation now proceeds by making use of the fact that L is self-adjoint. This gives

$$2\theta^* LE_z = 0 \tag{8}$$

Since θ^* is an arbitrary constant it cannot be equal to zero, so the only other possibility therefore coincides with $LE_z = 0$ as asserted.

19.3 Electric and magnetic energies in planar circuits

The eigenvalues or frequencies of any resonant circuit coincide with the condition for which the electric and magnetic energies stored in the circuit are equal:

$$W_\mathrm{m} = W_\mathrm{e} \tag{9}$$

There are in a planar circuit an infinite number of such solutions. To determine these it is convenient to form the following quantity:

$$I = W_\mathrm{e} - W_\mathrm{m} \tag{10}$$

The time average of the stored magnetic (W_m) and electric (W_e) energies in the resonator are obtained by using the Poynting vector. This gives

$$W_\mathrm{m} = \frac{\omega}{4}\,\mathrm{Re}\iiint_v \mathbf{H}\cdot[\mu]\mathbf{H}^*\,\mathrm{d}v \tag{11}$$

$$W_e = \frac{\omega}{4} \operatorname{Re} \iiint_v \mathbf{E} \cdot [\varepsilon] \mathbf{E}^* \, dv \tag{12}$$

Making use of the matrix relationship below:

$$\mathbf{A} \cdot \mathbf{B} = \mathbf{A}^T \mathbf{B}$$

indicates that W_m and W_e may also be written as

$$W_m = \frac{\omega}{4} \operatorname{Re} \iiint_v \mathbf{H}^T [\mu] \mathbf{H}^* \, dv \tag{13}$$

$$W_e = \frac{\omega}{4} \operatorname{Re} \iiint_v \mathbf{E}^T [\varepsilon] \mathbf{E}^* \, dv \tag{14}$$

The matrix operations inside the integral sign in the preceding equations have either the nature of a quadratic or Hermitian form. The expansion of either form produces a real scalar quantity. It may be separately demonstrated that a property of an energy function is that it is either one or the other of these two forms.

By constructing the integral I given in equation (10) in the case of a planar circuit a variational formulation can be derived. The minimisation of this functional satisfies both the eigenfunctions of the wave equation and the natural boundary condition on the side-walls of the circuit.

19.4 Electric and magnetic fields in planar circuits with top and bottom electric walls

To proceed with the evaluation of the functional it is usual to express H_x and H_y in terms of E_z prior to seeking a stationary value for E_z which satisfies the wave equation. The boundary of the circuit considered here is designated by a contour ξ along which two unit vectors are defined: a normal vector $\bar{n}$ and a tangential vector $\mathbf{t}$. The separation between the top and bottom walls of the problem region is arranged to be small with respect to the wavelength to ensure that higher order modes which vary in the z-direction are suppressed. These may either be electric or magnetic walls. This arrangement is illustrated in Figure 19.1.

The top and bottom walls, for the TM family of modes, are electric ones and

$$E_z \neq 0, \quad H_x \neq 0, \quad H_y \neq 0 \tag{15a}$$

$$E_x = 0, \quad E_y = 0, \quad H_z = 0 \tag{15b}$$

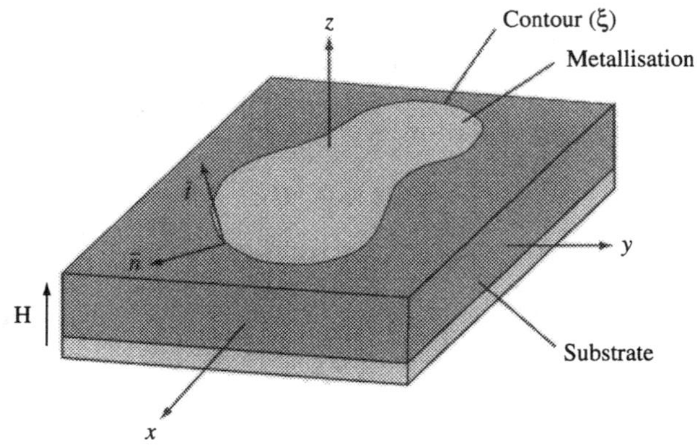

Figure 19.1 Schematic diagram of planar circuit with magnetic side-wall

In a scalar problem,

$$[\varepsilon] = \varepsilon_0 \begin{pmatrix} \varepsilon_f & 0 & 0 \\ 0 & \varepsilon_f & 0 \\ 0 & 0 & \varepsilon_f \end{pmatrix} \tag{16a}$$

$$[\mu] = \mu_0 \begin{pmatrix} 1 & 0 & 0 \\ 0 & 1 & 0 \\ 0 & 0 & 1 \end{pmatrix} \tag{16b}$$

Maxwell's equations then permits H_x and H_y to be written in terms of E_z only. The required result is

$$H_x = \frac{j\omega\varepsilon_0\varepsilon_f}{k_{eff}^2}\left(\frac{\partial E_z}{\partial y}\right) \tag{17a}$$

$$H_y = \frac{-j\omega\varepsilon_0\varepsilon_f}{k_{eff}^2}\left(\frac{\partial E_z}{\partial x}\right) \tag{17b}$$

k_{eff} is the wave number defined by

$$k_{eff} = k_0\sqrt{\varepsilon_f}$$

H_x and H_y may be determined from the preceding equations once E_z is established.

The total magnetic field at any point within the circuit is

$$\mathbf{H}_t = \frac{j\omega\varepsilon_0\varepsilon_f}{k_{eff}^2}\left[\mathbf{a}_x\left(\frac{\partial E_z}{\partial y}\right) + \mathbf{a}_y\left(-\frac{\partial E_z}{\partial x}\right)\right] \tag{18}$$

On the periphery of the circuit, it can be expressed in terms of components which are normal and tangential to its boundary:

$$\mathbf{H}_{\mathrm{t}} = \frac{j\omega\varepsilon_0\varepsilon_{\mathrm{f}}}{k_{\mathrm{eff}}^2}\left[\mathbf{n}\left(\frac{\partial E_z}{\partial t}\right) + \mathbf{t}\left(-\frac{\partial E_z}{\partial n}\right)\right] \tag{19}$$

If the boundary condition is taken as a magnetic wall one then the tangential component of the magnetic field must be zero there. The required boundary condition is then determined by

$$\frac{\partial E_z}{\partial n} = 0 \quad \text{on } \xi \tag{20}$$

An electric wall is established on the contour ξ of the circuit provided

$$E_z = 0 \tag{21}$$

The dual problem with top and bottom magnetic walls is easily understood. Its solution corresponds to the TE family of modes.

19.5 Derivation of functional for planar isotropic circuits

The magnetic energy (W_{m}) in the circuit defined by equation (13) may now be evaluated in terms of E_z only by using equations (17a) and (17b). The required result is

$$W_{\mathrm{m}} = \frac{\omega\varepsilon_0\varepsilon_{\mathrm{f}}}{4k_{\mathrm{eff}}^2}\,\mathrm{Re}\left\{\iiint_v |\nabla_t E_z|^2 \,\mathrm{d}v\right\} \tag{22}$$

where

$$\nabla_t = \mathbf{a}_x\frac{\partial}{\partial x} + \mathbf{a}_y\frac{\partial}{\partial y}$$

and

$$|\nabla_t A|^2 = (\nabla_t A)\cdot(\nabla_t A^*)$$

Since the fields in a planar circuit with a ground plane spacing H are independent of z the volume integral can be replaced by a surface integral over the area of the structure. Introducing this operation in the expression for the stored magnetic energy gives

$$W_{\mathrm{m}} = \frac{\omega\varepsilon_0\varepsilon_{\mathrm{f}}H}{4k_{\mathrm{eff}}^2}\,\mathrm{Re}\left\{\iint_S |\nabla_t E_z|^2 \,\mathrm{d}s\right\} \tag{23}$$

The corresponding result for the stored electric energy (W_e) in equation (14) is

$$W_e = \frac{\omega \varepsilon_0 \varepsilon_f H}{4} \mathrm{Re}\left\{ \int\int_S |E_z|^2 \, \mathrm{d}s \right\} \tag{24}$$

The energy functional I is then given by forming the difference between these two quantities,

$$I = W_e - W_m = \frac{\omega \varepsilon_0 \varepsilon_f H}{4k_{\mathrm{eff}}^2} \mathrm{Re}\left\{ \int\int_S [-|\nabla_t E_z|^2 + k_{\mathrm{eff}}^2 |E_z|^2] \, \mathrm{d}s \right\} \tag{25}$$

The minimisation of this quantity produces, in practice, an infinite number of modes which are described by

$$E_z = \phi_a \tag{26a}$$

$$k_{\mathrm{eff}} = k_a \tag{26b}$$

$$a = 0, 1, 2, \ldots \tag{26c}$$

Introducing this notation and missing out the common factors which are not significant (since only minimisation of the integral is required) gives the required variational formulation

$$I = \mathrm{Re}\left\{ \int\int_S [-|\nabla_t \phi_a|^2 + k_a^2 |\phi_a|^2] \, \mathrm{d}s \right\} \tag{27}$$

The zeros of this quantity satisfy both the wave equation and a magnetic wall boundary condition on the side-walls of the circuit.

$$(\nabla_t^2 + k_a^2)\phi_a = 0 \tag{28}$$

$$\frac{\partial \phi_a}{\partial n} = 0 \quad \text{on boundary } \xi \tag{29}$$

respectively.

The energy functional of the dual problem with magnetic top and bottom walls is

$$I = \mathrm{Re}\left\{ \int\int_S [-|\nabla_t H_z|^2 + k_{\mathrm{eff}}^2 |H_z|^2] \, \mathrm{d}s \right\} \tag{30}$$

It satisfies an electric wall boundary condition on the side-walls of the problem region.

The processes of minimisation and maximisation in variational calculus are often referred to as extremisation and the functional is said to be extremised or made stationary.

19.6 Rayleigh-Ritz procedure

The procedure for obtaining a solution of a Helmholtz differential equation by extremising a functional is known as a variational technique. One way this may be accomplished is by employing the Rayleigh-Ritz procedure. It starts by replacing the true solution by a trial function ϕ_a:

$$\phi_a = \sum_{i=1}^{n} \alpha_i(x, y) u_i \tag{31}$$

In the Rayleigh-Ritz approach the trial function is expanded in terms of a suitable set of real basis or shape functions $\alpha_i(x, y)$ which contain the spatial variation of the problem with complex coefficients u_i. In the finite element problem the u_i represent the values of the fields at the nodes of the elements. This step reduces the problem to a set of simultaneous equations of the form

$$\mathbf{U}^{\mathrm{T}}\{[A] - k_a^2[B]\}\mathbf{U} = 0 \tag{32}$$

This matrix equation retains the quadratic or Hermitian form of the original problem.

$[A]$ and $[B]$ are square matrices given by

$$A_{ij} = \iint_S (\nabla_t \alpha_i) \cdot (\nabla_t \alpha_j) \, \mathrm{d}s \tag{33}$$

and

$$B_{ij} = \iint_S (\alpha_i \alpha_j) \, \mathrm{d}s \tag{34}$$

where ∇_t has the meaning previously introduced and

$$(\nabla_t \alpha_i) \cdot (\nabla_t \alpha_j) = \left(\frac{\partial \alpha_i}{\partial x}\right)\left(\frac{\partial \alpha_j}{\partial x}\right) + \left(\frac{\partial \alpha_i}{\partial y}\right)\left(\frac{\partial \alpha_j}{\partial y}\right) \tag{35}$$

$$(\nabla_t \alpha_i) \cdot (\nabla_t \alpha_i) = \left(\frac{\partial \alpha_i}{\partial x}\right)^2 + \left(\frac{\partial \alpha_i}{\partial y}\right)^2 \tag{36}$$

$\mathbf{U}$ is a column vector containing the unknowns of the problem, and $\mathbf{U}^{\mathrm{T}}$ is its transpose.

To develop some detailed familiarity with the origin of the two matrices in equations (33) and (34) these will now be derived for a degree-2 approximation problem by way of an example. The assumed form for E_z is in this instance taken as

$$\phi_a = \alpha_1(x, y)u_1 + \alpha_2(x, y)u_2 \tag{37}$$

Introducing this approximation into the functional I in equation (27) gives

$$I = \iint_S \left\{ -\left[\frac{\partial(u_1\alpha_1 + u_2\alpha_2)}{\partial x}\right]^2 - \left[\frac{\partial(u_1\alpha_1 + u_2\alpha_2)}{\partial y}\right]^2 + k_a^2[u_1\alpha_1 + u_2\alpha_2]^2 \right\} \mathrm{d}s \tag{38}$$

Making use of the vector identities in equations (35) and (36) then gives

$$I = \iint_S \{-[u_1^2(\nabla_t\alpha_1)\cdot(\nabla_t\alpha_1) + u_2^2(\nabla_t\alpha_2)\cdot(\nabla_t\alpha_2)]$$

$$- 2u_1u_2(\nabla_t\alpha_1)\cdot(\nabla_t\alpha_2) + k_a^2(u_1\alpha_1 + u_2\alpha_2)^2\}\ \mathrm{d}s \tag{39}$$

The matrix notation in equation (32) is now recovered by extracting column vectors $\mathbf{U}^{\mathrm{T}}$ and $\mathbf{U}$ from the preceding relationship:

$$[u_1 u_2]\left\{\begin{bmatrix}(\nabla_t\alpha_1)\cdot(\nabla_t\alpha_1) & (\nabla_t\alpha_1)\cdot(\nabla_t\alpha_2)\\ (\nabla_t\alpha_1)\cdot(\nabla_t\alpha_2) & (\nabla_t\alpha_2)\cdot(\nabla_t\alpha_2)\end{bmatrix} - k_a^2\begin{bmatrix}\alpha_1\alpha_1 & \alpha_1\alpha_2\\ \alpha_1\alpha_2 & \alpha_2\alpha_2\end{bmatrix}\right\}\begin{bmatrix}u_1\\ u_2\end{bmatrix} \tag{40}$$

The final step proceeds by imposing the Rayleigh-Ritz condition. It amounts to taking the derivative of the functional of the problem at each node of the interpolation function

$$\frac{\partial f(\mathbf{U})}{\partial u_k} = 0 \tag{41}$$

This operation may be understood by recalling that $F(\mathbf{U})$ has the nature of a quadratic form so that each element in its expansion is a scalar quantity. The required operation leaves the matrix of the form in equation (32) unchanged:

$$\{[A] - k_a^2[B]\}\mathbf{U} = 0 \tag{42}$$

Once the basis functions are selected the eigenvalue equation will yield an eigenvalue k_a^2 and a column eigenvector $\mathbf{U}_a$ for each basis function included in the preceding equation. The derivation of this condition proceeds by differentiating the functional I with respect to u_1 and u_2 one at a time. Taking the first factor inside the integral sign in the original functional to start with gives

$$I_1 = u_1^2(\nabla_t\alpha_1)\cdot(\nabla_t\alpha_1) + u_2^2(\nabla_t\alpha_2)\cdot(\nabla_t\alpha_2) + 2u_1u_2(\nabla_t\alpha_1)\cdot(\nabla_t\alpha_2)$$

Its derivatives with respect to u_1 and u_2 are

$$\frac{\partial I_1}{\partial u_1} = 2u_1(\nabla_t\alpha_1)\cdot(\nabla_t\alpha_1) + 2u_2(\nabla_t\alpha_1)\cdot(\nabla_t\alpha_2)$$

$$\frac{\partial I_1}{\partial u_2} = 2u_1(\nabla_t\alpha_1)\cdot(\nabla_t\alpha_2) + 2u_2(\nabla_t\alpha_2)\cdot(\nabla_t\alpha_2)$$

respectively. The second factor in the functional is

$$I_2 = k_a^2[u_1^2\alpha_1^2 + u_2^2\alpha_2^2 + 2u_1u_2\alpha_1\alpha_2]$$

and its derivatives are

$$\frac{\partial I_2}{\partial u_1} = 2k_a^2[u_1\alpha_1^2 + u_2\alpha_1\alpha_2]$$

$$\frac{\partial I_2}{\partial u_2} = 2k_a^2[u_1\alpha_1\alpha_2 + u_2\alpha_2^2]$$

The required result in equation (42) is now readily constructed by forming

$$\frac{\partial I}{\partial u_1} = \frac{\partial}{\partial u_1}(-I_1 + I_2) = 0$$

$$\frac{\partial I}{\partial u_2} = \frac{\partial}{\partial u_2}(-I_1 + I_2) = 0$$

and once more identifying column vectors $(\mathbf{U})^{\mathrm{T}}$ and $\mathbf{U}$ in the ensuing relationship.

19.7 Field patterns

Once the eigenvalues of the matrix problem have been established the eigenvectors may be evaluated from a knowledge of the $[A]$ and $[B]$ matrices using standard numerical techniques. The eigenvector $\mathbf{U}$ contains the discrete nodal values of the field perpendicular to the top and bottom walls of the planar circuit. These values, which are in general complex, are distributed over the whole surface area in accordance with the finite element mesh. Each eigenvalue and eigenvector is associated with a unique eigensolution of the problem region. The eigenvectors are orthogonal and are normalised so that

$$\mathbf{U}_a^{\mathrm{T}}\mathbf{U}_a = 1 \tag{43}$$

The equipotential lines for each solution may be obtained by interpolating between the nodal values. Interpolation methods such as 'least squares' or 'weighted average' may be utilised here. The accuracy of the fields is dependent on the degree of the original approximation problem and the distribution of the finite elements in the problem region. At regions of rapid field variations low concentrations of finite elements or elements of low order will produce noticeable errors in both the eigenvalues and eigenvectors. This may produce discontinuities in the field plot which can only be removed by resolving the problem with an improved mesh.

19.8 Derivation of energy functional based on a mathematical technique

While it is usually possible to deduce an energy functional for any physical system whose stationary points satisfy its wave equation it would be preferable instead of deducing the energy explicitly to be able to start with the governing partial differential equation. Such a solution would then be considered a mathematical technique independent of the physics under consideration. Fortunately, such an approach has been developed. It is established by recognising that the functional whose stationary points satisfy the wave equation may be obtained by premultiplying it by E_z^* prior to integrating the ensuing quantity over the surface of the circuit:

$$I = \iint_S [E_z^*(\nabla_t^2 + k_c^2)E_z] \, ds \tag{44}$$

The functional I is again recognised as having the nature of a quadratic form which may be readily reduced to a scalar quantity.

The equivalence between this form of the functional and that previously deduced may be established by using one identity in vector algebra based on the divergence theorem

$$\iint_S \nabla_t \cdot \mathbf{A} \, dS = \int_\xi \mathbf{A} \cdot \mathbf{n} \, dt$$

The required identity is

$$\iint_S (A^* \nabla_t^2 A) \, dS = -\iint_S (\nabla_t A) \cdot (\nabla_t A^*) \, dS + \int_\xi A^* \frac{\partial A}{\partial t} \, dt$$

S is the surface of the planar circuit, ξ represents the periphery and t is the boundary tangent defined in a counter-clockwise direction. The Green theorem in a plane is readily obtained from this identity by forming a new one by replacing A by A^* and adding the two relationships.

The derivation now proceeds by introducing Green's theorem in a plane into the assumed form for I. This gives

$$I = \iint_S -[(\nabla_t E_z^*) \cdot (\nabla_t E_z)] \, \mathrm{d}S + \int_\xi E_z^* \frac{\partial E_z}{\partial t} \, \mathrm{d}t + \iint_S [k_e^2 |E_z|^2] \, \mathrm{d}s \qquad (45)$$

If the boundary condition on the edge of the circuit is a magnetic wall one then

$$\frac{\partial E_z}{\partial t} = 0$$

The assumed nature of the functional reduces, in this instance, to that previously derived by explicitly formulating the stored energy in the circuit,

$$I = \iint_S [-|\nabla_t E_z|^2 + k_e^2 |E_z|^2] \, \mathrm{d}s \qquad (46)$$

This development also indicates that a magnetic wall is a natural boundary condition of the functional. It need not therefore be separately catered for in any problem which is bounded by such a wall.

Bibliography

Chapter

AHMED, S., AND DALY, P.: 'Finite-element methods for inhomogeneous waveguides', *Proc. IEE*, 1969, **116**, pp. 1661–1664 7

AKAIWA, Y.: 'Bandwidth enlargement of a millimeter wave Y-circulator with half-wavelength line resonators', *IEEE Trans.*, 1974, **MTT-22**, pp. 1283–1286 11

ALLEN, P.J.: 'The turnstile circulator', *IRE Trans.*, 1956, **MTT-4**, pp. 223–227 1

AMARI, S., AND BORNEMANN, J.: 'Application of a pole-free model field-matching technique to ridged rectangular waveguides', *Microw. Opt. Technol. Lett.*, 1997, **14**, pp. 337–340 5

ANDERSON, T.N.: 'Directional coupler design nomograms', *Microw. J.*, May 1959 12

ANDERSON, W.W., AND HINES, M.E.: 'Wide-band resonance isolator', *IRE Trans.*, 1960, **MTT-8**, pp. 63–67 8

AULD, B.A.: 'The synthesis of symmetrical waveguide circulator', *IRE Trans.*, 1959, **MTT-7**, pp. 238–246 17

AYTER, S., AND AYASLI, Y.: 'The frequency behaviour of stripline junction circulator', *IEEE Trans.*, 1978, **MTT-26**, pp. 197–202 18

BALAJI, U., AND VAHLDIECK, R.: 'Radial mode matching analysis of ridged circular waveguides', *IEEE Trans.*, 1996, **MTT-44**, pp. 1183–1186 9

BALAJI, U., AND VAHLDIECK, R.: 'Mode matching analysis of circular-ridged waveguide discontinuities', *IEEE Trans.*, 1998, **MTT-46**, pp. 191–195 9

BERK, A.D.: 'Variational principles for electromagnetic resonators and waveguides', *IRE Trans.*, 1956, **AP-4**, pp. 104–111 10

BETHE, H.A.: 'Theory of diffraction by small holes', *Phys. Rev.*, 1944, **66**, pp. 163–182 12

BEUST, W., AND JOHNSTON, E.G.: 'High average power rotator', *Microw. J.*, May 1960 10

BEYER, A., AND SOLBACH, K.: 'A new fin-line ferrite isolator for integrated millimeter-wave circuits', *IEEE Trans.*, 1981, **MTT-29**, pp.1344–1348 16

BEYER, A., AND WOLFF, I.: 'Finline ferrite isolator and circulator for K-band'. Proc. 11th European microwave conference, 1981, pp. 321–326 16, 17

BHAT, B., AND KOUL, S.: 'Analysis, design and applications of fin-lines' (Artech House, Inc., Norwood, M.A., 1987) 16

BORNEMANN, J., AND ARNDT, F.: 'Modal-S-matrix design of optimum stepped ridged and finned waveguide transformers', *IEEE Trans.*, 1987, **MTT-35**, pp. 561–567 5, 14

BRAAS, M., AND SCHIEBLICH, C.: 'E-type circulator for finlines', *Electron. Lett.*, 1981, **17**, pp. 701–702 17

BUDIMIR, D.: 'Optimized E-plane filters with improved stopband performance', *IEEE Trans.*, 1997, **MTT-45**, pp. 212–220 14

CANATAN, F.: 'Cutoff wavenumbers of ridged circular waveguides via Ritz-Galerkin approach', *Electron. Lett.*, 1989, **25**, pp. 1036–1038 9

CASWELL, D.: 'Faraday rotation devices', *Microwaves*, May 1964, **3** 10

CHAIT, H., AND SAKIOTIS, G.: 'Broadband ferrite rotators using quadruply ridged circular waveguide', *IRE Trans.*, 1959, **MTT-7**, pp. 38–41 10

CHEN, C.C.: 'Quadruple ridge-loaded circular waveguide phased arrays', *IEEE Trans.*, 1974, **AP-22**, pp. 481–483 9

CHEN, M.H., TSANDOULAS, G.N., AND WILLWERTH, F.G.: 'Modal characteristics of quadruple-ridged circular and square waveguides', *IEEE Trans.*, 1974, **MTT-22**, pp. 801–804 9

CHEN, T.S.: 'Calculation of the parameters of ridge waveguide', *IRE Trans.*, 1957, **MTT-5**, pp. 12–17 1, 3, 5, 6

CHEN, T.S.: 'Nonreciprocal attenuation of ferrite in single ridge waveguide', *IRE Trans.*, 1960, **MTT-8**, pp. 247–248 15

DE KONING, *et al.*: 'Full band low loss continuous tracking circulation in K band', *IEEE Trans.*, 1977, **MTT-25**, p. 152 18

DILLION, B.M., AND GIBSON, A.A.P.: 'Applied field/frequency dependency of propagation in axially magnetised ferrite waveguides', *IEEE Trans.*, 1993, **MTT-41**, (12), pp. 2092–2098 10

DILLION, B.M., AND GIBSON, A.A.P.: 'Phase shift calculations in axially-magnetized ferrites using the finite element method'. IEEE MTT-S Digest, Atlanta, GA, 1993, pp. 469–472 10

DILLION, B.M., AND GIBSON, A.A.P.: 'Triple-ridged circular waveguides', *J. Electromagn. Waves Appl.*, 1995, **9**, pp. 145–156 9

DILLION, B.M., GIBSON, A.A.P., AND WEBB, J.P.: 'Cut-off and phase constants of partially filled, axially magnetized, gyromagnetic waveguides using finite elements', *IEEE Trans.*, 1993, **MTT-41**, pp. 803–808 10

FAO, J.-W., AND BAUDRAND, H.: 'Multi-modal variational analysis of uniaxial waveguide discontinuities', *IEEE Trans.*, 1991, **MTT-39**, pp. 506–516 11

FLERI, D., AND HANLEY, G.: 'Non-reciprocity in dielectric loaded TEM mode transmission lines', *IRE Trans.*, 1959, **MTT-7**, pp. 83–84 8, 15

FOX, A.G., MILLER, S.E., AND WEISS, M.T.: 'Behaviour and applications of ferrites in the microwave region', *Bell Syst. Tech. J.*, 1955, **34**, p.5 10

GARB, K., AND KASTNER, R.: 'Characteristic impedance of quadruple-ridged square waveguides', *Microw. Opt. Technol. Lett.*, 1995, **8**, (5), pp. 236–238 5, 9

GARB, K., AND KASTNER, R.: 'Characteristic impedance of a rectangular double-ridged TEM line', *IEEE Trans.*, 1997, **MTT-45**, pp. 554–557 5

GARDIOL, F.E.: 'Higher-order modes in dielectrically loaded rectangular waveguides', *IEEE Trans.*, 1968, **MTT-16**, pp. 919–924 7

GETSINGER, W.J.: 'Ridge waveguide field description and application to directional couplers', *IRE Trans.*, 1962, **MTT-10**, pp. 41–50 3, 4, 6, 8, 12

GIL, J.M., AND ZAPATA, J.: 'Efficient singular element for finite element analysis of quasi-TEM transmission lines and waveguides with sharp metal edges', *IEEE Trans.*, 1994, **MTT-42**, pp. 92–98 5, 8

GODTMANN, H.D., AND HAAS, W.: 'Magnetodynamic modes in axially magnetized ferrite rods between two parallel conducting sheets', *IEEE Trans.*, 1967, **MTT-15**, pp. 476–481 17

GOEBEL, U., AND SCHIEBLICH, C.: 'Broadband finline circulators'. Proc. IEEE symposium digest, 1982, pp. 249–251 17

GRIMES, E.S. JR., BARTHOLOMEW, D.D., SCOTT, D.C., AND SLOAN, S.C.: 'Broadband ridge waveguide ferrite devices', *IRE Trans.*, 1960, **MTT-8**, pp. 489–492 10, 15

HANO, M.: 'Finite element analysis of dielectric-loaded waveguides', *IEEE Trans.*, 1984, **MTT-32**, pp. 1275–1279 7

HASTINGS, W.I., AND HELSZAJN, J.: 'WRD 200 90° waveguide differential phase shift section', *IEE Proc., Microw. Antennas Propag.*, 1987, **134**, pp. 223–225 15

HAYASHI, Y., AND MITTRA, R.: 'Analytical investigation of fin-lines with magnetized ferrite substrate', *IEEE Trans.*, 1983, **MTT-31**, pp. 495–498 16

HAYATA, K., KOSHIBA, M., EQUICHI, M., AND SUZUKI, M.: 'Vectorial finite element method without any spurious solutions for dielectric waveguiding problems using transverse magnetic field component', *IEEE Trans.*, 1986, **MTT-34**, pp. 1120–1124 7

HELLER, G.S., AND LAX, B.: 'Use of perturbation theory for cavities and waveguides containing ferrite', *IRE Trans.*, 1957, **AP-4**, p. 588 17

HELSZAJN, J.: 'A symmetrical design of a short Faraday rotator', *Microw. J.*, August 1968, pp. 35–39 10

HELSZAJN, J.: 'Synthesis of quarter-wave coupled circulators with Chebyshev characteristics', *IEEE Trans.*, 1972, **MTT-20**, pp. 764–769 18

HELSZAJN, J.: 'Three-resonant mode adjustment of the waveguide circulator', *Radio Electron. Eng.*, 1972, **42**, pp. 1–4 17

HELSZAJN, J.: 'Scattering matrix of junction circulator with Chebyshev characteristics', *IEEE Trans.*, 1975, **MTT-23**, pp. 548–554 18

HELSZAJN, J.: 'Operation of tracking circulators', *IEEE Trans.*, 1981, **MTT-29**, pp. 700–707 18

HELSZAJN, J.: 'Complex gyrator circuit of an evanescent mode E-plane junction circulator using H-plane turnstile resonator', *IEEE Trans.*, 1987, **MTT-35**, pp. 797–806 17

HELSZAJN, J.: 'Ferrite phase shifters and control devices' (McGraw Hill, 1989) 15

HELSZAJN, J.: 'Synthesis of quarter-wave coupled semi-tracking octave band stripline junction circulators', *IEEE Trans.*, 1995, **MTT-43**, pp. 310–318 18

HELSZAJN, J.: 'Green's function, finite elements and microwave planar circuits' (John Wiley & Sons Ltd, Chichester, 1996)

HELSZAJN, J.: 'Characterization of discontinuity effects in single ridge waveguide', *Microw. Eng. Eur.*, March 2000, pp. 31, 32, 35, 36, 38, 41 11

HELSZAJN, J., AND CAPLIN, M.: 'Impedance and propagation in ridge waveguide', *Microw. Eng. Eur.*, May 1997, pp. 49–58 1, 3

HELSZAJN, J., AND CHENG, S.: 'Aspect ratio of open resonators in the design of evanescent mode E-plane circulators', *IEE Proc. H*, 1990, **137**, pp. 55–60 17

HELSZAJN, J., AND McDERMOTT, M.: 'Mode charts for E-plane circulators', *IEEE Trans.*, 1972, **MTT-20**, pp. 187–188 17

HELSZAJN, J., AND McKAY, M.: 'Voltage-current definition of impedance of double ridge waveguide using the finite element method', *IEE Proc., Microw. Antennas Propag.*, 1998, **145**, pp. 1–6 1, 4, 6

HELSZAJN, J., AND McSTAY, J.: 'External susceptibility tensor of magnetised ferrite ellipsoid in terms of uniform-mode ellipticity', *Proc. IEE*, 1969, **116**, pp. 2088–2093 16

HELSZAJN, J., AND SHRIMPTON, L.: 'Propagation constant of dielectric loaded square ridge waveguide', *Electron. Lett.*, 1996, **32**, (24), pp. 2250–2251 7, 9

HELSZAJN, J., AND THORPE, W.: '18–26 GHz finline resonance isolators using hexagonal ferrites', *IEE Proc. H*, 1985, **132**, pp. 73–76 16, 17

HELSZAJN, J., BAARS, R.D., AND NISBET, W.T.: 'Characteristics of circulators using planar triangular and disk resonators symmetrically loaded with magnetic ridges', *IEEE Trans.*, 1980, **MTT-28**, pp. 616–621 10

HELSZAJN, J., McDONALD, G., AND SUTHERLAND, D.: 'Design of finline circulators using turnstile junctions', *IEE Proc. H*, 1988, **135**, pp. 279–281 17

HIDEYA, G.: 'The Faraday rotation of waves in a circular waveguide', *J. Phys. Soc.*, 1953, **8**, pp. 176–182 10

HOEFER, W.J.R., AND BURTON, M.N.: 'Closed form expressions for parameters of finned and ridge waveguide', *IEEE Trans.*, 1982, **MTT-30**, pp. 2190–2194 — 3, 6, 16

HOFMANN, H.: 'Dispersion of planar waveguides for millimeter-wave applications', *Arch. Elektr. Uebertrag.*, 1977, **31**, pp. 40–44 — 16

HOGAN, C.L.: 'The ferromagnetic Faraday effect at microwave frequencies and its applications – the microwave gyrator', *Bell Syst. Tech. J.*, 1952, **31**, pp. 22–26 — 10

HOGAN, C.L.: 'The microwave gyrator', *Bell Syst. Tech. J.*, 1952, **31**, p. 1 — 15

HOPFER, S.: 'The design of ridged waveguides', *IRE Trans.*, 1955, **MTT-3,** pp. 20–29 — 1, 3, 5, 6, 12

IGARASHI, H., AND HONMA, T.: 'An analysis of electromagnetic waves by a complementary finite element method', *IEEE Trans. Magn.*, 1994, **30**, pp. 3104–3107 — 4, 6

ISRAEL, M., AND MINIOWITZ, R.: 'An efficient finite element method for non-convex waveguide based on Hermitian polynomials', *IEEE Trans.*, 1987, **MTT-35**, pp. 1019–1026 — 5, 6

ISRAEL, M., AND MINIOWITZ, R.: 'Hermitian finite-element method for inhomogeneous waveguides', *IEEE Trans.*, 1990, **MTT-38**, pp. 1319–1327 — 5, 6, 7

KALES, M.L.: 'Modes in waveguides containing ferrites', *J. Appl. Phys.*, 1953, **24**, pp. 604–608 — 10

KITAZAWA, T., AND MITTRA, R.: 'An investigation of strip lines and fin-lines with periodic stubs', *IEEE Trans.*, 1984, **MTT-32**, pp. 684–688 — 16

KITAZAWA, T., AND MITTRA, R.: 'Analysis of fin-line with finite metallization thickness', *IEEE Trans.*, 1984, **MTT-32**, pp. 1484–1487 — 16

KITTEL, C.: 'On the theory of ferromagnetic resonance absorption', *Phys. Rev.*, 1948, **73**, pp. 155–161 — 15, 16

KNORR, J.R., AND SHAYDA, P.M.: 'Millimeter wave fin-line characteristics', *IEEE Trans.*, 1980, **MTT-28**, pp. 737–743

KONISHI, Y., AND MATSUMURA, H.: 'Short end effect of ridge guide with planar circuit mounted in a waveguide', *IEEE Trans.*, 1979, **MTT-27**, pp. 168–170 — 11

KONRAD, A.: 'Vector variational formulation of electromagnetic fields in anisotropic media', *IEEE Trans.*, 1976, **MTT-24**, pp. 553–559 — 7, 10

KOSHIBA, M., HAYATA, K., AND SUZUKI, M.: 'Improved finite-element formulation in terms of the magnetic field vector for dielectric waveguides', *IEEE Trans.*, 1985, **MTT-33**, pp. 227–233 7

KOSHIBA, M., HAYATA, K., AND SUZUKI, M.: 'Finite element formulation in terms of the electric-field vector for electromagnetic waveguide problems', *IEEE Trans.*, 1985, **MTT-33**, pp. 900–906 7

LAX, B., AND BUTTON, K.: 'Microwave ferrites and ferrimagnetics (McGraw-Hill, New York, 1962) 15, 16

LEE, J.F., SUN, D.K., AND CENDES, Z.J.: 'Full-wave analysis of dielectric waveguides using tangential vector finite elements', *IEEE Trans.*, 1991, **MTT-39**, pp. 1262–1271 7, 10

LEE, J.F., WILKINS, G.M., AND MITTRA, R.: 'Finite-element analysis of axisymmetric cavity resonator using a hybrid edge element technique', *IEEE Trans.*, 1993, **MTT-41**, pp. 1981–1986 7

LEVY, R., AND HELSZAJN, J.: 'Specific equations for one and two section quarter-wave matching networks for Stub-resistor loads', *IEEE Trans.*, 1982, **MTT-30**, pp. 55–68 18

LEVY, R., AND HELSZAJN, J.: 'Synthesis of series STUB-R complex gyrator circuits using halfwave long impedance transformers', *IEE Proc., Microw. Antennas Propag.*, 1993, **141**, pp. 426–432 11

LYON, R.W., AND HELSZAJN, J.: 'A finite element analysis of planar circulators using arbitrarily shaped resonators', *IEEE Trans.*, 1982, **MTT-30**, (11), pp. 1967–1974 19

MABAYA, N., LAYASSE, P.E., AND VANDENBULCKE, P.: 'Finite element analysis of optical waveguides', *IEEE Trans.*, 1981, **MTT-29**, pp. 600–605 7

MAGERL, G.: 'Ridged waveguides with inhomogeneous dielectric slab loading', *IEEE Trans.*, 1978, **MTT-26**, pp. 413–416 7, 8, 15

MANSOUR, R.R., TONG, R.S.K., AND MacPHIE, R.H.: 'Simplified description of the field distribution in fin lines and ridge waveguides and its application to the analysis of E-plane discontinuities', *IEEE Trans.*, 1988, **MTT-36**, (12), pp. 1825–1832 4, 6, 8, 12, 14, 16

MARCUVITZ, N.: 'Waveguide handbook', MIT Radiation Lab. Series, No. 10 (Boston Technical Publishers Inc., 1964) 3, 5, 11

McKAY, M., AND HELSZAJN, J.: 'Characterisation of dielectric loaded double ridge waveguides using hybrid and vector finite element solvers', *Microw. Eng. Eur.*, Feb. 2001, not yet published 7

McKAY, M., ROBICHAUD, R., FRENA, J., AND HELSZAJN, J.: 'The cross-guide directional coupler in ridge waveguide', *Microw. Eng. Eur.*, Feb. 1999, pp. 23–24, 27–28, 31–32, 34 12

MEIER, P.J.: 'Equivalent relative permittivity and unloaded Q factor of integrated fin-line', *Electron. Lett.*, 1973, **9**, (7), pp. 162–163 16

MEIER, P.J.: 'Integrated fin line millimeter components', *IEEE Trans.*, 1974, **MTT-22**, pp. 1209–1216 16

MEINEL, H., AND REMBOLD, B.: 'New millimeter-wave fin line attenuators and switches'. 1979 IEEE-MTT-S international microwave symposium digest, Orlando, FL, pp. 249–251 16

MEIXNER, J.: 'The behaviour of electromagnetic fields at edges', *IEEE Trans.*, 1972, **AP-20**, pp. 442–446 4

MIHRAN, T.G.: 'Closed and open-ridged waveguide', *Proc. IRE*, 1949, **37**, pp. 640–644 1, 3, 6

MIKHLIN, S.G.: 'Variational methods in mathematical physics' (Pergamon Press, 1964 and MacMillan, 1974), Chap. 3 19

MIRSHEKAR-SYAHKAL, D., AND DAVIES, J.B.: 'An accurate, unified solution to various fin-line structures, of phase constant, characteristic impedance and attenuation,' *IEEE Trans.*, 1982, **MTT-30**, pp. 1854–1861 16

MONTGOMERY, J.P.: 'On the complete eigenvalue solution of ridged waveguide', *IEEE Trans.*, 1971, **MTT-19**, pp. 547–555 3, 4, 6, 8, 12, 14

MORISHITA, K., AND KUMAGAI, N.: 'Unified approach to the derivation of variational expression for electromagnetic fields', *IEEE Trans.*, 1977, **MTT-25**, (1), pp. 34–40 19

MUMFORD, W.W.: 'Directional couplers', *Proc. IRE*, 1947, **35**, pp. 160–166 12

NG, K.T., AND CHAN, C.H.: 'Unified solution of various dielectric-loaded ridge waveguides with a mixed spectral-domain method', *IEEE Trans.*, 1989, **MTT-37**, pp. 2080–2085 7

OHM, O.E.: 'A broadband microwave circulator', *IRE Trans.*, 1956, **MTT-4**, p. 210 — 10

OKONIEUSKI, M.: 'Vector wave equation 2–D-FDTD method for guided wave problems', *IEEE Microw. Guid. Wave Lett.*, 1993, **3**, pp. 307–309 — 7

OMAR, S., AND SCHUNEMANN, K.: 'Transmission matrix representation of finline discontinuities', *IEEE Trans.*, 1985, **MTT-33**, pp. 765–770 — 14

OMORI, M.: 'An improved E-plane waveguide circulator'. G-MTT international microwave symposium digest, 1968, p. 228 — 17

OWEN, B., AND BARNES, C.E.: 'The compact turnstile circulator', *IEEE Trans.*, 1970, **MTT-18**, pp. 1096–1100 — 17

PANTIC, Z., AND MITTRA, R.: 'Quasi-TEM analysis of microwave transmission lines by the finite element method', *IEEE Trans.*, 1986, **MTT-34**, pp. 1096–1103 — 5

PATZELT, H., AND ARNDT, F.: 'Double plane steps in rectangular waveguides and their application for transformers, irises and filters', *IEEE Trans.*, 1982, **MTT-30**, pp. 771–776 — 14

PINCHERLE, L.: 'Electromagnetic waves in metal tubes filled longitudinally with two dielectrics', *Phys. Rev.*, 1944, **66**, pp. 118–130 — 7

POSTOYALIKO, V., AND BUDIMIR, D.S.: 'Design of waveguide E-plane filters with all-metal inserts by equal ripple optimization', *IEEE Trans.*, 1994, **MTT-42**, pp. 217–222 — 14

PRAMANICK, P., AND BHARTIA, P.: 'Accurate analysis equations and synthesis techniques for unilateral finlines', *IEEE Trans.*, 1985, **MTT-33**, pp. 24–30 — 16

PYLE, J.R. : 'The cut-off wavelength of the TE_{10} mode in ridged rectangular waveguide of any aspect ratio', *IEEE Trans.*, 1966, **MTT-14**, pp. 175–183 — 1, 3, 5

RAHMAN, B.M.A., AND DAVIES, J.B.: 'Finite element analysis of optical and microwave waveguide problems', *IEEE Trans.*, 1984, **MTT-32**, pp. 20–28 — 7

RAHMAN, B.M.A., AND DAVIES, J.B.: 'Penalty function improvement of waveguide solution by finite elements', *IEEE Trans.*, 1984, **MTT-32**, pp. 922–928 — 7

RAHMAN, B.M.A., FERNANDEZ, F.A., AND DAVIES, J.B.: 'Review of finite element methods for microwave and optical waveguides', *Proc. IEEE,* 1991, **79**, pp. 1442–1448 — 10

RIBLET, G.: 'The measurement of the equivalent admittance of 3–port circulators via an automated measurement system', *IEEE Trans.*, 1977, **MTT-25**, pp. 401–405 18

RIBLET, H.J., AND SAAD, T.S.: 'A new type of waveguide directional coupler', *Proc. IRE*, 1948, **36**, pp. 61–65 12

RIZZI, P.A.: 'High-power ferrite circulators', *IRE Trans.*, 1957, **MTT-5**, pp. 230–237 10

RIZZI, P.A.: 'Microwave properties and applications of ferrite rotators', *Microw. J.*, Nov/Dec 1958, **1** 10

RODRIGUE, G.P., PIPPIN, J.E., AND WALLACE, M.E.: 'Hexagonal ferrites for use at x to v-band frequencies', *J. Appl. Phys.*, 1962, **33**, (Suppl.), pp. 1366–1368 16

RONG, Y.: 'The bandwidth characteristics of ridged circular waveguide', *Microw. Opt. Tech. Lett.*, 1990, **3**, pp. 347–350 9

RONG, Y., DOU, W.-B., AND LI, S.F.: 'Generalized field theory of irregular waveguides partially-filled with longitudinally magnetized ferrite', *IEE Proc. H*, 1991, **138**, (5), pp. 412–416 10

ROZZI, T.E.: 'Network analysis of strongly coupled transverse apertures in waveguide', *Int. J. Circuit Theory Appl.*, 1972, **1**, pp. 161–170 14

ROZZI, T.E.: 'The variational treatment of thick interacting inductive irises', *IEEE Trans.*, 1973, **MTT-21**, pp. 82–88 14

ROZZI, T.E., AND MECKLENBRAUKER, W.F.G.: ' Wide-band network modeling of interacting inductive irises and steps', *IEEE Trans.*, 1975, **MTT-23**, pp. 235–245 14

SAAD, A.M.K., AND BEGEMANN, G.: 'Electrical performance of fin-lines of various configurations', *Microw. Opt. Antennas*, 1977, **1**, (2), pp. 81–88 16

SAAD, A.M.K., AND SCHUENEMANN, K.: 'A simple method for analyzing fin-line structures', *IEEE Trans.*, 1978, **MTT-26**, (12), pp. 1002–1007 16

SAFAVI-NAINI, R., AND MacPHIE, R.H.: 'Scattering at rectangular-to-rectangular waveguide junctions', *IEEE Trans.*, 1982, **MTT-30**, (6), pp. 2060–2063 14

SANGSTER, A.J., AND LYON, R.W.: 'Analysis of nonresonant slots in a notched-wall dual-mode rectangular waveguide', *Microw. Opt. Acoust.*, 1978, **2**, pp. 141–146 12

SCHAUG-PATTERSON, T.: 'Novel design of a 3–port circulator'. Norwegian Defence Research Establishment, January 1958 17

SCHIERLICH, C.: 'Mode charts for magnetized ferrite cylinders', *IEEE Trans.*, 1989, **MTT-37**, pp. 1555–1561 17

SCHIFF, B.: 'Eigenvalues for ridged and other waveguides containing corners of angle $3\pi/2$ or 2π by the finite element method', *IEEE Trans.*, 1991, **MTT-39**, (6), pp. 1034–1039 5, 6

SCHMIDT, L.P., AND ITOH, T.: 'Spectral domain analysis of dominant and higher order modes in fin-lines', *IEEE Trans.*, 1980, **MTT-28**, pp. 981–985 16

SCHMIDT, L.P., ITOH, T., AND HOFMANN, H.: 'Characteristics of unilateral fin-line structures with arbitrarily located slots'. IEEE MTT-S international microwave symposium digest, 1980, pp. 255–257 16

SHARMA, A.K., AND HOEFER, W.J.R.: 'Empirical expressions for fin line design', *IEEE Trans.*, 1983, **MTT-31**, pp. 350–356 16

SHIH, Y.C., AND ITOH, T.: 'E-plane filters with finite-thickness septa', *IEEE Trans.*, 1983, **MTT-31**, pp. 1009–1012 14

SHIH, Y.C., ITOH, T., AND BUI, L.Q.: 'Computer-aided design of millimetre wave E-plane filters', *IEEE Trans.*, 1983, **MTT-31**, pp. 135–141 14

SHIMIZU, J.K.: 'Octave-bandwidth feed horn for paraboloid', *IRE Trans.*, 1961, **AP-9**, pp. 223–224 9

SHU, S., GOGGANS, P.M., AND KISHK, A.A.: 'Computation of cut-off wavenumbers for partially filled waveguides of arbitrary cross section using surface integral formulations and the method of moments', *IEEE Trans.*, 1993, **MTT-41**, pp. 1111–1118 7

SICH, E.M., AND MacPHIE, R.H.: 'The conservation of complex power technique and E-plane step-diaphram junction discontinuities', *IEEE Trans.*, 1982, **MTT-30**, pp. 198–201 14

SILVESTER, P., AND FERRARI, R.L.: 'Finite elements for electrical engineers' (Cambridge University Press, New York, 1990, 2nd edn.) 4, 5, 6, 19

SIMON, J.W.: 'Broadband strip-transmission line Y junction circulators', *IEEE Trans.*, 1965, **MTT-13**, pp. 335–345 18

SINGER, J.R.: 'Nonreciprocal attenuation of ferrites in single-ridge waveguide', *IRE Trans.*, 1960, **MTT-8**, pp. 247–249 15

SOLBACH, K.: 'Status of printed millimeter-wave E-plane circuits', *IEEE Trans.*, 1983, **MTT-31**, pp. 107–121 16

SOOHOO, R.F.: 'Theory and application of ferrites' (Prentice-Hall, 1960) 7, 15

SUBRAMANISAM, S., DEVENDRAN, V., RATNAJEEVAN, S., AND HOOLE, H.: 'Optimization of ridged waveguide', *J. Appl. Phys.*, 1993, **73**, pp. 6790–6792 5

SUHL, H., AND WALKER, L.R.: 'Topics in guided wave propagation through gyromagnetic media', *Bell Syst. Tech. J.*, 1954, **33**, pp. 579–659, 939–986, 1133–1194 10

SUN, W., AND BALANIS, C.A.: 'MFIE analysis and design of ridged waveguides', *IEEE Trans.*, 1993, **MTT-41**, pp. 1965–1971 4, 6, 9

SUN, W., AND BALANIS, C.A.: 'Analysis and design of quadruple-ridged waveguides', *IEEE Trans.*, 1994, **MTT-42**, pp. 2201–2207 9

TAFT, D.R., HARRISON, G.R., AND HODGES, L.R.: 'Millimeter resonance isolators utilizing hexagonal ferrites', *IEEE Trans.*, 1963, **MTT-11**, pp. 346–350 16

TAN, J., AND PAN, G.: 'A new edge elements analysis of dispersive waveguiding structures', *IEEE Trans.*, 1995, **MTT-43**, pp. 2600–2607 7

TAO, J.W., AND BAUDRAND, H.: 'Rigorous analysis of triple-ridge waveguide', *Electron. Lett.*, 1988, **24**, pp. 820–821 14

TAO, J.W., AND BAUDRAND, H.: 'New design procedure for evanescent-mode ridged waveguide filters'. Proc. 1989 URSI EM theory symposium, pp. 458–460 14

TAO, J.W., AND BAUDRAND, H.: 'Multimodal variational analysis of uniaxial waveguide discontinuities', *IEEE Trans.*, 1991, **MTT-39**, pp. 506–516 14

TAO, J.W., AND BAUDRAND, H.: 'Variational analysis and synthesis of stepped ridged waveguide transformers', *Microw. Opt. Technol. Lett.*, 1989, **2**, pp. 103–110 14

TEDJINI, S., AND PIC, E.: 'Investigation of nonreciprocal fin-line devices', *IEE Proc. H*, 1984, **131**, (1), pp. 61–63 16

UTSUMI, Y.: 'Variational analysis of ridged waveguide modes', *IEEE Trans.*, 1985, **MTT-33**, (2), pp. 111–120 5, 6

VAHLDIECK, R.: 'Accurate hybrid-mode analysis of various fin-line configurations including multilayered dielectrics, finite metallization thickness and substrate holding grooves', *IEEE Trans.*, 1984, **MTT-32**, pp. 1454–1460 16

VAHLDIECK, R., AND HOEFER, W.J.R.: 'The influence of metallization thickness and mounting grooves on the characteristics of fin-lines'. IEEE MTT-S international microwave symposium digest, 1985, pp. 143–144 16

VAHLDIECK, R., BORNEMANN, J., ARNDT, F., AND GRAUERHOLTZ, D.: 'Optimized waveguide E-plane metal insert filters for millimeter-wave applications', *IEEE Trans.*, 1983, **MTT-31**, pp. 65–69 14

VAN BLADEL, J.: 'Small-hole coupling of resonant cavities and waveguides', *Proc. IEE*, 1970, **117**, pp. 1098–1104 12

VAN BLADEL, J.: 'Field singularities at metal-dielectric wedges', *IEEE Trans.*, 1985, **AP-33**, pp. 450–455 4

VAN TRIER, A.A.: 'Guided electromagnetic waves in anisotropic media', *Appl. Sci. Res.*, 1953, **3**, pp. 305–371 10

VOSS, W.A.G.: 'Optimized crossed slot directional coupler', *Electron. Radio Eng.*, May 1963, pp. 83–87 12

WALDRON, R.A.: 'Electromagnetic wave propagation in cylindrical waveguides containing gyromagnetic media', *J. Br. Inst. Radio Eng.*, 1958, **18**, pp. 597, 677, 733 10

WEBB, J.P.: 'Finite element analysis of dispersion in waveguides with sharp metal edges', *IEEE Trans.*, 1988, **MTT-36**, pp. 1819–1824 5

WEISS, M.T., AND DUNN, F.A.: 'A 5-mm resonance isolator', *IRE Trans.*, 1958, **MTT-6**, p. 331 16

WELT, D., AND WEBB, J.: 'Finite element analysis of dielectric waveguides with curved boundaries', *IEEE Trans.*, 1985, **MTT-33**, pp. 576–585 7

WESCLER, G.: 'Solution of waveguide discontinuities of modal analysis', *IEEE Trans.*, 1967, **MTT-15**, pp. 508–517 14

WU, Y.S., AND ROSENBAUM, F.J.: 'Wideband operation of microstrip circulators', *IEEE Trans.*, 1974, **MTT-22**, pp. 849–856 18

ZARGANO, G.F., LYAPIN, V.P., MICHALEVSKY, V.S., SINELNIKOV, Y.M., SINYAVSKY, G.P., AND CHEKRIGINA, I.M.: 'Waveguides of complex cross-section'. Radio and Communication, Moscow, 1986 (in Russian) 9

ZIENIUTYCZ, W.: 'Nonreciprocal phase shift in fin-line with layered substrate containing magnetised ferrite', *Electron. Lett.*, 1984, **20**, (4), pp. 183–185 16

Index